American Energy Policy in the 1970s

American Energy Policy in the 1970s

Edited by

Robert Lifset

University of Oklahoma Press : Norman

Library of Congress Cataloging-in-Publication Data

American energy policy in the 1970s / edited by Robert Lifset.
    pages cm
Includes index.
ISBN 978-0-8061-4450-4 (pbk. : alk. paper) 1. Energy policy—United
States—History—20th century. 2. Energy consumption—United States—
History—20th century. I. Lifset, Robert, 1974–
HD9502.U52A429 2014
333.790973'09047—dc23

                                                              2013038520

The paper in this book meets the guidelines for permanence and durability of
the Committee on Production Guidelines for Book Longevity of the Council
on Library Resources, Inc. ∞

Contents

Acknowledgments   vii

List of Abbreviations   ix

Introduction   3
    Robert Lifset

Part I. Political Leadership
    Chapter 1. "The Toughest Thing": Gerald Ford's Struggle
    with Congress over Energy Policy   19
    *Yanek Mieczkowsk*i

    Chapter 2. Conflict or Consensus? The Roots
    of Jimmy Carter's Energy Policies   47
    Jay E. Hakes

Part II. Foreign Policy
    Chapter 3. From the Nixon Doctrine to the Carter Doctrine:
    Iran and the Geopolitics of Oil in the 1970s   61
    David S. Painter

Chapter 4. The United States, Iran, and the "Oil Weapon":
From Truman to George W. Bush   93
Steve Marsh

Part III. Supply
Chapter 5. Diving into Deep Water: Shell Oil and
the Reform of Federal Offshore Oil Leasing   123
Tyler Priest

Chapter 6. The U.S. Strategic Petroleum Reserve and
Energy Security Lessons of the 1970s   163
Bruce Beaubouef

Chapter 7. The Development and Demise of the Agrifuels
Ethanol Plant, 1978–1988: A Case Study in U.S. Energy Policy   184
Jason P. Theriot

Chapter 8. Pipe Dreams for Powering Paradise:
Solar Power Satellites and the Energy Crisis   203
Jeff Womack

Chapter 9. The Nuclear Power Debate of the 1970s   221
J. Samuel Walker

Part IV. Demand
Chapter 10. The Consumer's Hand Made Visible: Consumer
Culture in American Petroleum Consumption of the 1970s   257
Brian Black

Chapter 11. Environmentalism and the Electrical
Energy Crisis   283
Robert Lifset

List of Contributors   303
Index   307

Acknowledgments

This volume is the product of a conference held at the University of Houston in November 2007. Sponsored by the Center for Public History and the University of Houston Department of History and made possible by the efforts of Joseph Pratt and Martin Melosi, "Energy in Historical Perspective: American Energy Policy in the 1970s" brought together historians working on these issues since the early 1980s with a group of younger scholars. The Center for Public History at the University of Houston has long served as a vital site for the advancement of energy history. Professors Pratt and Melosi have worked tirelessly, through their teaching and scholarship, to advance the field. The conference and this book would not have been possible without their dedication, hard work, and energy.

Abbreviations

AEC	U.S. Atomic Energy Commission
AIOC	Anglo-Iranian Oil Company
API	American Petroleum Institute
BLM	U.S. Bureau of Land Management
CAFE	Corporate Average Fuel Economy
DOD	U.S. Department of Defense
DOE	U.S. Department of Energy
DOT	U.S. Department of Transportation
ECPA	Energy Conservation and Production Act of 1976
EIA	U.S. Energy Information Administration
EPA	U.S. Environmental Protection Agency
EPAA	Emergency Petroleum Allocation Act of 1973
EPCA	Energy Policy Conservation Act of 1975
ERDA	Energy Research Development Administration

FEA	Federal Energy Administration
FPC	U.S. Federal Power Commission
IEA	International Energy Agency
MAFLA	Mississippi, Alabama, Florida region
MMS	U.S. Minerals Management Service
NASA	National Aeronautics and Space Administration
NEPA	National Environmental Policy Act of 1969
NHTSA	U.S. National Highway Transportation Safety Administration
NRC	U.S. Nuclear Regulatory Commission
NSC	U.S. National Security Council
OAPEC	Organization of Arab Petroleum Exporting Countries
OCS	Outer Continental Shelf
OPEC	Organization of the Petroleum Exporting Countries
SALT	Strategic Arms Limitation Treaty
Shell E&P	Shell Oil Exploration and Production division
SPR	Strategic Petroleum Reserve
SPS	solar power satellite
UCS	Union of Concerned Scientists
USDA	U.S. Department of Agriculture

American Energy Policy in the 1970s

Introduction

Robert Lifset

O n a cold January morning in 1974, Mary Korechoff swung her delivery van in front of Frank Knight's blue paneled truck. They were waiting on a gas line at a Mobil station on Atlantic Avenue in Brooklyn, New York. Mr. Knight, who used his truck to deliver rugs, was two hours behind schedule and had been waiting in line for forty-five minutes.

"The hell with her," he said as he swung his truck onto the sidewalk and after maneuvering between some parked cars cut back into his rightful place in line.

Ms. Korechoff, a carpenter also making deliveries, was unapologetic. "In my case it's either gas or welfare. I need the gas and I need the time."[1]

For Mary Korechoff, Frank Knight, and many other Americans in the 1970s, there was a deep-seated anger and frustration at their inability to secure adequate supplies of a commodity that many had never expected to become scarce. Mary Korechoff and Frank Knight were experiencing the consequences of America's dependence on oil.

A volume on energy policy in the 1970s needs to acknowledge that the central effort of energy policymaking in that decade was an effort to address the energy crisis. For many Americans the long gas lines that formed during the winter of 1973/74 and the spring and summer of 1979 defined the energy crisis of the 1970s as an oil crisis. In both of these cases, events in the Middle East curtailed the world oil supply while price controls in the United States interfered with the markets' ability to supply consumers.[2]

American consumers also found themselves at the mercy of events in the Middle East because of a large and significant shift in the domestic oil market. In 1950 the United States produced 5.9 million barrels of oil per day (bpd). Twenty years later U.S. oil production peaked at 11.6 million bpd. Over that same period of time, U.S. oil consumption rose from 6.4 million to 14.6 million bpd. In 1950 the United States imported 5.5 percent of the oil it consumed; twenty years later that figure was 21.5 percent. During that period, oil production doubled, but consumption nearly tripled.[3]

The rise of suburbs in the decades after World War II had a profound impact on the demand for oil. This suburbanization produced a growing reliance on gasoline-guzzling automobiles and energy-consuming single-family homes. Petroleum used for transportation grew between 1950 and 1970 by 131 percent. Significant growth of oil consumption in commercial (85 percent), industrial (109 percent), and residential (114 percent) use, coupled with a peaking of domestic oil production in 1970, fueled America's growing reliance on foreign oil.[4]

A crisis within the utility sector of the energy economy complicates our understanding of the energy crisis of the 1970s as an oil crisis. Since the early twentieth century the utility industry relied upon a business model in which increasing economies of scale and improving thermal efficiency allowed companies to meet electrical demand while lowering prices (thermal efficiency is the percentage of a fuel's energy content actually converted into electricity). Thermodynamic theory limits thermal efficiency in a steam-generated power plant to 48 percent. In the 1970s the nation reached a technological barrier as manufacturers began to discover that improving thermal efficiency began to produce diminishing returns, with metallurgical problems appearing at around 40 percent. Manufacturers and utilities learned that less efficient plants could be run more reliably. Hoping to overcome the decline of thermal efficiency improvements and meet increasing demand, utility companies tried to build larger power plants. Lacking the time to test and slowly introduce larger turbines, manufacturers extrapolated from existing designs and produced equipment that frequently broke down. In addition to these problems, the amount of oil used by the utility industry had been slowly growing in the postwar years such that by 1970 oil was used to produce 13 percent of America's electrical generation.[5] As a result, higher oil prices played a role in fostering a crisis within the utility sector (this is discussed in greater length in chapter 11).

American policymakers and journalists were aware of the trends within the oil market years before the first gas lines materialized. The Nixon administration focused its attention on energy issues when, in 1969, it began to consider abolishing the oil import quota. President Eisenhower had established this quota in 1959 to protect the domestic oil industry from international competition. Nixon's labor secretary, George Schultz, released a report in early 1970 that found that the quota, which blocked access to cheap foreign supplies, was costing Americans $5 billion a year.[6] But the report noted that dropping the quota would allow oil imports to reach 51 percent of total oil consumption by 1980. Shortly after Nixon's reelection in the fall of 1972, energy analysts in the Interior Department warned the White House that if it took no action there would be widespread gasoline shortages by the following summer.[7] In response, Nixon ended the import quota in the spring of 1973, but the embargo of world oil supplies by the Organization of Arab Petroleum Exporting Countries (OAPEC) in the fall resulted in the gasoline shortages that Mary Korechoff and Frank Knight experienced in January 1974.

A good deal of energy was expended attempting to address this energy crisis. In an effort to understand these efforts, this volume is divided into four parts. It begins with an examination of political leadership, followed by foreign policy issues, before turning its attention to a series of essays focusing on efforts to increase supply and reduce demand.

Part I: Political Leadership

Three presidents and five congresses crafted energy policy in the 1970s. Overall, those policies focused on increasing the energy supply while reducing demand. However, the complex and tangled nature of energy politics complicated the policymaking process, pitting energy-producing regions against energy-consuming regions and Republicans against Democrats. It also involved significant implications for both foreign policy and the domestic economy. Energy bewildered and confused many Americans. They had come to believe that cheap and plentiful energy was a birthright.[8] By the 1970s, however, the energy crisis challenged this deeply held belief. This challenge took place at a time when many Americans were losing faith in key institutions in society, including their political leaders. This profound skepticism shaped the way they viewed the energy crisis. A majority of Americans were unaware that the country imported oil. When prices rose and oil company profits soared, a Roper poll conducted in 1974 revealed that 73 percent of respondents believed that there was no shortage of oil. Many Americans believed the energy crisis was a fraud perpetuated by the oil companies to manipulate and increase prices.[9]

In November 1973, President Nixon famously called for the United States to become energy independent by the end of the decade. Though many of his advisors found this implausible, nearly every president since has either renewed that pledge or called for reducing America's dependence on foreign oil imports. It should be noted that since the United States has never imported a significant amount of electricity, coal, or natural gas, calling for energy independence is really a call for eliminating the nation's dependence on foreign supplies of oil. To fulfill his pledge Nixon increased funding for energy research and development (independence was to be achieved through new technology), supported some conservation measures (e.g., reducing highway speed limits, asking Americans to use mass transit and car pools, reducing air traffic, lowering thermostats), and leaned on Saudi Arabia to temper the more radical demands of its OPEC partners (i.e., to resist calls for larger production cuts and switch Arab financial reserves out of dollars) while seeking a diplomatic solution to the embargo.

Although Nixon supported legislation exempting the Alaskan pipeline from environmental litigation that had prevented its construction, he also charted a role for the federal government in the energy economy more invasive than anything since the early New Deal.[10] In 1971, Nixon imposed price controls on oil and gas. Two years later his administration made those controls permanent with the passage of the Emergency Petroleum Allocation Act of 1973. Despite the market-distorting effect of price controls, the impact of the embargo was real. By February 1974, U.S. oil imports had declined by 1.2 million bpd in just five months—a 19 percent reduction. It took both of these developments (price controls and reduced imports due to the embargo) to produce the oil crisis experienced by Mary Korechoff and Frank Knight.[11]

In the fall of 1974, President Ford brought a new approach to the White House energy policy. Ford continued Nixon's policy of increased funding for energy research but rejected his predecessor's reliance on price controls. He was convinced that the energy crisis could be resolved only if prices were allowed to rise to their natural market-driven price. In chapter 1 of this volume, "'The Toughest Thing': Gerald Ford's Struggle with Congress over Energy Policy," Yanek Mieczkowski provides a detailed examination of the Ford administration's efforts to end price controls. Ford pressed for free-market solutions—policies that Carter and Reagan would implement. But Ford also supported and initiated policies that sought to advance energy conservation, most notably the right-on-red laws and Corporate Average Fuel Economy (CAFE) standards (CAFE standards are the sales-weighted average fuel economy of a manufacturer's fleet of current model year passenger cars or light trucks, manufactured for sale in the United States). Indeed, CAFE standards and the strategic petroleum reserve (discussed in chapter 6), both signed into law by President Ford, are among the more lasting and influential energy policies of the era. Mieczkowski argues that Ford worked hard (but with limited success) to play a consensus-making role to reduce the bitter partisanship in Washington generated by Watergate and Vietnam.

President Carter's energy policies placed a greater emphasis on conservation and support for alternative forms of energy. Carter proposed tax breaks to encourage alternative energy (a case study of the effort to support ethanol is discussed in chapter 7, and the support for a particular form of solar power is discussed in chapter 8). He proposed increasing CAFE standards above 27.5 miles per gallon after 1985 and establishing new standards for light trucks. Like Ford, Carter supported the expansion of nuclear power (discussed in chapter 9) and a new tax on gasoline. Unlike Ford, Carter put off the deregulation of oil and gas prices. He felt doing so would allow OPEC to set domestic oil prices, allow the oil industry to garner windfall profits, and increase inflation. Carter agreed with Ford that prices were too low, but his solution called for regulation that would allow for a gradual increase in prices.

Carter's policies found expression in the National Energy Act of 1978, which provided tax incentives for alternative energy sources and gasohol, allowed for higher natural gas prices, created incentives for conservation, and empowered the energy secretary to order power plants to switch from using oil when feasible. It did not include new, tougher CAFE standards after 1985 or a new tax on gasoline.[12]

Although sharp ideological distinctions might describe American energy policy in the early twenty-first century, Jay E. Hakes argues in chapter 2, "Conflict or Consensus? The Roots of Jimmy Carter's Energy Policies," that there was a great deal of consensus and continuity between Nixon, Ford, Carter, and the Congresses they worked with on energy policy. In examining the overlapping personnel responsible for making energy policy, the creation of the Department of Energy, debates over price controls and windfall profits, nuclear policy, and the support for alternative energy technologies, Hakes finds broad support in Washington for an activist government. To suggest otherwise is to see the 1970s through the prism of the changed ideological assumptions of the 1980s.

Hakes essay underscores the revolutionary nature of the 1980 election that brought Reagan to the White House. Ultimately, an accurate assessment of the energy policies formulated in the 1970s must account for the impact of so many of those policies (support for the development of alternative forms of energy, support for conservation and efficiency measures, the connections between energy and foreign policy) reversed and ridiculed in the following decade.

Part II: Foreign Policy

A meteoric rise in oil consumption coupled with a peaking in domestic production produced the oil crisis; and a foreign cartel of oil-producing nations was responsible for the price spikes and reduction in available supply. How American foreign policy leaders dealt with this crisis is the subject of chapter 3, David S. Painter's "From the Nixon Doctrine to the Carter Doctrine: Iran and the Geopolitics of Oil in the 1970s." In this essay, Painter argues that diplomatic historians writing about the postwar period have largely ignored oil, despite the fact that most of the major postwar foreign policy doctrines (those of Truman, Eisenhower, Nixon, and Carter) relate to the Middle East and its oil. Painter analyzes the role of oil in world politics by focusing on three issues: (1) the withdrawal of the British from the Middle East, and the American decision under the Nixon Doctrine to view Iran as the guardian of its interests in the region; (2) how changes in the world oil economy influenced changes in the global balance of power, with the Middle East emerging as an important source of the world's oil supply at the moment the United States became highly dependent on foreign oil; and (3), how concerns about cold war oil supply shaped the Carter Doctrine. According to Painter, the Iranian revolution highlighted a central weakness of the Nixon Doctrine (in which a local ally tasked with promoting stability suddenly adopts a different agenda). The loss of Iran as an ally coupled with the Soviet invasion of Afghanistan pushed the United States to assume a more active military posture in the region. Painter argues that the strategic and economic benefits of controlling the world oil supply deeply affected perceptions of Soviet intentions in the region and shaped U.S. foreign policy in the 1970s.

Writing oil into American foreign relations highlights the strands of continuity in pre–and post–cold war foreign policy. It also does much to explain how in the early twenty-first century Americans find themselves with large and long-term military commitments in the Middle East.

Steve Marsh also highlights how the United States has used military force to secure Middle Eastern energy supplies in chapter 4, "The United States, Iran, and the 'Oil Weapon': From Truman to George W Bush." This military dependence is a sign of growing U.S. weakness in the global economy, as the nation has become more vulnerable to Iran's strategic use of oil. Marsh argues that this vulnerability is the product of processes begun or accelerated in the 1970s, including the weakening position of the international oil companies, the fewer options available to Western nations to isolate and coerce oil producers into line, and the U.S. dependence on market forces to deal with oil shocks. Coupled with the failure to develop a policy to address energy supply and type (and the demand side of the equation), embracing market forces served to normalize energy price volatility. Marsh concludes that these developments serve to constrain America's power and options in meeting the foreign policy challenges posed by Iran in the twenty-first century.

Part III: Supply

The resourcefulness and ability of the oil industry to find new supplies of oil was tested in the 1970s like never before. The results were nothing less than spectacular. The oil industry's aggressive response to the energy crisis displayed, in the words of President Nixon, "the spirit of Apollo, with the determination of the Manhattan Project." Large investments, advances in technology, and sustained effort most noticeably transformed the Gulf of Mexico into a world-class oil-producing region.

In chapter 5, "Diving into Deep Water: Shell Oil and the Reform of Federal Offshore Oil Leasing," Tyler Priest traces the longer history of this project through a case study of the Shell Oil Company's pioneering efforts in offshore exploration in the postwar period. Shell led the industry in extending the art of the possible in offshore drilling. In the late 1950s, deep offshore consisted of drilling in 300 feet of water; twenty-five years later Shell was conducting experimental drilling in over 6,000 feet of water. The rising cost of oil produced by the energy crisis encouraged Shell and others to venture farther offshore into the Gulf of Mexico and to explore the coast of Alaska. Priest also describes the technological and political challenges the company faced while drilling in the Atlantic and Pacific. It is useful to note that, although the company spent billions acquiring leases and drilling, with the exception of one project off the coast of Los Angeles it came up with nothing but dry holes.

The Gulf of Mexico was another story. Priest details the technological advances necessary for the company (and industry) to begin producing in several thousand feet of water. The federal government aided the effort to expand offshore exploration and production. Even before the OAPEC embargo, the Nixon administration conducted large auctions of offshore acreage. In the early 1980s, the Reagan administration reformed the leasing system, fundamentally altering the economics of the offshore business. Priest presents the company's perspective on the Reagan administration's controversial "area-wide leasing" program. His essay provides insight into the government's efforts to increase domestic oil supply.

In chapter 6, "The U.S. Strategic Petroleum Reserve and Energy Security Lessons of the 1970s," Bruce Beaubouef describes the historical roots of the preeminent emergency energy policy tool to survive the 1970s: the strategic petroleum reserve (SPR). The SPR is a wholly government-owned reserve of crude oil stored in large salt caverns along the Gulf of Mexico in Louisiana and Texas. The reserve (at 695 million barrels as of April 2013) is designed to serve as a defense against a sudden disruption in supply.[13]

Beaubouef argues that stockpiling oil was one of several policies the federal government pursued in response to the energy crisis of the 1970s. However, unlike the efforts to accelerate domestic production and promote conservation, the SPR has succeeded because it is a passive, supply-side energy tool. The SPR enjoyed support across the political spectrum and became one of the few 1970s energy policies that subsequent administrations of both parties enthusiastically supported.

Beaubouef argues that the reserve is a policy tool designed to protect the consumers and purchasers of petroleum. As such, the SPR is a product of a profound shift in political power from oil producers to consumers. It remains one of the enduring ironies of this era that the very moment the larger culture became fascinated by conspiracy theories focusing on the secretive power of the oil industry is also the moment when that industry suffered a profound loss of political capital, two developments that were clearly connected.

In the 1970s, federal and state governments began to explore, and in some cases subsidize, the production of alternative forms of energy. In chapter 7, "The Development and Demise of the Agrifuels Ethanol Plant, 1978–1988: A Case Study in U.S. Energy Policy," Jason P. Theriot examines the experience of one Louisiana company that attempted to produce ethanol (or gasohol, as it was called at the time) from sugarcane by-products and sweet sorghum. The project had much to recommend it and received generous subsidies from the State of Louisiana and the federal government. Yet the declining price of oil in the 1980s and the uneven, uncertain, and declining political support eventually drove the company into bankruptcy.

This story serves to underscore how the dominance of oil and problems faced by alternatives are directly related to the wild swings in oil prices over the past several decades. Arguments pointing to the decline of an earlier generation of ethanol as a natural outcome of market forces must consider that there is nothing "natural" or "free" about a market influenced by a large and powerful international oil cartel such as OPEC. The inconsistency of state and federal energy policy was a reaction to these declining prices. Scrapping gasohol subsidies in the 1980s might have made sense in the short term, but the recent past suggests that an energy policy wholly reliant on an idealized free market places tremendous burdens on our national security apparatus to guarantee the security of supply.

There once existed an effort to confront the energy challenges of the 1970s so fantastic it could only be sketched by a science fiction writer. Indeed, Jeffrey Womack credits Isaac Asimov with the idea of satellites beaming solar-generated power back to the planet. In chapter 8, "Pipe Dreams for Powering Paradise: Solar Power Satellites and the Energy Crisis," Womack describes how this plan attracted the interest of the federal government and why advocates of solar power rejected it. Womack argues that the fate of this program and its periodic resurgence in the popular press provide important insights into the nation's perception of solar power and technological optimism. The technological, economic, and environmental problems with this program serve to highlight how the hidden (or in this case nearly hidden) costs can serve to derail plans for any energy transition.

In chapter 9, "The Nuclear Power Debate of the 1970s," J. Samuel Walker takes us back inside the controversies that helped to explain why a new nuclear power plant was not licensed between 1978 and 2012.[14] Walker argues that the energy crisis dimmed the prospects of nuclear power by simultaneously draining utilities of cash (because of increased fuel costs) and reducing demand (because the price of electricity increased). Rising inflation and interest rates presented the industry with the nearly insurmountable problem of financing these capital-intensive power plants. Even without these problems the industry was running into new headwinds generated by a determined and growing anti-nuclear movement. Walker examines how and why nuclear power became controversial. In so doing, he reveals that this story is about more than simply a rising environmentalism flexing its muscles. Rather, we see here a growing debate about the safety of nuclear power that took on emotional and religious overtones because the issues (the effects of low-level radiation, the probability and consequences of a severe accident, the dangers of radioactive waste disposal, the level of threat posed by terrorists, the cost and reliability of nuclear power) were themselves the subjects of dispute among experts.

Part IV: Demand

Brian Black's "The Consumer's Hand Made Visible: Consumer Culture in American Petroleum Consumption of the 1970s" (chapter 10) argues that the United States began, ever so slightly, to shift away from a petroleum-based economy in the 1970s. That decade brought a new level of scrutiny to how we acquire and use energy. Part of the impetus for this shift came from the environmental movement, which provided the intellectual ammunition for changing consumption patterns through conservation and increased efficiency.

Black traces this shift by exploring the recent history of automobiles. The law establishing CAFE standards created the loophole that allowed automobile manufacturers to build the light trucks and sport utility vehicles (SUVs) that by the early twenty-first century would nearly dominate the market. By increasing the weight of their vehicles, and thereby reducing their efficiency, manufacturers could escape the intent of the CAFE standards. However, at this very moment normal-size and smaller cars were becoming increasingly fuel efficient (and dominated by foreign automakers), creating a truly bifurcated automobile market. Detroit built the larger vehicles because people wanted to drive them. Yet Black argues that this consumer preference was itself a form of denial—an unwillingness to recognize the increasing problems of sustaining the petroleum age. Black finds that, since the economic collapse of 2008, American automobiles have begun once again to reflect an ethic first found in the 1970s. In 2011, 70 percent of hybrid automobiles sold globally were made in the United States; between 2005 and 2012, hybrid and electric car sales grew by an estimated 268 percent. Consumer preferences for vehicle muscle and size that reemerged between 1990 and 2005 appear to be fading.

In chapter 11, "Environmentalism and the Electrical Energy Crisis," I examine both the roots of the crisis within the utility sector of the energy economy and the impact of environmentalism on the industry. In the 1960s, environmentalists gained new tools. One of these new tools was gaining access to the federal courts. The ability of environmentalists to delay projects by using the courts drew the power to effect land-use decisions down to the grassroots level, challenging the utility sector and simultaneously changing the nature of environmentalism. One response of the utility industry to this challenge was to accept the critique long made by environmentalists and adopt conservation measures to reduce the demand for electricity.

. . . .

This volume contains an eclectic collection of essays that revisit the last period in which energy issues rose to the forefront of the nation's political discourse. Exploring this historical record provides us with valuable contextualization and perspective that should enlighten the discussion of contemporary energy issues.

Broad synthetic surveys of American energy history include Martin Melosi's *Coping with Abundance*, Richard H. K. Vietor's *American Energy Policy since 1945*, and David Nye's *Consuming Power*. Fiona Venn's *The Oil Crisis* examines the OAPEC oil embargo from the perspective of diplomatic history and does not provide an in-depth study of how the crisis impacted the domestic policy and politics of Western nations. Many of the contributors to this volume (Beaubouef, Hakes, Mieczkowski, Priest, and Walker) have recently published books that address some aspect of the energy crisis, but there remains no single volume history of the energy crisis of the 1970s.[15]

Several themes emerge from this collection of essays. First, energy policy-making has not been immune to the law of unintended consequences. Price controls, though only partially designed to address an energy concern, did discourage production and conservation. Loopholes in the CAFE standards did help generate a bifurcated automobile market. The overthrow of Mohammed Mossadegh and the later selection of Iran as the United States' chief political ally in the Arab Middle East did contribute to rising producer nationalism and U.S. vulnerability to the "oil weapon."

Second, efforts to produce a significant change in the country's energy supply required a sustained political commitment. Ethanol and solar power did not achieve this, whereas nuclear power, offshore oil, and the SPR benefited from consistent support in high and low price environments.

Third, the 1970s witnessed a significant increase in the political power of energy consumers. None of the changes implemented—including CAFE standards (a conservation measure that inconvenienced manufacturers and not consumers), the SPR, the growing power of environmentalists to influence the siting of power plants, right-on-red laws, and gasoline additives—called for consumers to sacrifice. Policies that did call for sacrifice, such as the removal of price controls, the expansion of nuclear power, and expanded offshore oil exploration and development in the Atlantic and Pacific Oceans, all met stiff public resistance.

Finally, this volume supports the view that there was far more consensus in 1970s energy policymaking than conventional wisdom suggests. Certainly

from the perspective of the Oval Office there is nothing inherently contra-
dictory in seeking to increase domestic supply (through nuclear power or
offshore oil), subsidizing alternatives (solar, gasohol), and ramping up conservation and efficiency efforts while seeking to defend American influence
at the site of significant world oil production (Carter Doctrine). Of course,
this interpretation downplays the centrality of price controls and the impact
the struggle over this issue had on energy policymaking. The history offered
in this volume suggests that it is a mistake to allow this single issue to distract
us from recognizing the broad outlines of a consensus.

Notes

1. "Drivers Waiting to Fill Empty Tanks Pour Out Their Many Tales of Woe," *New York Times,* Jan. 4, 1974.

2. There exists a common misperception that the oil embargo in response to the Yom Kippur war was organized by OPEC. In fact it was the Arab members of OPEC—OAPEC—that deliberately disrupted the international supply of oil in the service of political and economic objectives. Many OPEC members, such as Iran and Venezuela, did not participate in the embargo. On the role that events in the Middle East played in sparking an OAPEC oil embargo , see Fiona Venn, *The Oil Crisis* (London: Longman, 2002), 7–32; on the role played by price controls, see Daniel Yergin, *The Prize, The Epic Quest for Oil, Money and Power* (New York: Free Press, 1991), 692; a far more detailed examination of this issue can be found in Philip K. Verlerger et al., "The U.S. Petroleum Crisis of 1979," *Brookings Papers on Economic Activity* 1979, no. 2 (1979), 463–76.

3. Table 5.1a: Petroleum and Other Liquids Overview, Selected Years, 1949–2010, *Annual Energy Review 2010* (Washington, D.C.: U.S. Energy Information Administration, 2011). In 2010 the United States imported 49 percent of oil it consumed.

4. Table 5.13a: Petroleum Consumption Estimates: Residential and Commercial Sectors, Selected Years, 1949–2010, and Table 5.13c: Petroleum Consumption Estimates: Transportation Sector, Selected Years, 1949–2010, *Annual Energy Review 2010.*

5. Table 8.4b: Consumption for Electricity Generation by Energy Source: Electric Power Sector, 1949–2010, *Annual Energy Review 2010.*

6. See United States Cabinet Task Force on Oil Import Control, *The Oil Import Question: A Report on the Relationship by Oil Imports to the National Security* (Washington, D.C.: Government Printing Office, 1970).

7. Jay Hakes, *A Declaration of Energy Independence: How Freedom from Foreign Oil Can Improve National Security, Our Economy, and the Environment* (Hoboken, N.J.: Wiley, 2008), 21.

8. On this point, see David Nye, *Consuming Power: A Social History of American Energies* (Cambridge: MIT Press, 1999).

9. See Eric R. A. N. Smith, *Energy, the Environment, and Public Opinion* (Plymouth, U.K.: Rowman and Littlefield, 2002), 25.

10. On the Alaskan pipeline, see Peter A. Coates, *Trans-Alaskan Pipeline Controversy: Technology, Conservation, and the Frontier* (Fairbanks: University of Alaska Press, 1993); on the increasing involvement of the federal government in the energy economy in the 1920s and '30s, see Gerald D. Nash, *United States Oil Policy, 1890–1964: Business and Government in Twentieth Century America* (Westport, Conn.: Greenwood, 1968); John G. Clark, *Energy and the Federal Government: Fossil Fuel Policies, 1900–1946* (Champaign: University of Illinois Press, 1986); Steve Isser, *The Economics and Politics of the United States Oil Industry, 1920–1990* (New York: Garland, 1996), 3–60.

11. Hakes, *Declaration of Energy Independence*, 35.

12. The National Energy Act of 1978 included many more policies than those listed here and consisted of the following statues: Public Utility Regulatory Policies Act (Pub.L. 95–617), Energy Tax Act (Pub.L. 95–618), National Energy Conservation Policy Act (Pub.L. 95–619), Power Plant and Industrial Fuel Use Act and the Natural Gas Policy Act. (Pub.L. 95–620), and Natural Gas Policy Act (Pub.L. 95–621).

13. Data on the Strategic Petroleum Reserve can be found in Table 5.17: Strategic Petroleum Reserve, 1977–2011, *Annual Energy Review 2012*. Due to the lag time involved in publishing the *Annual Energy Review*, the most up to date information on the strategic petroleum reserve can be found here: www.eia.gov/dnav/pet/pet_stoc_typ_d_nus_sas_mbbl_m.htm.

14. Although there was a long pause in the issuing of construction licenses for new nuclear power plants, fifty-one of the nation's 104 nuclear reactors became operational since 1980. Roughly half of the nation's nuclear power plants were initially licensed in the 1970s but built in the 1980s and '90s.

15. Hakes, *Declaration of Energy Independence*; Yanek Mieczkowski, *Gerald Ford and the Challenges of the 1970s* (Lexington: University Press of Kentucky, 2005); Tyler Priest, *The Offshore Imperative: Shell Oil's Search for Petroleum in Postwar America* (College Station: Texas A&M Press, 2009); J. Samuel Walker, *Three Mile Island: A Nuclear Crisis in Historical Perspective* (Berkeley: University of California Press, 2006).

Part I

Political Leadership

Chapter 1

"The Toughest Thing"

Gerald Ford's Struggle with Congress over Energy Policy

Yanek Mieczkowski

On July 22, 1975, the presidential yacht *Sequoia* took an evening cruise on the Potomac River. Aboard were President Gerald Ford and First Lady Betty Ford, Vice President Nelson Rockefeller and his wife Happy, top administration advisors, and the most powerful men in Congress and their wives, including Senate Majority Leader Mike Mansfield, Minority Leader John Rhodes, House Speaker Carl Albert, and House Majority Leader Thomas P. "Tip" O'Neill.[1]

Aboard the 108-foot vessel, the gregarious Ford enjoyed drinks, conversation, and the company of his former congressional colleagues. But he had a firm purpose. After a deadlock on energy policy that had dragged on for half a year, Ford took the congressmen out to exert friendly persuasion on formulating national energy policy. Ford had pressed Congress on the issue, and he indicated publicly that he was "ready to go halfway" in seeking compromise.[2]

Among the administration members who accompanied the president that evening was Frank Zarb, a New Yorker who, as a young man, had once planned to become an aircraft mechanic. He ended up working in investment banking instead and then joined the government in 1971. In late 1974 he became head of the Federal Energy Administration (FEA).

Zarb recalled that in energy policy the Democratic members of Congress "all agreed that what we [the Ford White House] were proposing was intelligent, but politically dynamite within their own party. . . . We did try the *Sequoia* [cruise] to try to persuade them. I don't think we made any progress that night."[3]

When Ford took office in August 1974, memories of energy short-ages and the Arab oil embargo were raw. "There was no question that in 1974–75, we faced a serious energy problem," President Ford recalled. "Nuclear energy was just beginning to be a factor. Hydroelectric power was at its peak. . . . Oil and gas were having troubles. Domestically, we were so much more dependent on foreign oil. We had a real crisis, so it was important that we make some headway, not only in conservation but in more production."[4]

To forge viable policy, Ford had to secure cooperation from Capitol Hill. His friendships there were warm and mutual, and Ford also deeply revered the institution of Congress. Representing Michigan's Fifth District for twenty-five years, he became House minority leader in 1965 and served in that post until President Richard Nixon picked him as vice president in 1973 following Spiro Agnew's resignation. Though Ford still adored the legislative branch, tussles like that over energy policy changed his perspective. A year after he took office, a reporter asked him, "What is the toughest thing that you are doing in Washington?" Ford replied that it was trying to get Congress to pass an energy program. The struggle proved frustrating, but it enabled Ford to lay the groundwork for national energy policy and rebuild executive-legislative ties that the Nixon administration had ruptured.

Ford, Congress, and the 1970s Energy Crisis

One of Gerald Ford's great legacies as president was mending relations between the White House and Capitol Hill. The mercurial Nixon believed he could shut Congress out of governing the nation and often refused to see lawmakers. Legislative aide Max Friedersdorf recalled that Nixon "much preferred that [the congressional liaison staff] wrap it up in a memo . . . and he would act on the memo without ever seeing anybody. I think if we left him to his own devices he would never, ever see a congressman or senator."[5]

Ford provided the consummate contrast. He had enjoyed tremendous popularity in the House, and members of Congress spoke fondly of dinners at "Jerry and Betty's" house in Alexandria, Virginia, and admired the handsome brood of four children that Ford and his wife raised. In 1965 he won election as House minority leader largely because his colleagues liked him so much. After Nixon nominated him as vice president, Ford won confirmation by overwhelming majorities (387–35 in the House, 92–3 in the Senate). As vice president he focused on his greatest task, saying, "Working with Congress will be my major responsibility, as I see it."[6]

Ford continued that work after becoming president. Democratic congressman James Hanley of New York remembered Ford as "a very open person who . . . tried hard to maintain an open door and enjoy good relations with the members [of Congress]."[7] After Ford retired from public life, the House published a flattering book of tributes from his former colleagues. The praise—and his record—suggests Ford might have enjoyed the best relations between a president and Congress in the post–World War II era.

Ford's congenial congressional relations made the energy policy fight remarkable. At times it became a heated battle, with harsh words flying between the Oval Office and Capitol Hill. House majority leader O'Neill, a Massachusetts Democrat who spent many friendly hours golfing with Ford, "used to give me hell" on energy proposals, Ford recalled. The sharp clash reflected the wide chasm between Ford and Democrats on energy policy.

The fight had its roots in conflicting economic philosophies. Growing up in Grand Rapids, Ford was raised in an environment that stressed almost Calvinistic values, including frugality. His parents told him of a "peacestone" that frontier families placed above the fireplace mantle when they paid off their home mortgages, indicating that a house was truly theirs.[8] Ford's stepfather owned and operated a small business, the Ford Paint and Varnish Company, and during the Great Depression the older Ford struggled to operate in the black. Imbued with the ideal of fiscal integrity at the household and business levels, Ford tried to practice it in government, arguing for lower federal spending and balanced budgets. Married to these concepts was the idea of small government. He feared the growth of government, arguing that it smothered individual initiative and warped the economy. Ford often warned, "A government big enough to do everything for you is a government big enough to take everything away from you."[9]

Democrats were less averse to government intervention. During the 1970s, they held the upper hand in Congress, especially after the 1974 midterm elections, when voters disgusted with Watergate vented their frustrations on Republicans. Democrats gained 43 House seats and three more in the Senate. Brimming with post-Watergate confidence and new political power, Democratic members of Congress stood ready to oppose Ford, however much they liked him personally. Heirs to the liberal tradition of Franklin Roosevelt and Lyndon Johnson, many Democrats believed in humane government programs to solve America's problems.

Energy policy furnished a textbook case of the political fight between President Ford and his Capitol Hill friends. In 1971, Nixon had slapped price controls on oil as part of his overall program of wage and price controls designed to slow inflation. Energy markets—which are extremely price sensitive—reacted adversely. By the winter of 1972/73, media nationwide wrote of an "energy crisis." Various localities reported power brownouts, and fuel shortages forced some school districts, such as those in Denver and Des Moines, to shut down for days during the winter season.

The emerging crisis forced the Nixon administration to act. It did, but unwisely. During the summer of 1973 it began a voluntary program to allocate crude oil and refinery products. The intention of these measures was to share scarcity nationwide, but as Nixon's first energy advisor, Charles DiBona, warned in opposing the program, it would pave the way for mandatory allocation, which would be a nightmare.[10]

That nightmare came in October 1973, when the administration announced mandatory allocation of propane, heating oil, jet fuel, and other middle-distillate fuels. The program caused market distortions throughout the country, as the Federal Energy Office, headed by William Simon, a former bond trader who had no experience in the energy field, tried to mete out fuel based on 1972 usage levels. The federal government attempted to reformulate fuel supply chains that had taken decades to establish. The result, in Simon's own words, was "a disaster."[11] Some regions suffered severe shortages, including long lines at gas stations, while others received adequate supplies.

Amid this calamity, the Arab oil embargo struck. In October 1973, when the United States sent aid to Israel during the Yom Kippur War, the Arab nations of the Organization of Petroleum Exporting Countries retaliated by withholding oil from the United States. The sudden shut-off of Middle Eastern oil exacerbated existing shortages and illustrated how much America depended on oil imports from an unstable region.

Panic led to bad legislation. In November 1973, as America floundered during the oil embargo, Congress approved and Nixon signed the Emergency Petroleum Allocation Act (EPAA), which established two tiers for domestically produced oil to hold down prices. One tier, for "old" oil (defined as oil produced before 1973), set a price cap of $5.25 per barrel. "New" oil, made after 1973, was uncontrolled, a move designed to encourage oil companies to explore for oil by profiting more from new discoveries.[12]

The allocation system and the EPAA proved economic catastrophes. Allocations represented rationing at the producer level, and when gas stations exhausted their monthly gasoline rations motorists grew anxious. Their worries led them to store gasoline and top off their tanks even when unnecessary, and they waited in long lines at gas stations—which became the most visible, doleful symbol of the 1970s energy crisis. The EPAA bred shortages of domestically produced oil because oil companies felt less incentive to produce oil, since its price was limited. Fears of energy shortages and long gas station lines added to the decade's woes of high inflation, retreat from Vietnam, and Watergate. These problems made the era the most economically troubled since the Great Depression.

Ford's Energy Program

An economics major at the University of Michigan, Gerald Ford—unlike many presidents—enjoyed delving into the nitty-gritty of economic policy. Alan Greenspan, Ford's chair of the Council of Economic Advisors, wrote that the president "had a sophisticated and consistent outlook on economic policy."[13] An exponent of free markets, Ford knew the deleterious effects of government controls on an economy. By limiting prices and hence profits, controls discouraged producers from supplying goods—as the law of supply and demand predicted. Ford described himself as "an absolute opponent" of wage and price controls. Earlier in his life he had witnessed controls' effects on the economy. "When I was in Congress the first term," he remembered, "we imposed wage and price controls during the Korean War. They were a mistake. And we suffered until we got rid of them."

Despite cries from some members of Congress and the public to reinstate wage and price controls to combat inflation, President Ford ruled them out. He also wanted to eliminate the remaining price controls on oil, later recalling that "in the back of my mind I knew [controls] were wrong, and we just had to find the right time to get rid of them."

Yet what Ford planned was politically unpopular on Capitol Hill. Many Democrats supported oil price controls as a means both to contain inflation and to protect consumers from allegedly greedy oil companies—and thus appear as champions of the people. The stage was set for political confrontation.

As both House minority leader and president, Ford displayed a skill at finding or promoting talented public servants. This ability allowed Ford protégés to leave a mark that transcended his 895 days in the White House. In the House, Ford advanced the careers of Barber Conable, who later became World Bank president, and Bob Michel, who served as minority leader during the 1980s. Ford's White House alumni included Dick Cheney, Alan Greenspan, James Lynn, Paul O'Neill, Donald Rumsfeld, Brent Scowcroft, and William Seidman, all of whom went on to prominent careers in government or industry, or both. Another person whom Ford tapped to serve him was Frank Zarb.

To confront the energy crisis, Ford needed to end the revolving-door pattern characterizing the top spot at the FEA and establish stability at the agency. Nixon's energy advisors included Charles DiBona, John Love, and then Simon, all in quick succession; in late October 1974, Ford fired FEA administrator John Sawhill because he continued to support a gasoline tax despite the president's opposition to it. Ford nominated Zarb as new FEA chief. Zarb, who had good relations with reporters and members of Congress, received quick confirmation and assumed his post in late November 1974. Serving the balance of Ford's term in office, he enjoyed direct access to the president and provided the leadership continuity that the agency needed.

Ford's first stab at energy policy, in an October 8, 1974, address to Congress, was embedded in a numbingly long thirty-one-point program, the nub of which aimed to reduce inflation. These initiatives had little impact. The energy measures, entombed within the anti-inflation proposals, got lost. Although including some concrete proposals like natural gas deregulation, the measures centered mostly on conservation. Ford introduced these initiatives as part of his voluntary anti-inflation campaign designed to spur citizen participation, "Whip Inflation Now" (WIN). He asked Americans to "drive less, heat less," take carpools, and strive toward a national goal of reducing oil consumption by one million barrels a day.[14]

President for just two months, Ford had scrambled to create these new proposals, which helped to explain their tentative nature. Given the time constraints administration members worked under, they fell back on the one safe and sure policy goal—conservation. Congress and the public could easily understand this imperative, and the president could use his office as a splendid (or "bully," the adjective of Theodore Roosevelt's day) pulpit to exhort it on them. "In the short term, conservation is about the only thing you can do," observed Glenn Schleede, the associate director for energy and science in Ford's Domestic Council.[15]

But as Ford grew more ensconced in office, he looked at the long term. In energy, he wanted tougher steps. By late 1974 his advisors raced to reverse policy and stimulate rather than slow down the economy—which had suddenly plunged into a recession—with a tax cut. Ford also formulated a comprehensive energy program that proposed much more than conservation. "The public cooperation effort [to save energy] has not achieved all we feel is necessary," he said. "So there will be stronger measures."[16]

On January 15, 1975, in a nationally televised State of the Union address, Ford announced his tougher stand. He proposed decontrolling oil—that is, removing price controls—by April 1. That meant in less than ninety days. It was a dramatic move, but his advisors estimated that it would yield savings of 850,000 barrels of oil per day by 1977.[17] The price of a barrel of oil could shoot up to the market level of $11 per barrel, spurring inflation and eliciting howls of protest from consumers, yet higher prices would also encourage conservation—and over the long term, prices would stabilize. More important, the incentive of higher prices would stimulate oil companies to explore for and produce more oil. As Ford explained it, his program's objective was "to allow the prices of oil and gas to move higher—high enough to discourage wasteful consumption and encourage the development of new energy sources."[18]

Ford's plan involved a complicated system of taxes and fees. He proposed increasing the national oil import fee from just $0.18 a barrel to $2 to encourage conservation. At the same time, he wanted to impose an excise tax of $2 a barrel, which would raise the price of domestic oil and discourage companies from raising prices to earn more profits. Because the oil import fee would hit the East and West Coasts hardest, since they relied more on imported oil, the excise tax would also spread the burden of cost increases by making regions such as the South pay more for domestically produced oil.[19]

Knowing Congress's sluggishness, Ford had a plan to get his former colleagues to act quickly. He announced that he would impose a dollar-a-barrel fee on imported oil at the beginning of each month—on February 1, March 1, and April 1, 1975—until Congress passed legislation to decontrol oil. It was a clever move that used Ford's knowledge of the legislative process. He later explained that "when that [fee] was imposed, we didn't call it a 'tax.' If it had been a tax, we would have had to get authority from Congress. Because we called it a 'fee,' involving national security," Ford reasoned, he could circumvent congressional approval and use his power as president to impose it.

Altogether, the new taxes would raise an estimated $30 billion. Yet philosophically Ford opposed higher taxes; moreover, taking so much money out of the economy threatened to deepen the economic recession. Thus Ford proposed returning the revenue to the economy through permanent tax cuts and federal entitlement programs.[20]

Ford's State of the Union energy program contained other measures to alleviate the energy crisis. He asked Congress to open special oil fields under government control—the naval petroleum reserves—to more development. To take advantage of coal, the country's most abundant fuel, he proposed converting power plants from oil- to coal-burning. He wished to spur development of alternative energy sources, such as nuclear. His plans to increase conservation included labeling appliances to guide consumers to select energy-efficient models.

Driving Ford's proposals was his fear that America would remain vulnerable as long as it depended on Middle Eastern oil. He later explained that "we were more and more dependent on the Middle East supply of oil. We were being held hostage by the Middle East oil barons." In a more candid moment, he told a reporter, "The Saudis have us by the balls."[21] He announced the ambitious goal of reducing oil imports by a million barrels a day by the end of 1975 and by 2 million barrels a day by the end of 1977. Although the objectives might have been lofty, Ford wanted

Americans to aim high and accomplish something substantive in striving to attain them.[22]

Ford's was a thoughtful program designed both to increase domestic oil production and to encourage conservation. *Business Week* called it "the first comprehensive plan for patching and refilling the nation's leaky oil drum."[23] Indeed, it was the most thorough energy initiative a president had ever proposed. And as a veteran of Congress, Ford understood that if members "don't like what the White House recommends—under our coequal, coordinate branch concept—they have an obligation to come up with something they say is better." Congress had to act on Ford's energy program—or develop its own.

Stalemate

Many members of Congress resisted Ford's program. They objected to his energy taxes, which lacked progressive scaling and would hit lower-income groups harder than wealthier Americans. Not all legislators had backgrounds in economics, and some members of Congress seemed to doubt the principle of supply and demand, arguing that higher prices would fail as an incentive for oil companies to produce more oil. Democratic congressman Bob Eckhardt of Texas revealed his confusion between economic and physical limitations when he said, "A person can pay me to jump 2½ feet for $5, but he cannot get me to jump 6 feet for $15. I just cannot do it. Such is analogous to the theory that you can ring in new oil quickly by merely raising the prices."[24]

FEA deputy administrator John Hill recalled another important reason for congressional opposition to Ford's program: inflation. In 1974 the inflation rate hit the dreaded "double-digit" rate of 11 percent, and rising prices dominated Americans' fears. "Inflation was such a big part of the backdrop here," Hill said. "It really did shape people's views as to whether we should have immediate decontrol or not."[25] Members of Congress feared that, once controls ended, the price of domestic oil would shoot up and send ripples across the inflation-plagued economy. House Speaker Albert predicted that Ford's program would have "an astounding inflationary impact."[26]

The dollar-a-barrel oil import fee also stuck a broomstick into a beehive. Members of Congress were "apoplectic" about the fee, recalled

Robert Nordhaus, a counsel to the House Commerce Committee.[27] After Watergate, legislators were sensitive to any hint of presidential abuse of power and reacted indignantly to actions they considered beyond the pale. The fee represented "unbridled discretionary authority," charged Democratic senator Henry Jackson of Washington. By overwhelming margins, the House and Senate approved a measure to suspend the president's power to levy the fee.[28]

Ford backed down. Fearing an override if he vetoed the legislation blocking the fee, he decided to postpone imposing the March 1 and April 1 fees. The breathing space would clear the air and allow Congress to act on Ford's program or develop an alternative. Meanwhile, though, eight governors of northeastern states filed legal action to deny the president's authority to impose the fee, arguing that it invaded a congressional prerogative, the power to tax.

The negative reactions frustrated Ford. After he unveiled a comprehensive, 167-page energy program, Congress only drafted four pages of legislation to deny his power to impose a dollar-a-barrel fee. "Now, it seems to me that the American people want something that is a plan for forward-moving action rather than a four-page bill to move backward," he commented.[29]

Ford postponed the next dollar-a-barrel import fee, scheduled to take effect May 1. But at the end of that month *Business Week* reported that a "massive amount of work still remains to be done by Congress on energy legislation."[30] Ford's frustrations reached a boiling point. On May 27, he gave a televised national address from the Oval Office in which he chastised legislators. Ripping pages off a large calendar next to his desk, he asked, "Now, what did Congress do in February about energy? Congress did nothing—nothing, that is, except to rush through legislation suspending for 90 days my authority to impose any import fees on foreign oil." Telling viewers he delayed the oil import fee three times to give Congress time to work, he asked, "What did the Congress do in March? What did the Congress do in April about energy? Congress did nothing." After pulling the page for May off the calendar, Ford ruefully noted that for four months Congress just dawdled on energy policy. He announced that on June 1 he would impose the second dollar-a-barrel oil import fee.[31]

Thus, as the summer of 1975 began, energy policy in the nation's capital seemed stalemated.

Congressional Sluggishness

Perhaps the confrontation over energy was predictable. House Speaker Albert of Oklahoma commented that "the so-called energy crisis is the most difficult, controversial and emotion-laden problem I've faced since I've been in Congress."[32]

Northeasterners had different concerns from Americans in oil-producing states in the Midwest. House members and senators from the latter regions wielded considerable power, including Democratic senator Russell Long of Louisiana, chair of the Finance Committee. Whereas the Northeast and upper Midwest imported most of their oil, states along the Gulf Coast supplied their own, and the latter chided the former for Yankee wastefulness. "Let the bastards freeze in the dark" was a popular bumper sticker in southern oil-producing states. Democratic majority leader O'Neill rued energy as "the most parochial issue ever to hit the [House] floor."[33]

Factors besides geography explained the energy stalemate. The nature of the 94th Congress, which began its term in January 1975, worked against the speedy action that Ford desired. Disorganization and delay often mark the first months of a new congressional session. Max Friedersdorf observed that legislators get new committee assignments and agendas, and "when you have a new Congress like [the 94th], and they have to reorganize and get their ducks in a row, they're not going to do much for the first six months."[34] The 94th Congress welcomed the largest freshman class since 1949, and the newcomers muscled their way into Capitol Hill aiming to challenge existing power structures. The new legislators had a rebellious streak to them, and their leaders faced difficulty in making them toe the party line. Democratic congressman John Dingell of Michigan, who chaired an energy subcommittee, said that the new representatives often proved intractable.[35]

The parceling of Ford's program added to delays. During the 1970s, a proliferation of committees and subcommittees sprung up, blurring jurisdiction on issues. House minority leader John Rhodes complained of "jurisdictional confusion on energy committees" and "too many cooks in the kitchen," which hindered progress on energy. President Ford quipped that, when his energy proposals hit Capitol Hill, they got chopped up into "535 energy programs."[36]

The confusion led to sluggishness. A slow-moving body to begin with, Congress was even more slothlike on energy. This issue was new and involved economic principles and an understanding of energy markets; legislators had trouble grasping its complexities and reaching consensus. Ford noted, for example, that Congress took four years to approve Alaska oil pipeline legislation, one of the most vital energy bills of the twentieth century.[37]

From an economic standpoint, price controls brought out the worst in politicians. When Democratic senator Birch Bayh of Indiana ran for reelection, his campaign literature touted his votes "to roll back the price of petroleum products. He believes it is wrong for the large multi-national oil companies to make record profits while we consumers pay ever higher prices for gasoline, heating oil, plastic, synthetics, and other petroleum-based products."[38] That was the classic pose of the politician acting as public defender, battling venal corporations, whereas in reality price controls proved economically disastrous, contributing to an energy crisis.

Yet some members of Congress invoked the populist portrait. Senator Henry Jackson thundered against oil companies' "obscene profits," and some Americans believed that was the issue.[39] An Oregon woman complained to her congressman, "Do the gas companies want more money? Do they want the Alaska pipeline bad enough to make everyone miserable by forcing people to stop fighting it? Are they wanting to force the independent dealers out of business?" *Washington Post* columnist Joseph Kraft derided Ford's proposals as "just a cozy arrangement for lining the pockets of the big oil companies."[40]

Ford respected market forces and the law of supply and demand, and he foresaw that oil decontrol would ameliorate energy shortages. Although higher oil prices could result, they were a necessary incentive for companies to produce more oil. Yet distrust of oil companies, dating back to John D. Rockefeller's ruthless reign over the Standard Oil Company, made Ford's program unpopular among members of Congress, who feared economic dislocations—especially higher inflation—and a public backlash. If Ford's program meant higher prices, Congress wanted none of it.

If some legislators disliked or distrusted Ford's proposals, others were apathetic. By 1975 the shock of the fuel shortages during the winters of 1972/73 and 1973/74 had worn off, and fuel prices had stabilized. The sense of crisis faded, and congressional attitudes reflected it. Zarb recalled asking Henry Jackson what could be done to instill a sense of urgency in Congress. Jackson laughed and replied, "Do you know how to make another embargo?" Zarb lamented, "The implication was that, in the absence of a crisis, we weren't going to get public policy moving."

Some congressional Democrats faulted their own behavior. Congressman Henry Reuss of Wisconsin said, "We're saying nuts to the Ford [energy] program, and we should, but we don't have anything of our own." Observers agreed. The London *Economist* commented that on energy "Congress has indeed not done much," and David Broder of the *Washington Post* noted that "there are a few large issues—e.g., energy—where the conscientious members know they have to lead their districts, even if it entails a degree of political risk." Months into the congressional session, Broder wrote, "So far, there is little sign the juniors in Congress recognize that distinction."[41]

Ford was willing to take actions that benefited the nation even if they hurt him politically, a quality uncommon among politicians. The most controversial act of his presidency—pardoning former president Nixon—emanated from his belief that the country needed to move past Watergate and concentrate on urgent problems such as energy and inflation. When faced with difficult decisions, Ford cast polls aside and instead asked himself the litmus test question of what would help America most over the long term. "A politician is interested in the next election," Ford once wrote. "A statesman is concerned about the next generation." Arizona's John McCain, who became a Republican senator a decade after Ford left office, praised him for offering America "the kind of leadership we most needed at the time, the selfless kind."[42]

But in the summer of 1975, most members of Congress thought otherwise. The differing perspectives made energy the tar baby of 1970s policy issues. At the White House, Ford—a star athlete who had played football center in high school and at the University of Michigan—gravitated toward the center of political issues, trying to find common ground. He began to ponder new strategies on energy.

Gradual Decontrol

On Capitol Hill, alternatives to Ford's program failed to win support. A House Ways and Means Committee bill proposed an increase of $0.23 per gallon in the gasoline tax, but the House defeated it. In the Senate, Henry Jackson sponsored a bill to create a byzantine, four-tier system of controls plus mandatory energy conservation. Democratic members of Congress lurched toward the increased government intrusion into the energy marketplace that Ford wanted to avoid.

During the summer of 1975, to encourage a compromise with Congress, the Ford administration began considering phased—rather than immediate—decontrol of oil. The idea appealed to Zarb, and Congress preferred trying gradual decontrol to going cold turkey. Zarb recalled that it "gave the Congress a political way out. It gave the economy a soft landing. It gave the system a chance to adjust. For some of us, it was not the best answer, but it was as good an answer as any."

Thus on July 14—which happened to be Ford's sixty-second birthday—he unveiled a compromise proposal to decontrol oil over a thirty-month period, with controls eased 3.3 percent per month. Just eight days later, the House voted down the proposal. Disappointed, Ford quickly offered a new plan. This version extended decontrol over thirty-nine months; Congress rejected it as well. Zarb remembered, "I was astounded that Congress rejected the thirty-month and thirty-nine-month decontrol [proposals]. It was so logical that they [vote for] it. It was shocking that they didn't do it."[43]

As Congress continued to tarry, a troublesome deadline passed. On August 30, all price controls expired. Members of Congress worried that oil companies would hike prices, sending disastrous ripples across the economy and dampening consumer spending, which could cause another recession. Capitol Hill passed a bill to extend price controls for six months; Ford vetoed it.

The president continued to work toward compromise. He met with Zarb, Senate majority leader Mansfield, and House Speaker Albert, and the four men forged an agreement whereby the two congressional leaders were to secure passage of a thirty-nine-month decontrol program while the president extended controls for forty-five days. Yet Congress refused to support Mansfield and Albert. "I sat in the Oval Office and we all shook hands, Carl Albert and Mike Mansfield," Zarb remembered. "And they went back and they could not deliver. The most astounding public policy lesson I have ever learned. The leadership went back and couldn't control their own people." The incident illustrated the fractiousness of Congress on energy and the need for support far beyond that of just two legislative leaders.

In a show of good faith, Ford agreed to extend price controls until November 15. In that instance, he tried too hard to find common ground with legislators. Members of Congress painted near-doomsday scenarios of a country without oil controls. John Dingell warned of "economic disaster" if controls expired, with higher inflation and an abrupt end to recovery.[44] The media had predicted that crude oil prices could shoot up to $14 a barrel (especially after Ford's import fees took effect).[45] Those Cassandra-like warnings proved false, but in the mid-1970s they were believable because inflation loomed large and the economy wobbled. These were the very controls that Ford had sought to eliminate, yet he agreed to extend them. He feared the economic dislocation and political damage that would result from their absence; still, the best action would have been to let controls expire and end a disastrous chapter in America's economic history.[46]

The Energy Policy and Conservation Act of 1975

On November 12, 1975, Congress finally developed an answer to Ford's energy program when a congressional conference committee produced an omnibus bill, the Energy Policy and Conservation Act of 1975 (EPCA). The legislation differed from Ford's plan in a dramatic and unfortunate way. Initially it would implement more controls on oil by rolling back the price of all domestically produced oil, pulling down the price of "new" oil from $12.50 to $11.28 a barrel and keeping "old" oil at $5.25 a barrel. Oil prices would rise over forty months (with the increases not to exceed 10 percent per year), allowing all oil to be free of controls by May 1979. The president could permit annual increases greater than 10 percent unless either chamber of Congress disapproved.[47]

The bill contained promising conservation measures. One of the most important was the right-on-red traffic law, designed to save the gasoline that motorists wasted while idling at red lights. Since 1937, California had permitted right-on-red turns, and that provision inspired Democratic senator Dale Bumpers of Arkansas, newly elected in 1974, to introduce the measure into the bill. When the White House learned of the measure it initially balked, fearing harm to pedestrians, but it soon warmed to the idea.[48] Zarb recalled that right-on-red would help drive home the point of conservation because it was "something that was visible and that the

American people could touch." To pressure states to enact right-on-red laws, EPCA tied their adoption to federal aid, stipulating that states draw up conservation plans with the new traffic law plus thermal efficiency standards for federal buildings, limits on building lights, and the promotion of public transportation and carpools.[49]

Another significant conservation measure involved pressure on America's marquee industry, automobile manufacturing. A Michigan native sensitive to the industry's concerns, Ford supported a voluntary 40 percent increase in fuel efficiency by 1980. Automakers had resisted mandatory federal standards, arguing that they represented onerous regulation. Moreover, industry officials warned that they lacked the technology to make rapid gains in fuel efficiency within a prescribed time period. Besides, until the 1970s, American consumers seemed to prefer muscular cars with powerful engines. But after the energy crisis, fuel-stingy vehicles came into vogue—especially Japanese imports such as Datsun (later Nissan), Honda, and Toyota—and Congress rejected domestic automakers' protests. Many lawmakers had no faith that Detroit would voluntarily meet efficiency standards, so the House-Senate conference committee included mandatory fuel efficiency standards in the final bill. Whereas average fuel economy for American cars was just 14 miles per gallon in 1975, the new EPCA standards stipulated that corporate average fuel economy must reach 20 mpg by 1980 and 27.5 mpg by 1985.[50]

As an omnibus bill, EPCA contained other provisions, including some of Ford's State of the Union proposals. For example, the United States would develop a strategic petroleum reserve to store oil for emergency use, with the ultimate aim of depositing one billion barrels of oil there. The president would also have enhanced authority to encourage power plants to convert from oil to coal.

Though pleased at these provisions, Ford felt gravely disappointed at the bill's plan to roll back "new" oil prices and continue all controls. EPCA also left out proposals from his State of the Union address such as natural gas deregulation and increased production from the naval petroleum reserves. The bill's shortcomings made Ford's decision about signing it a difficult one.

In a perverse irony, the man who helped to engineer the oil allocation scheme that caused much of the energy crisis, William Simon, now treasury secretary, opposed controls with indignation. He drafted an eight-page memo urging the president to veto the bill and even called Ford at the White House just before midnight to press his case. Simon was not alone in his concerns. Alan Greenspan worried that continued controls on the oil industry might encourage their spread into other energy industries, such as coal and nuclear power. Some Republicans in Congress also objected to EPCA. Barber Conable recommended a veto, writing, "It is one thing to be reasonable and to try to achieve a compromise, and another to appear to capitulate." The popular mail count ran 10,381 for vetoing the bill and just 70 for signing it.[51]

Ford's precarious political position weighed heavily, too. In November 1975, former California governor Ronald Reagan announced that he would challenge Ford for the GOP presidential nomination, and the first primary in New Hampshire loomed just two months ahead. After the Nixon pardon, Ford's popularity had been mired below 50 percent, and his campaign staff worried about Reagan pulling an upset victory in New Hampshire. The challenger denounced EPCA and called on the president to veto it, a stand popular in Texas, site of a key primary. Yet New Hampshire, which produced no oil, relied heavily on imports and suffered through freezing winters. There the idea of capping oil industry profits remained popular, and a veto of EPCA could have hurt Ford in that critical first primary (he ended up winning New Hampshire—but by a dime-thin margin, just 1,317 votes).

A veto might have harmed Ford's political image, too. He had spent nearly a year imploring Congress to act on energy policy, and a veto would have given Ford nothing to show for the long struggle. It might have also hurt his image as a president who promoted healing between the executive and legislative branches. "This war with Congress was tense, and it was hot, it was nasty," recalled John Hill. "Zarb really believed we had to end this fight. It was not healthy. And I think Ford, being a [former] member of Congress, really appreciated that argument."

Zarb's opinion held heavy sway over Ford, who said that "his influence was significant" in signing the bill. The energy czar conceded that EPCA was "flawed" and would result in "a major reduction in incentives for investment in new, high-cost oil production."[52] Yet Zarb worried that Congress might override a veto of EPCA, and he believed that the bill could help Ford enough politically that he could win the 1976 presidential election and during a full term, when he would enjoy a mandate, he could forge energy policy more on his terms. Zarb recalled advising the president, "If you sign it, we've got enough tools to work with to get you reelected, and we can still get our objectives done, and there was enough in [EPCA] to do it with. If you don't sign it, we have immediate chaos in the country. And we're getting ready for the primary in New Hampshire, who knows what's going to happen."

Economic advisor William Seidman, a Grand Rapids native who shared Ford's political and economic philosophy, agreed, remarking that EPCA "sets a national energy policy that, while delayed, is a sound step in the right direction." Ultimately, Ford believed that "half a loaf was better than none."[53] Over the past year he had witnessed the futility of getting Congress to pass immediate decontrol. He, too, worried about its inflationary effects, especially since he regarded the fight to control inflation "the basic theme of my administration."[54] Hill remembered, "I think the thing that drove most of the policymakers, ultimately, toward signing this bill was fear of inflation, or what . . . immediate deregulation might do to inflation." Also lurking over Ford was the threat of a veto override; he recalled worrying that, if he pushed through legislation for immediate decontrol, Congress "probably would have rammed through something that, if I had vetoed [it], I probably would have been overridden." Gradual decontrol seemed the only viable option.

Just before Christmas, on December 22, 1975, Ford signed EPCA. His signing statement cited three main reasons for accepting the bill. First, it embodied important elements of his State of the Union energy program. Second, it called for gradual decontrol. Last, Ford remarked, "I am also persuaded that this legislation represents the most constructive bill we are likely to work out at this time. If I were to veto this bill, the debates of the past year would almost surely continue throughout the election year and beyond. The temptation to politicize the debate would be powerful, and the Nation could become further divided."[55]

It was one of the most difficult decisions of his presidency. EPCA, the result of a year-long ideological and political struggle, was a letdown. One cartoonist showed Ford sitting in children's pajamas by a Christmas fireplace, holding an unprepossessing, unplugged Mickey Mouse doll labeled "Energy Bill." With a tear in his eye, the Ford character asks, "Honest, Santa? It really took you a whole year to build this?"[56]

Signing a bill that Ford considered "half-good and half-bad" and only "marginally signable" was no choice he took lightly. He later termed the decision to approve EPCA "a very close call." He remembered, "I signed it primarily after listening to the arguments of Frank Zarb. . . . Zarb convinced me that we had a timetable for decontrol [in EPCA], we ought to accept it; there was no way we could get decontrol [immediately]. So I thought it was the best package we could get."

The divisiveness that energy policy generated in Congress found a correlate within the White House. Some administration members felt embittered over the president's decision and believed that Zarb had exaggerated reasons for signing the bill, warning of "mad dogs loose" on Capitol Hill were Ford to veto it.[57] Indeed, Zarb's depiction of "immediate chaos" following a veto painted too graphic a picture.

The bill's opponents had a good point. One economist later calculated that the 1970s oil controls dropped oil production by 1.4 million barrels a day.[58] Mapco, a Tulsa, Oklahoma, oil company, expressed its anger with EPCA by taking legal action to prevent the price rollback on new oil, arguing that the controls would force it to sell new oil at a loss.[59] Ford rued the economic havoc that controls created, and if not for their support in Congress, the history of the 1970s energy crisis would have been much different. "If I could [have acted] unilaterally, on my own, I probably would have decontrolled period, immediately," Ford reflected. "There would have been a little turmoil for a short period of time. But we would have opened the markets and the markets would have corrected things rather quickly. But that was not the real world. The real world was that I had to deal with Congress."

Ford's support for EPCA helped to cut the Gordian knot tying up Washington on energy policy. Although he had little enthusiasm for the bill, he signaled his political moderation and willingness to work with Congress. "Signing the bill took so much heat out of the partisan debate," Hill reflected, "and that enabled Ford to calm Washington down for the first time since Nixon had left."

Immediate decontrol had whipped up resentment in Congress, and Zarb felt that he had learned a lesson. "If I had to do it over again, knowing everything I know, I think I would have backed off sooner and gone incremental," he reflected, meaning that he would have proposed gradual decontrol at the outset.

Yet Ford's ambitious proposal of immediate decontrol served a purpose. Congress often opts for languid solutions; that tendency was one reason

Zarb opposed an increase in the gasoline tax as a substantive part of Ford's energy policy. Congress would "put a nickel on a gallon of gas and then go home, saying we don't need any more energy policy," Zarb believed. Had Ford proposed gradual decontrol at the outset, it might have failed to grab Congress's attention as dramatically as immediate control had, and lawmakers might have watered down Ford's proposals by stretching out the decontrol period far longer than he wanted.

Ford's program fought a growing complacency among members of Congress and their constituents. By 1975, Zarb recalled, "the people lost the need to drive down oil imports after the embargo ended [in March 1974]. The sense of urgency started to dissipate." By early 1976, Republican congressman Charles Grassley of Iowa said, "Talk about energy has pretty much faded away. There is even an occasional gas war."[60]

Ford's proposals kept energy in the headlines. With EPCA, he hoped to set a foundation for a more far-reaching program during a full term. His method was to continue the strategy forged in 1975—keep on the offensive, introduce new ideas, force Congress to respond to his initiatives. "He wanted to get the Congress to react to him," Zarb recalled. That desire helped Ford to enact more of his energy policy before he left office.

Final Victories in Energy Policy

The singular aspect of Ford's last full year as president was that he found time to implement policy at all. Reagan's challenge for the GOP nomination forced Ford to devote time and effort to campaigning and almost dashed his hopes of becoming the Republican standard bearer. The two combatants ran neck-and-neck for virtually the entire primary season. Although the president won the nomination, once the contest was over administration members felt like spent rockets, drained and exhausted. They had little time to recover, as their candidate soon faced the Democratic nominee, former Georgia governor Jimmy Carter, in the general election.

Yet Ford still pursued energy policy. In April 1976, one of his original State of the Union proposals became law when Ford signed the Naval Petroleum Reserves Production Act, authorizing increased production from two federally owned oil fields, in Elk Hills, California, and Teapot

Dome, Wyoming. In August, Ford signed the Energy Conservation and Production Act (ECPA), which freed so-called stripper wells (those producing fewer than 10 barrels a day) from price controls. Collectively, oil from stripper wells contributed 12 percent of domestic production. The Ford White House estimated that decontrolling stripper wells would increase domestic oil production by 450,000 barrels annually while costing consumers just a penny more per gallon of gasoline.[61]

Buried within EPCA was a provision that allowed Ford to take an important step toward oil price decontrol. The president could propose decontrol of finished petroleum products, which Congress could block with a one-house majority vote. Absent the latter, decontrol would proceed, giving Ford a way to dismantle a part of the cumbersome structure of price controls and allocation that the Nixon administration had erected. Controls on crude oil would remain, but by removing controls on finished products the EPCA provision would ultimately allow retailers to offer lower prices to consumers.

On June 1, the FEA secured decontrol of residual fuel oil (the heavy, least refined products of a crude oil barrel, such as heating oil used in large buildings and factories). The next month, the agency decontrolled distillate oils, which included kerosene and diesel. In September, the FEA ended controls on specialty oils such as benzene, naphtha, and lubricants. On his last day as president, Ford proposed decontrolling the most critical fuel, gasoline; unfortunately, President Carter rescinded the proposal after taking office.[62]

The decontrol of finished products constituted one of Ford's final victories in energy policy. As he had done in proposing immediate decontrol, he pushed ahead with initiatives that caught Congress off-guard. Legislators "thought they put energy behind them [with EPCA], and all of a sudden we were consuming their time on these deregulation issues," John Hill recalled. All told, the Ford administration decontrolled 50 percent of the finished products from a barrel of crude oil.

Ford's hopes for scoring similar victories in alternative fuels fell short. As early as October 1974, he had asked to decontrol natural gas prices, but legislators refused. Ford aspired to increase nuclear power's role; he wanted it to furnish 20 percent of the nation's power, up from 9 percent in 1975. But environmental concerns over plant locations and waste disposal stymied progress in nuclear energy throughout Ford's presidency.[63]

Coal, America's most abundant fuel, was a different story. The country held one-third of the world's proven coal reserves, and Ford was an out-spoken exponent of its mining and use. Yet drawbacks were clear. Thick, dark fumes rose from coal-burning power plants, a visible sign of the fuel's damage to the air. Environmentalists especially fumed over strip mining. Strip mining (or euphemistically, "surface mining") accounted for half of all the coal extracted in the United States. Although it was the most efficient means of extracting coal near the earth's surface, it ravaged the landscape, as giant machines cleared out acres of land to gain access to the coal seams beneath. Congress passed two bills, in late 1974 and in the spring of 1975, to restrict strip mining. Ford vetoed both.

The outcry was earsplitting; the vetoes were among Ford's most con-troversial actions as president. Democratic congressman Morris Udall of Arizona, an environmentalist whose brother Stewart served as interior secretary during the Kennedy administration, charged Ford with cham-pioning the strip mining industry's "avarice" and "greed." The *New York Times* lambasted the president for his "remarkable indifference to the harm that such mining inflicts upon the nation's land and water."[64]

But Ford viewed coal as "America's ace in the hole" for reducing foreign oil imports. The congressional bills would cut coal production by 20 per-cent and cost 40,000 coal workers their jobs. Moreover, the administration believed that existing environmental safeguards guarded against "raping the landscape," as Udall characterized it. Hill recalled the White House arguing that "we already had enough laws on the books to deal with strip [mining]. It was not that we were saying don't regulate it; we were saying don't shut it down."

To Ford's satisfaction, Congress sustained his vetoes. During his final year as president, he achieved an even more substantive victory by encour-aging energy production using coal. EPCA and ECPA provided $750 million in loan guarantees (whereby the federal government promised to pay part or all of the principal and interest to a lender in case the borrower defaulted) to companies that mined new coal sites. In 1976 coal production shot up to 670 million tons, a new record, and the FEA ordered almost eighty power plants to burn coal rather than other fuels. By the early 1980s, coal produced more energy than oil in America and has led the country's energy production ever since. Ford's actions allowed America's most plentiful fuel to take a larger role in the nation's energy profile.[65]

Ford's Energy Legacy

During Gerald Ford's long retirement, as he lived to age ninety-three and set a record for the nation's oldest former president, he observed, "Historians generally rank Presidents by what they complete while in office. Another way of assessing them is by what they begin."[66] In energy—as in other aspects of Ford's presidency—that paradigm provides a realistic way to judge his legacy.

Ford faced many obstacles to setting sound energy policy. He had only two and a half years to work on an issue that required decades to study and address. He faced a feisty Congress that resisted presidential authority in general and his energy proposals in particular. Public apathy on energy grew as gas station lines disappeared and the issue lost urgency. Yet by the end of Ford's term America's dependence on foreign oil had grown, and the country imported 46 percent of its oil, up from 37 percent two years earlier.

Ford recognized the problem and proposed a solution. The price controls on oil were a bitter legacy of the Nixon presidency, with effects even more tangible than those of Watergate—energy shortages that affected Americans daily. "The most egregious mistake in economic policy during that era was President Nixon's decision to go for wage and price controls," wrote Paul McCracken, Nixon's first Council of Economic Advisors chair, who also advised Ford on economic policy. "The result was predictable. The President would be popular temporarily, but soon the action would have been found to be somewhere between ineffectual and harmful."[67]

Ford's role in history, noted press secretary Ron Nessen, "was to clean up other people's messes." Nessen drew a vivid metaphor, using an incident that he witnessed during Christmas with the Ford family. The president's dog had an accident on a rug, and a White House steward tried to take care of the result. Ford interceded, telling the aide he would clean it himself. "No man should have to clean up after another man's dog," Ford explained. On a national level, Nessen noted, Ford had to clean up messes such as "the shattered faith" of Americans after Watergate and "a sick economy" plagued by high inflation.[68] Two other messes that Ford tried to clean up were oil price controls and the resulting energy crisis and the poisonous relations between the president and Congress.

Ford was "absolutely right" to advocate decontrol, McCracken said.[69] Decontrolling oil prices was essential to increasing domestic oil production. In January 1981, when Reagan removed the last remaining controls on oil as one of his first presidential actions, oil companies responded as Ford had predicted, increasing production. Between 1979 and 1985, domestic oil production increased by 400 million barrels a day.[70]

Ford foresaw the benefits of decontrol and began the policy debate to initiate it. By 1976, Zarb noted that the "working environment on energy has changed. A year ago, if the President sent me up to the Hill to talk about deregulating the price of home heating oil, I would have been tarred and feathered."[71] Even Democrats acknowledged that Ford, by keeping Congress's feet to the fire, helped to pressure legislators to act on national energy policy. In mid-1975, Mike Mansfield wrote to Ford, "I am frank to say that it has been your effort that has provided the primary impetus to the energy issue and to the need to develop a comprehensive energy policy for the Nation." The Senate majority leader gave Ford a remarkable tribute: "Because of your effort, much has been done to shape and implement [energy] policy; more, in fact, in the past 6 months than ever before in the Nation's history."[72]

Half a year after Mansfield wrote those words, the president induced a reluctant Congress to produce an omnibus energy bill. He had gotten lawmakers to act through a combination of coercion, cajolery, and compromise. Though EPCA was far from perfect, Ford accepted it partly to further a spirit of cooperation with Capitol Hill. He subsequently used EPCA to push for decontrol of finished petroleum products during his last year in office, and after he left the White House EPCA's provisions promoting coal usage made this alternative fuel America's number-one energy source.

Some other EPCA measures have proved sensible. All fifty states have adopted the right-on-red traffic law, which saves motorists time and gasoline. The Strategic Petroleum Reserve remains an emergency cushion that the president can draw upon during emergencies. For example, in August 2005, after Hurricane Katrina battered the Gulf Coast and destroyed oil rigs there, President George W. Bush ordered oil released from the reserve and loaned to oil refineries. In December 2007, Bush signed legislation to increase corporate average fuel economy (CAFE) standards to 35 miles per gallon by 2020. Automakers continue to protest CAFE requirements, and critics have argued that the lighter vehicles they entail have reduced motorist safety. Still, a 2002 National Academy of Sciences study concluded, "The

CAFE program has clearly contributed to increased fuel economy of the nation's light-duty vehicle fleet during the past 22 years."[73]

In signing EPCA, Ford demonstrated the collegial politics that he practiced throughout his career. EPCA had the economically baneful effect of continuing controls longer than Ford wanted. But it had the politically helpful effect of mending the rift between the president and Congress and producing energy policy measures that still benefit America.

Notes

I thank former Ford administration members, members of Congress, and energy experts whom I interviewed for my book *Gerald Ford and the Challenges of the 1970s* (University Press of Kentucky, 2005), which provided a basis for this essay. They include former President Gerald R. Ford, Glenn Schleede, Dale Bumpers, Charles DiBona, Max Friedersdorf, Ken Glozer, Paul MacAvoy, Paul McCracken, Robert Nordhaus, and Frank Zarb. Geir Gundersen and the archivists at the Gerald R. Ford Presidential Library provided help and vital information. Dr. Michael S. Green of the College of Southern Nevada and Dr. Bogdan and Seiko Mieczkowski caught mistakes and offered critiques during proofreading. I gratefully acknowledge research assistant Carolyn Badalucco of Dowling College and the Dowling College Faculty Development and Curriculum Committee for the Social Sciences, which provided released time that facilitated this writing.

1. E-mail from Geir Gunderson, Gerald R. Ford Presidential Library archivist, to author, Oct. 25, 2007.

2. *Washington Post,* July 1, 1975, A6.

3. Recollections from Frank Zarb are from his personal interviews with author, Sept. 26, 1994, and Feb. 6, 2003.

4. Here and in the following discussion, quoted passages from President Ford are from his personal interview with author, Mar. 30, 1994, unless otherwise noted.

5. Max Friedersdorf, personal interview with author, Aug. 14, 2001.

6. *U.S. News and World Report,* Dec. 17, 1973, 25.

7. James Hanley, telephone interview with author, June 14, 1994.

8. Gerald R. Ford, *A Time to Heal: The Autobiography of Gerald R. Ford* (New York: Reader's Digest, 1979), 125.

9. Jerald terHorst, *Gerald Ford and the Future of the Presidency* (New York: Third Press, 1974), 212–13.

10. Charles DiBona, personal interview with author, Jan. 23, 2003.

11. William Simon, *A Time for Truth* (New York: Reader's Digest, 1978), 57.

12. Paul MacAvoy, *Federal Energy Administration Regulation: Report of the Presidential Task Force* (Washington, D.C.: American Enterprise Institute, 1977), 7–8; *New York Times,* Apr. 11 and July 4, 1975; *Washington Post,* Jan. 19, 1975.

13. Alan Greenspan, *The Age of Turbulence: Adventures in a New World* (New York: Penguin Press, 2007), 65.

14. Gerald R. Ford, *Public Papers of the President, 1975* (Washington, D.C.: Government Printing Office, 1976), 1:237.

15. Glenn R. Schleede, personal interview with author, Oct. 20, 1994.

16. Yanek Mieczkowski, *Gerald Ford and the Challenges of the 1970s* (Lexington: University Press of Kentucky, 2005), 219.

17. President's Energy Briefing Book, Dec. 19, 1974, Box 50, Presidential Handwriting Files, Gerald R. Ford Presidential Library, Ann Arbor, Mich.

18. Ford, *Time to Heal*, 242.

19. Excise Tax and Import Fees on Crude Oil, Jan. 13, 1975, Box 34, Folder: Qs & As—Taxes: 1975 (Jan. 17), William Simon Papers, David Bishop Skillman Library, Lafayette College, Easton, Pa. (hereafter Simon Papers); *New York Times*, Jan. 3, 1975.

20. *New York Times*, Jan. 15, 1975; President's Economic and Energy Proposals: A Factual Summary, Box 13, Folder: Economic and Energy Program, 1/75 (1), John Marsh Files, Ford Library.

21. Ron Nessen, *It Sure Looks Different from the Inside* (New York: Playboy Books, 1978), 89.

22. Mieczkowski, *Gerald Ford*, 225.

23. *Business Week*, Jan. 27, 1975, 47.

24. *Congressional Quarterly Almanac 1975* (Washington, D.C.: Congressional Quarterly, 1976), 246.

25. Here and below, recollections of John Hill from his personal interview with author, Sept. 19, 2002.

26. *Business Week*, Feb. 3, 1975, 16.

27. Robert Nordhaus, personal interview with author, Sept. 18, 2003.

28. Mieczkowski, *Gerald Ford*, 231.

29. *Public Papers of the President, 1975*, 1:217.

30. *Business Week*, May 26, 1976, 26.

31. *Public Papers of the President, 1975*, 1:730–31.

32. Mieczkowski, *Gerald Ford*, 245.

33. Ibid., 243.

34. Max Friedersdorf, personal interview with author, Aug. 14, 2001.

35. John Dingell, telephone interview with author, Feb. 14, 1994.

36. *Washington Post*, May 23, 25, 1975; *Public Papers of the President, 1975*, 1:801.

37. Ford, *Time to Heal*, 243.

38. Campaign Materials 1974, Box 1092, Folder 2, Birch Bayh Papers, Lilly Library, University of Indiana, Bloomington.

39. Daniel Yergin, *The Prize: The Epic Quest for Oil, Money and Power* (New York: Touchstone, 1992), 657.

40. Letter from Lowell Zentner, May 29, 1073, Simon Papers; *Washington Post*, Jan. 15, Aug. 14, 1975.

41. Mieczkowski, *Gerald Ford*, 242, 244, 245.

42. Gerald Ford, "My Definition of a Statesman as Opposed to a Politician," materials from the writing of *A Time to Heal*, Ford Library; John McCain with Mark Salter, *Hard*

Call: Great Decisions and the Extraordinary People Who Made Them (New York: Twelve Books, 2007), 350.

43. Mieczkowski, *Gerald Ford*, 248.

44. Statement of Hon. John Dingell, July 2, 1975, Box 8, Folder: EPCA (3), Loen and Leppert Files, Ford Library.

45. *Business Week*, Jan. 27, 1975, 44.

46. Mieczkowski, *Gerald Ford*, 250.

47. "Summary," Box 12, Folder: Energy 1975: Omnibus Energy Bill (S. 622), Oct.–Nov. (2), Glenn Schleede Files; "Summary of the Adverse Effects of the Energy Policy and Conservation Act of 1975," Box 44, Folder: Energy-Legislation (1), Alan Greenspan Files; and White House Fact Sheet: Energy Policy and Conservation Act (S. 622), Dec. 22, Box 9, Folder: Energy-General (2), Ron Nessen Papers, all in Ford Library; *Time*, Nov. 24, 1975, 91.

48. Dale Bumpers, personal interview with author, June 17, 2003.

49. "Significant Problems with the Non-Pricing Provisions of Conference Bill," Box 12, Folder: Energy, 1975: Omnibus Energy Bill (S. 622), Oct.–Nov. (2), Schleede Files, Ford Library; *Washington Post*, Jan. 16, 1976.

50. Steven Engler, "The CAFE Standards and the Politics of Energy in the 1970s" (senior essay at Yale University, Apr. 9, 2007).

51. Mieczkowski, *Gerald Ford*, 252–53.

52. Ibid., 253, 254.

53. Ibid., 255.

54. A. James Reichley, *Conservatives in an Age of Change: The Nixon and Ford Administrations* (Washington, D.C.: Brookings Institution, 1981), 383–84.

55. *Public Papers of the President, 1975*, 1:741–42.

56. *Business Week*, Jan. 12, 1976, 23.

57. Paul MacAvoy, personal interview with author, May 2, 2002.

58. Joseph Kalt, *The Economics and Politics of Oil Price Regulation: Federal Policy in the Post-embargo Era* (Cambridge, Mass.: MIT Press, 1981), 91, 207.

59. *Business Week*, Jan. 12, 1976, 23.

60. Mieczkowski, *Gerald Ford*, 258, 259.

61. *New York Times*, Aug. 11, 15, 1976; *Washington Post*, Aug. 11, 1976.

62. Mieczkowski, *Gerald Ford*, 254–55.

63. Ibid., 265–67.

64. Ibid., 268.

65. Ibid., 267–69.

66 *Parade*, July 5, 1998, 20.

67. Paul McCracken, letter to author, Nov. 22, 2005.

68. Nessen, *It Sure Looks Different*, xiv.

69. Paul McCracken, personal interview with author, Feb. 8, 1994.

70. Mieczkowski, *Gerald Ford*, 354.

71. *New York Times*, Oct. 18, 1976.

72. Letter to the president from Mike Mansfield, Aug. 29, 1975, Box 10, Folder: Energy-Oil Decontrol, Nessen Papers, Ford Library.

73. National Research Council, National Academy of Sciences, *Effectiveness and Impact of the Corporate Average Fuel Economy (CAFE) Standards* (Washington, D.C.: National Academies Press, 2002), 3.

Chapter 2

Conflict or Consensus?
The Roots of Jimmy Carter's Energy Policies

Jay E. Hakes

T he energy policies of Jimmy Carter became a signature component of his presidency. Since he left office in 1981, these policies have been both widely condemned and praised. Critics have
painted him as the creator of an ineffective energy bureaucracy (the Department of Energy) and a propagator of a "limits on growth" philosophy
with too much emphasis on the environment and fuel conservation. Those
more favorably disposed toward Carter say that if we had adopted his bold
goals on conservation and alternative sources of energy, we would not have
become so dependent on foreign oil.

Both sides of this debate seem to agree on one thing: Carter's ideas were
out of step with his time, by being either stuck in the past (the view of
conservatives that he was a big-government Democrat) or ahead of his
time (the view of liberals that the country was not yet ready to accept his
progressive ideas). A widespread perception that little substantive action
was taken to deal with rampant energy problems is reflected in the 2006
assessment by Barack Obama that, when confronted with the problem of
gasoline lines, "the best Jimmy Carter could suggest was turning down the
thermostat."[1]

All of this conventional wisdom distorts the impact of the Carter presidency on energy policy and the dominant role that energy issues played in American politics during the 1970s. A detailed review of the period offers persuasive evidence for a different perspective.[2] The Arab oil embargo (1973–74), in fact, produced a substantial consensus for about six years among federal officials on what to do about energy and a series of major legislative results. Although there were definitely heated disputes over energy policy, Carter and many members of Congress in both parties reflected this consensus and agreed enough on energy matters to pass significant legislation. Carter stood out for his willingness to back the most comprehensive approach to energy; his "moral equivalent of war" announced in April 1977 included 113 proposals, and in 1979, after the Iranian revolution (1978–79), he recommended many additional steps. But even if Carter pushed harder for energy solutions and tilted more than some others toward conservation, his general sense of direction was widely accepted among elected officials as what was needed to avoid escalating dependence on Persian Gulf oil.

Several factors have contributed to the perception of a lack of consensus or any substantive results on energy. Major battles where consensus was elusive included bitter fights throughout much of the 1970s over what to do about price controls on natural gas and oil, which pitted the energy-producing states against the consuming regions of the country, particularly the Northeast, where heating oil was a winter necessity. Proposals for energy taxes, though they had support from some Republicans and Democrats, always created a hornets' nest and mostly failed to gain much traction in the legislative process. News coverage of these contentious issues overwhelmed frequent agreement on other matters. It is necessary to go behind the daily headlines and the public statements to find the areas of agreement that led to significant (if temporary) policy agreement.

In addition, presidential politics tended to obscure areas of consensus among Republicans and Democrats at the time. President Gerald Ford's run for reelection in 1976 required fending off a stiff challenge from former California governor Ronald Reagan, who espoused an anti-government philosophy. To accommodate the Reagan wing of the Republican Party, Ford avoided mention during the campaign of his activist views on energy in 1974 and 1975. Carter took advantage of Ford's silence on energy by charging in the first presidential debate in sixteen years on September 23,

1976, that "almost every other developed nation in the world has an energy policy except us," a position repeated in his presidential memoirs.[3]

Ford returned fire during Carter's campaign to win reelection in 1980, stating, "We still lack a comprehensive energy policy, and, in my opinion, the Carter Administration and the Democratic Congress must shoulder much of the blame." He blistered Carter on energy initiatives that were not very different from those Ford had pursued when he worked in the Oval Office.[4] The political rhetoric of Ford and Carter tended to obscure the energy issues on which they agreed and the substantial energy legislation passed during both administrations.

In this essay, five topics illustrate a general consensus on energy in the post-embargo 1970s that crossed party lines and the branches of government: the overlapping personnel between Republican and Democratic administrations, creation of the Department of Energy, debates over price controls and windfall profits, nuclear policy, and support for alternative energy technologies.

The Apolitical Energy Experts

Carter's energy proposals, particularly those announced in 1977, built on several analytical studies produced after the Arab oil embargo. Much of the work was done by the Nixon and Ford administrations, and few experts at the time doubted that bold action was needed to avoid another oil embargo and the transfer of huge sums of money to the volatile Persian Gulf even without another interruption in supplies.

Other than the president himself, the six most influential members of the Carter team on energy were his two energy secretaries (James Schlesinger and Charles Duncan), the department's two deputies (Jack O'Leary and John Sawhill), David Freeman (behind-the-scenes idea man for the 1977 energy plan), and Stuart Eizenstat (key policy advisor during development of new plans in 1979 and 1980). Four had worked previously for Republican presidents.

Schlesinger, Sawhill, and O'Leary held positions in the administrations of both Richard Nixon and Gerald Ford. Freeman wrote most of Nixon's broad-based energy message of 1971 before leaving the White House to head the Ford Foundation's *A Time to Choose: America's Energy Future*.

Sawhill, as the Nixon-appointed head of the Federal Energy Administration, was in charge of the 1974 *Project Independence* report, which was issued by the White House in the early months of the Ford presidency and like the Ford Foundation study contained many ideas that showed up in Carter's 1977 energy plan.[5] Business executive Duncan had no government service before becoming deputy secretary of defense in 1977 but was widely admired on both sides of the aisle for his management skills. He was confirmed as energy secretary by a vote of 95 to 1 in July 1979, with the only negative vote coming from Democratic senator William Proxmire of Wisconsin.

Those serving under Republican presidents had been considered dissidents to varying degree, and none stayed with Ford into the 1976 election year. Yet others at the next rung of influence remained at least for a while into the Carter term and did not have to alter their public statements greatly on a wide variety of energy issues. This continuity within the executive branch would not have been possible if there had not been some general sense of direction on national energy policy that crossed party lines.

Creating a New Department

President Ford was not as keen on reorganizing the energy functions of government into a single agency as his predecessor, Richard Nixon—a constant tinkerer with federal organization charts. Responding to a mandate from Congress, however, Ford late in 1976 presented a plan for creating a department of energy. Ford advisors were surprised to find that their effort produced a plan almost identical to the one proposed by Carter during the presidential campaign.

Like Ford, Carter wanted to combine the Energy Research and Development Administration (which included the research and development responsibilities of the old Atomic Energy Commission) and the Federal Energy Administration (which handled energy analysis and regulation). Henry Jackson (D-Wash.), chairman of the Senate Energy Committee and the legislator with the greatest influence on the matter, wanted a larger energy operation that included the natural resource functions of the Interior Department, but he deferred to the Carter position in order to expedite action. His agreement eased the way for the plan favored by Ford and Carter.

The remarkably uncontroversial bill creating the Department of Energy sped through Congress, receiving final passage in August 1977. The House agreed 353–57, the Senate 76–14. The breadth of support in 1977 for the newborn agency was further demonstrated by Senate confirmation of Schlesinger as its secretary by acclamation on the very day of its establishment.

After his election, Ronald Reagan initially proposed abolishing the department but eventually settled for draconian cuts in the budget for its nondefense functions. (Most of the department's budget goes to nuclear weapons.) Reagan's attacks on the department left memories of it in later years as a source of controversy. But at the time of its creation, the agency has a lot of wind at its back, and Carter was working well within the dominant thinking in Washington.[6]

Price Controls and Windfall Profits

The major energy studies after the Arab oil embargo—from diverse sources such as the Ford Foundation, the American Enterprise Institute, and the White House's *Project Independence*—agreed that price controls on oil would need to be jettisoned to reduce dependence on foreign supplies.[7] This consensus among energy experts viewed as liberal and conservative confronted a prevalent view among the general populous that energy fuels were necessities of life and government should prevent increases in prices.

Nixon, Ford, and Carter all straddled the fence to some extent on what to do about price controls. Nixon announced a wage and price freeze for the entire economy with great fanfare in August 1971. It was wildly popular from Wall Street to Main Street and with both political parties. Nixon was so convinced of the political popularity of price controls (which continued on even after the original freeze ended) during his first term that he ordered his staff not to discuss deregulation even in private. Nixon challenged the idea of price controls rhetorically (arguing that his controls would prevent the Democrats from imposing even worse controls) and, after being safely reelected, publicly called (not very loud) for decontrol of natural gas prices. But even after reelection Nixon, fearful of an additional loss of popularity during the Watergate investigation, refused to consider ending price caps on crude oil and oil products. Of the three presidents, Nixon was clearly the most supportive of price regulation.

Ford spoke out strongly for decontrol of oil and gas prices. He did not want to go cold turkey, however. Phasing out regulated prices would carry less risk of economic turmoil and political backlash. After the oil embargo, presidents could not end energy price controls without the consent of Congress. As Ford fought with legislators over the issue, he decided not to take advantage of several opportunities to end controls immediately by allowing the authorizing legislation to lapse. He opted to keep controls rather than end them precipitously.

During the 1976 presidential campaign, Carter committed to work toward deregulation of natural gas but avoided doing so until a complex plan for partial deregulation emerged in 1978. When his first opportunity to end price controls on crude oil by executive action came in the spring of 1979, Carter, acting against the advice of his political advisors, used his authority under the Energy Policy and Conservation Act of 1975 to begin a bold phase-out. The substantial easing of government controls contributed to a rising energy prices and inflation, a sharp drop in support from his Democratic base, and the lowest poll numbers of his time in office.

Carter showed the greatest political courage on price controls (though Ford may well have done the same thing). He has, however, not received as much credit as might be expected from free-market advocates. More attention has gone to Ronald Reagan's expedited and final termination of controls in 1981.

It was assumed by Nixon, Ford, and Carter that the termination of price controls would have to be accompanied by new taxes on the profits of oil companies to offset the sudden jump in the price for domestic oil as it reached world levels. Decontrol would also be a harder political sell to the public if the companies were not taxed on their sudden "windfall." Carter's proposal for a windfall profits tax in 1979 was not without controversy, but final passage of a compromise version in 1980 came with ample margins of support. The 66–31 tally in the Senate included the two Democratic senators from oil-producing Louisiana (Russell Long and Bennett Johnston) and seventeen of forty Republicans voting aye. Carter had to work tirelessly to pass the largest tax on an industry in history. But with such broad support for it in the Congress, it is hard to argue that he was out of step with the general political consensus of the time.

Nuclear Policy

The 1970s was a pivotal time for American nuclear power, which started the decade as the wave of the future and ended it with cloudy prospects. Carter is viewed by many as an anti-nuclear president. He called nuclear power "a last resort," fought hard to stop funding for building a nuclear breeder reactor plant in Tennessee, and held office during the Three Mile Island accident, often viewed as the incident that stopped the momentum of an emerging technology. This common interpretation appears rather improbable upon close examination. Carter, after all, worked with (later Admiral) Hyman Rickover while in the U.S. Navy to develop the first light-water reactors for American submarines, a technology that formed the foundation of civilian nuclear power. As president, Carter often defended the need for nuclear power.

The nuclear industry was facing a trend of cancellations for conventional plants far outnumbering orders for new ones even before Carter took office. The electric companies had discovered that coal generation was less expensive than nuclear, and the performance of nuclear plants fell far below what had been advertized. The success of light-water plants was adversely affected by environmental concerns and delays in licensing, but the decline resulted mainly from the economics of the electric industry.[8]

Before Carter took office, the tide was also turning against the breeder reactor, an advanced technology (and Nixon's favorite energy technology) that produced as much nuclear fuel as it consumed, thus protecting the country against any future shortages of uranium. Very expensive to construct and much riskier from a security standpoint than light-water reactors, the breeders no longer made sense economically, since there was no imminent shortage of uranium. The White House budget office led the fight against the breeder as early as the Nixon years, despite the president's personal support.

After the Three Mile Island accident in March 1979, Carter visited the site to demonstrate that any health hazards were under control. Later he fought calls for a moratorium on the completion of more nuclear plants. In fact, forty-two operable nuclear units (net) were added in the decade after the accident.

Although Carter was a supporter of the light-water reactor and willing to fund research that might find better alternative breeder technologies, his opposition to moving ahead on construction of a breeder reactor with the then available technology and under then current economics bred hostility to Carter within some segments of the nuclear industry. Yet it was not just environmental organizations that fought the breeder. Conservative think tanks like the American Enterprise Institute found the economic arguments for the breeder unconvincing.[9] The strongest support in Congress for Carter's high-profile stand against the breeder reactor came from Michigan's Republican representative, David Stockman, who went on to be Reagan's first budget director. The final termination of breeder funding occurred under Reagan through a conservative coalition in the House led by Newt Gingrich, Jack Kemp, and Trent Lott.

Nuclear power had strong advocates for and against it during the Carter presidency, as it does today. Yet among electric industry executives, the Carter White House, and Congress, there was general agreement that the country needed nuclear power and that it could be produced safely with appropriate safeguards. On the other hand, nuclear power was expected to compete with other sources of electric power on the basis of cost.

Alternative Fuels

Momentum for big increases in the budgets for research and development on renewable and other forms of energy started well before Carter came to Washington. Nixon, urged on by Democrats like Jackson, proposed major increases in funding for energy research even before the Arab oil embargo. After the embargo, Nixon, Ford, and a Democratic Congress all supported more money for virtually all forms of alternative energy, ranging from coal chemically altered into oil and gas to renewable fuels like solar and wind power. Carter pushed successfully for even more funding in this area. During his last two years in office, energy research reached (controlled for inflation) its highest level in American history, either before or after.

It is useful to note that Carter was not facing a reluctant Congress on his support for alternative supplies of energy. A bipartisan Solar Caucus lobbied Carter in 1979 to set the bold goal he later adopted of 20 percent of all energy coming from renewables by the year 2000.[10] Although Carter is remembered for putting solar collectors on the roof of the White House, a more significant contribution was ramping up spending on photovoltaic cell research, technology that was more strategically important and had already gained momentum during the Ford years.

The most novel development on renewables under Carter was the emergence of the "gasohol" (later called ethanol) industry. Carter, prodded by members of Congress from corn-producing states, was the first president to mention the subject. A substantial tax break for ethanol was passed in the 1978 energy package. A brutal fight over natural gas pricing delayed and almost scuttled the omnibus energy legislation, but the ethanol tax preference (like most items in the package) was not controversial.[11] The break has continued without interruption into the current era and constitutes one of the major unique legacies of the Carter presidency.

In this period, the Carter presidency represented the greatest commitment of American resources to the development of alternative fuels in American history. It was, however, a logical growth of trends already in place when he took office. The level of support seems so high in historical perspective, in part, because of the sharp decline after he left office.[12]

Waves of Legislation

Because of the substantial consensus on energy policy after the Arab oil embargo, more significant energy legislation passed under Ford and Carter than is generally recognized. In 1975, Congress passed and Ford signed the Energy Policy and Conservation Act, which for the first time established the Strategic Petroleum Reserve and mileage efficiency standards for automobiles. It also provided for the eventual elimination for oil price controls, the authority later used by Ford, Carter, and Reagan.

Major energy packages passed in 1978 and 1980 contained many, though far from all, of the Carter recommendations to Congress. They contained generous tax breaks for building insulation and renewable energy like solar and alcohol fuels, a "gas guzzler" tax, mandates on backing oil out of generation of electricity, and massive spending on alternative energy research, development, and deployment.

Political momentum for this historic flurry of energy policy innovation was built on fears that American national security was threatened by dependence on Middle Eastern oil. The Arab oil embargo, the Iranian revolution, and the Russian invasion of Afghanistan (1979) were utilized to generate political support for presidential energy initiatives. Carter also argued to good effect in 1979 and 1980 that heavy dependence on expensive foreign oil weakened the U.S. dollar. At the time, few officials in Washington disputed that strong government initiatives were needed to counter these threats.

Conclusion

The 1980s saw a great reversal in public and official views on energy policy. What had generally been agreed on just a few years earlier—the need for an activist government forcing new directions in energy markets—suddenly became an objective of ridicule. The idea that government should reduce its role in the business of energy held great sway in the 1980s and, in many respects, continued to prevail into the next century.

In the 1970s the role of an activist government in energy was widely presumed to be a good thing—by the public and by officials of both parties. Carter came to believe that government caps on oil prices were counterproductive, but he sought a wide of array of policy carrots and sticks to reverse the prevailing trends in energy. On most of these issues he had plenty of company. In the pre-Reagan era, he was able to find considerable support for many of his views on both sides of the aisle.

Notes

1. Barack Obama, *The Audacity of Hope: Thoughts on Reclaiming the American Dream* (New York: Crown, 2006), 29.

2. This chapter is based on a review of the energy records at the Nixon Materials Project, the Ford Presidential Library, the Carter Presidential Library, the William Simon Papers at Lafayette College, a fresh analysis of congressional debates and votes, and a survey of historical energy data from the period. Additional information on U.S. energy policy during the Carter years can be found in Jay Hakes, *A Declaration of Energy Independence: How Freedom*

from Foreign Oil Can Improve National Security, Our Economy, and the Environment (Hoboken, N.J.: Wiley, 2008), chap. 2.

3. Public Papers of the Presidents, Sept. 23, 1976, 2297 (U.S. National Archives and Records Administration); Jimmy Carter, *Keeping Faith: Memoirs of a President* (New York: Bantam Books, 1982), 91.

4. Gerald R. Ford, *A Time to Heal*, rev. ed. (New York: Berkley Books, 1980), xxviii–xxx.

5. Energy Policy Project of the Ford Foundation, *A Time to Choose: America's Energy Future* (Cambridge, Mass.: Ballinger, 1974), and Federal Energy Administration, *Project Independence* (Washington, D.C.: Government Printing Office, 1974).

6. In retrospect, it can be argued that energy policy received stronger direction with an influential advisor to the president like Schlesinger at the White House rather than in the cabinet and with the functions of the old Energy Research and Development Administration and Federal Energy Administration kept separate. Schlesinger himself later expressed some sympathy for this view; see interview with James Schlesinger, Miller Center Interviews, July 19, 1984, 24, University of Virginia.

7. The American Enterprise Institute issued many studies during the period to make the case for deregulating oil and gas prices. These were conducted under the general direction of two veterans of the Nixon administration, Melvin Laird and Paul McCracken. Ironically, McCracken, as the chairman of Nixon's Council of Economic Advisors, privately supported wage and price controls when they were implemented in 1971. Key studies published by the Institute include Paul W. McCracken and Thomas Gale Moore, *Competition and Market Concentration in the American Economy* (1973); Robert B. Helms, *Natural Gas Regulation: An Evaluation of PFC Price Controls* (1974); and Edward J. Mitchell, ed., *Perspective on U.S. Energy Policy: A Critique of Regulation* (1976). Although the American Enterprise Institute and the Ford Foundation were often viewed as antagonists on energy matters, the gulf was narrow enough that some analysts worked in both camps.

8. See J. Samuel Walker, *Three Miles Island: A Nuclear Crisis in Historical Perspective* (Berkeley: University of California Press, 2004), 7–43.

9. Brian G. Chow, *The Liquid Metal Fast Breeder Reactor: An Economic Analysis* (Washington, D.C.: American Enterprise Institute, 1975).

10. The congressional Solar Caucus included Republicans, such as Jim Jeffords (Vt.) in the House and Charles Percy (Ill.) and Pete Domenici (N.Mex.) in the Senate, and many current and future Democratic leaders on energy such as Al Gore, Jr. (Tenn.), Richard Gephardt (Mo.), and Norm Mineta (Calif.) in the House and Gary Hart (Colo.) and Paul Tsongas (Mass.) in the Senate.

11. The support was controversial within the White House. The Council of Economic Advisors argued that corn-based ethanol did not provide much benefit for the considerable subsidies being provided. The economists urged that the emphasis be place on research on cellulosic ethanol, which produced more benefits but was not yet commercially available.

12. Many other instances of bipartisan consensus on energy policy during the Carter presidency could also be cited. It was widely assumed, at the time, that offshore oil and gas resources outside the Gulf of Mexico ought to be developed. There was also broad support for the Strategic Petroleum Reserve that crossed party lines.

Part II

Foreign Policy

Chapter 3

From the Nixon Doctrine to the Carter Doctrine

Iran and the Geopolitics of Oil in the 1970s

David S. Painter

O il was central to world power in the twentieth century. As Daniel Yergin points out in his Pulitzer Prize–winning study, oil was
"an essential element in national power, a major factor in world
economics, a critical focus for war and conflict, and a decisive force in international affairs." Thus, it is strange that many histories of international
relations neglect oil. Although Paul Kennedy focuses on the importance
of economic power factors in international relations in his best-selling *Rise
and Fall of the Great Powers*, he ignores the importance of oil to "economic
change and military conflict." Likewise, even though most of the major
postwar doctrines of U.S. foreign policy—the Truman, Eisenhower, Nixon, and Carter doctrines—relate, either directly or indirectly, to the Middle East and its oil, John Lewis Gaddis barely mentions oil in his popular
survey *Strategies of Containment*.[1]

Similarly, most histories of international oil neglect the larger geopolitical context. *The Prize* is a partial exception. Yergin's emphasis on the
strategic importance of oil during the period from World War I through
the 1956 Suez Crisis sets this work apart from most studies of the oil industry, but this aspect of his analysis is largely absent from the period after
Suez. Apart from Yergin, Fiona Venn's pioneering study *Oil Diplomacy in*

the Twentieth Century and Simon Bromley's interpretative survey *American Hegemony and World Oil* are among the few works that discuss the role of oil in world politics in any depth.[2]

This chapter, which is part of a larger study of oil and world power, analyzes the role of oil in world politics from the announcement of the Nixon Doctrine in July 1969 through the oil crises of the 1970s to the proclamation of the Carter Doctrine in January 1980. Although Henry Kissinger wrote in his memoirs that the energy crisis "altered irrevocably the world as it has grown up in the postwar period," neither Kissinger nor most studies of the 1970s examine the interaction of the decade's oil crises and the course of the cold war.[3] Oil is mentioned only twice in the 1,180 pages of Raymond Garthoff's *Détente and Confrontation*, and there is surprisingly little on oil in Odd Arne Westad's prize-winning *Global Cold War*, despite its focus on the cold war in the third world in the 1970s. Although William Bundy discusses oil, Iran, and the Nixon Doctrine, he does not connect this discussion to the course of the cold war, and there is nothing on oil in a recent collection of essays on Nixon's foreign policy. Although he does not deal with oil in depth, Melvyn Leffler points to the importance of American fears about loss of access to oil as a factor in the demise of détente. Most of the essays in a recent collection on the 1970s by Niall Ferguson and colleagues mention the oil crises, though the volume lacks an essay devoted this important topic.[4]

Similarly, although there are many fine studies of the energy crises of the 1970s, most do not take the larger geopolitical context into account. Neither *The Prize* nor such fine studies of the oil crises as Raymond Vernon's edited volume or Fiona Venn's survey address cold war issues.[5]

Preliminary research in the records of the Nixon, Ford, and Carter administrations and other primary sources suggests that students of the 1970s can gain new insights into both topics by examining the ways they interacted.[6] Rather than try to cover the whole decade, this essay focuses on three issues that illustrate the relationship between oil and geopolitics in the 1970s. First, it examines how the withdrawal of British military forces from "east of Suez" and the growing importance of Middle Eastern oil to the world economy led the United States, under the rubric of the Nixon Doctrine, to look to Iran as the guardian of Western interests in the Persian Gulf. Second, it looks briefly at how changes in the world oil economy interacted with changes in the global balance of power and with internal developments in Iran. Third, it analyzes the interaction of cold war and oil supply concerns with regional instability in the origins of the Carter Doctrine.

The Nixon Doctrine and the Middle East

On July 25, 1969, during a stopover in Guam, President Richard M. Nixon told reporters that the United States would in the future refrain from sending U.S. troops to help countries facing internal subversion but rather would provide them with "assistance in helping them solve their own problems."[7] Initially intended to reassure its Asian allies that the United States would not abandon them while convincing the American public and European allies that the United States would avoid future Vietnams, Nixon and his national security advisor Henry Kissinger gradually expanded the president's "informal and ambiguous" remarks into a more comprehensive statement of policy that became known as the Nixon Doctrine.[8] In his State of the Union address on January 22, 1970, Nixon announced: "Neither the defense nor the development of other nations can be exclusively or primarily an American undertaking. The nations of each part of the world should assume the primary responsibility for their own well being; and they themselves should determine the terms of that well-being."[9]

Despite its origins in Nixon's Vietnam strategy, the Nixon Doctrine had a profound impact on U.S. policy in the Middle East. In January 1968, in the midst of the Vietnam War, the British government announced that it intended to end Britain's military commitments "east of Suez" by the end of 1971. Caught up in the Vietnam War, the Johnson administration protested but failed to convince the British to change their minds.[10]

British military forces in the Persian Gulf, which numbered around 8,400 in 1968, had historically excluded unfriendly major powers from the Gulf, suppressed radical groups, and prevented long-standing Arab-Iranian and inter-Arab antagonisms from erupting into open conflict. In addition to ending their responsibilities for the defense and foreign relations of the nine Arab sheikhdoms of the lower Gulf and for the defense of Kuwait, British withdrawal from the Far East eliminated carriers and bases available to support or relieve forces in the Gulf.[11]

By the end of the 1960s, around 40 percent of Britain's oil supplies and 50 percent of Western Europe's came from the Persian Gulf. Around 40 percent of the Gulf's oil was controlled by British companies, constituting (in 1968) an investment of around £1 billion and contributing around £1.2 billion a year to Britain's balance of payments. Its oil interests in the area also allowed Britain to pay for over half of its oil imports in sterling, furthering bolstering the pound. Persian Gulf oil was also important to the U.S. balance of payments; the net gain to the U.S. balance of payments from the Persian Gulf by the late 1960s amounted to about $1.5 billion, which covered about 40 percent of the U.S. worldwide balance-of-payments deficit.[12]

A few months after the British announcement, the Soviets began to deploy naval forces to the Indian Ocean. According to military analyst Michael MccGwire, the Indian Ocean was "the best area in the world from which to target both Russia and China with submarine-launched ballistic missiles," and the initial Soviet deployment in June 1968 was clearly intended to counter the threat from U.S. guided-missile-carrying submarines, though the Soviets probably also hoped that "showing the flag" would bolster regional revolutionary movements. In addition, the Soviets were undoubtedly concerned about the establishment of the U.S. base on the Indian Ocean island of Diego Garcia, which they saw as threatening their sea communications with East Asia, and suspected that the United States planned to interfere with Soviet aid to North Vietnam. By the end of the decade, the Soviets were also anxious about defending sea routes to the Far East in event of a war with the People's Republic of China.[13]

Although the Soviet deployments were small, the United States feared that the Soviet presence could endanger Western access to Persian Gulf oil. To make matters worse, by the early 1970s the Soviet Union had achieved rough strategic parity with the United States, which raised the risks involved in U.S. intervention in the region.[14]

At the same time, U.S. relations with the Arab oil producers, including Saudi Arabia, were becoming increasingly strained because of U.S. support for Israel. The 1967 Arab-Israeli war left the Sinai, the West Bank, East Jerusalem, and the Golan Heights under Israeli control, and this outcome pushed radical Arab states further toward the Soviets and made it more difficult for conservative Arab states to maintain their ties with the United States. In addition to existing radical regimes in Syria, Iraq, and South

Yemen, Libya, which had been seen as an alternative source of oil outside the Persian Gulf, came under the control of a radical regime under Muammar Qaddafi in September 1969. On the other hand, the end of the war of attrition in August 1970 followed by the death of Egyptian leader Gamal Nasser in September 1970 and his replacement by Anwar Sadat seemed to promise a waning of radical Arab nationalism.[15]

After a review of the situation by an interdepartmental group led by the National Security Council (NSC), the United States adopted a strategy of looking to Iran and, to a lesser extent, Saudi Arabia to protect Western interests in the Persian Gulf. In early November 1970, President Nixon "approved a general strategy for the near term of promoting cooperation between Iran and Saudi Arabia . . . while recognizing the preponderance of Iranian power" to promote stability in the Persian Gulf. The strategy, outlined in National Security Decision Memorandum 92, also called for the United States to develop relations with the states of the lower Gulf and to maintain the existing U.S. naval presence in the region.[16]

This "twin pillar" strategy reflected the Nixon administration's belief that the United States was politically unable to assume Britain's role as "guardian of the gulf" because of the Vietnam War and widespread opposition at home and abroad to the United States acting as world policeman. The strategy recognized that Iran was the most powerful state in the area, with a comparatively strong and modern military and an active intelligence organization with assets in most of the smaller states. As Douglas MacArthur II, the American ambassador to Iran, noted in a conversation with President Nixon in April 1971, "Between Japan, and NATO, and Europe, it's the only building block we've got that is strong, that is sound, that is aggressive, and that above all regards us as just about its firmest friend." Nixon agreed: "Iran is the only thing there," he noted, and it was important to "have them strong."[17]

For the other pillar of its policy, the United States looked to Saudi Arabia, which had a much smaller population and weaker military forces than Iran. Moreover, the Saudis were committed to the Arab struggle against Israel, which greatly complicated, in U.S. eyes, their ability to play a stabilizing role in the region. U.S. intelligence also had concerns about Saudi Arabia's long-term stability, a view shared by the British. Still, Saudi Arabia was a key source of oil, especially for Western Europe and Japan, and the United States looked to the Saudis to take over the U.S. role of maintaining spare production capacity that could be used to supply oil in an emergency and help moderate prices.[18]

The shah of Iran was eager and determined to assume responsibility for the maintenance of order and security in the Gulf after the British left. In addition to his desire for Iran to play a larger role in the world, he was concerned about the spread of radical Arab nationalism and its possible impact on the large Arab minority in southern Iran. His determination to build up Iran's military may also have been influenced by his reading of the 1965 India-Pakistan war, during which Pakistan did not receive assistance from the United States or the Central Treaty Organization.[19]

Iran's economy had been booming since the mid-1960s, with a real rate of growth of 9 percent a year between 1962 and 1970. The rapid economic expansion coupled with high military expenditures led to serious balance-of-payments problems by the end of the decade. Oil revenues led Iranian economic growth as Iranian oil exports climbed from 1.9 million barrels per day (bpd) ($500 million) in 1965 to 3.8 million bpd ($1.1 billion) in 1970.[20] Nevertheless, to continue economic growth and to finance the expansion of Iran's military power that the shah desired, Iran needed more money.

The shah had a long history of pressuring the consortium that ran Iran's oil industry for increased production, but the consortium members had interests in other countries in the region and around the world, and they were limited in the extent to which they could accede to his wishes. He tried unsuccessfully to convince the United States to grant Iran privileged access to the U.S. oil market. In November 1970, the shah was able to convince the consortium to agree to an increase in the tax rate on net income from 50 to 55 percent, but this would not produce sufficient revenue to meet the his needs.[21]

The World Oil Economy in Transition

Geological and economic factors reinforced the political and strategic changes that made the Middle East more important. By 1972 world oil consumption had risen to 52.7 bpd, with oil and gas accounting for around 64 percent of total energy consumption. The tremendous increase in world oil consumption after World War II had dried up spare productive capacity outside the Middle East. Proven oil reserves in the continental United States peaked at 39.9 billion barrels in 1968, and U.S. oil production peaked in 1970 at 11.3 million bpd. Alaskan oil, though promising, would not be available until the mid-1970s. North Sea oil was also not yet available. Although Venezuelan production increased during the 1960s, Venezuelan reserves declined as the oil companies shifted the focus of their operations to the Middle East. In the 1970s, Venezuelan production began a gradual decline. In contrast, oil reserves in the Middle East (including North Africa) had increased from 126.2 billion barrels in 1955 to 433.7 billion barrels in 1972, around two-thirds of the world total. Middle Eastern production reached 21.7 million bpd in 1972, half of all noncommunist world production.[22]

As a result of these changes, Middle Eastern oil had come to play a crucial role in the world energy situation. In 1972, Middle Eastern oil accounted for 47 percent of total Western European energy consumption and 57 percent of Japanese energy consumption; around 80 percent of Western Europe's and Japan's oil imports came from the Middle East and North Africa. Middle Eastern oil constituted a much smaller share of U.S. energy consumption—only 2 percent in 1972—but the high absolute level of U.S. oil consumption and the key role of oil in transportation made even this small amount significant in the face of supply difficulties. U.S. annual per-capita oil consumption increased from 19.6 barrels in 1960 to 28.6 barrels in 1972. Moreover, U.S. oil imports were increasing, rising from 18.6 percent of U.S. oil consumption in 1960 to 23.2 percent in 1971. In addition, 85 percent of the oil used by U.S. forces in Southeast Asia came from the Persian Gulf.[23]

The shift in the geography of world oil provided Middle Eastern oil producers with leverage to gain higher prices. In January 1970 the Libyan government demanded that the oil companies pay higher prices for Libyan oil. Libyan oil was more favorably located in relation to oil markets in Western Europe and the United States and had a higher API gravity and lower sulfur content than most crude from the Persian Gulf. The Libyans argued that they, and not the oil companies, should receive the rents resulting from the transportation cost and quality differentials. Colonel Qaddafi warned that his country could afford to live without oil revenues while it trained its own technicians if necessary.[24]

Qaddafi made his move at an opportune time. Nigerian oil, which competed with Libyan oil in Western Europe and the United States, was still recovering from the disruption caused by civil war. Iraqi production, part of which moved to the Mediterranean by pipeline, was down because of the ongoing dispute between the Iraqi government and the Iraq Petroleum Company. In May 1970 an accident in Syria shut down the Trans-Arabian Pipeline, which carried around 500,000 bpd of oil from Saudi Arabia to the Mediterranean. In addition, the Suez Canal was still closed as a result of the 1967 war, forcing tankers to sail around Africa to reach Western Europe. In these circumstances tanker rates tripled, giving Libyan oil an even greater transportation advantage.[25]

The structure of the Libyan oil industry also proved to be to Qaddafi's advantage. To avoid domination by the major oil companies, the Libyan government in the 1950s had invited many small oil companies as well the majors to seek concessions. The result was that by 1970 independent oil companies accounted for around 55 percent of Libyan production. Determined to carve out markets for their oil and unconcerned about the impact of their actions on prices, the independents cut prices in order to sell their oil. Libyan oil production soared from 180,000 bpd in 1962 to 3.3 million bpd by 1970.[26]

The largest producer in Libya was Occidental Petroleum Company, an independent U.S. oil company totally dependent on Libyan oil to supply its Western European refineries. Charging that the companies had not followed sound conservation policies, the Libyan government ordered a drastic cut in production. Occidental's production fell from 800,000 to 400,000 bpd, and other companies suffered similar but smaller cuts. After failing to convince Exxon and Shell to supply his company with oil at cost so he could resist Qaddafi's demands, Occidental head Armand Hammer agreed in early September 1970 to a price increase of 30 cents per barrel and an increase in the tax rate on company profits from 50 to 58 percent. Unable to agree on a common strategy, the other oil companies soon agreed to similar price and tax increases.[27]

Realizing their bargaining power, the other OPEC members decided to press the oil companies for price and tax increases equal to those won by Libya. In response, the oil companies decided to present a united front to OPEC. Under the guidance of oil company counsel John J. McCloy, and in cooperation with the U.S. Departments of State and Justice, the companies drafted a message informing OPEC that they would negotiate only if the settlement covered all companies and all countries. The Justice

Department issued business review letters permitting U.S. companies to take collective action without violating the antitrust laws, and the companies operating in Libya signed an agreement pledging them to support each other in case of cutbacks by Libya. Negotiations were set to begin on January 19, 1971, in Tehran.[28]

Analyses prepared by the NSC concluded that U.S. options were limited. It was not clear if the companies could win a confrontation with the producing countries, and adopting a strong position in favor of the companies could further erode the U.S. position in the Middle East. Moreover, to an extent, higher oil prices were in the U.S. economic interest. The United States was a major oil producer and home to five of the seven great international oil companies (the so-called Seven Sisters), so higher prices benefited some sectors of the U.S. economy. Indeed, as far as U.S. companies with production in OPEC countries were concerned, higher oil prices were not a problem as long as they applied to all companies. Higher oil prices for Western Europe and Japan would end the economic advantage over the United States they had enjoyed as a result of low oil prices. Finally, the companies were not opposed to higher prices as long as they could achieve some assurance of stability.[29]

Taking a strong stand against higher oil prices also risked a disruption in oil supplies. Kissinger warned Nixon that, if the companies held the line against Libyan demands, Libya would probably cut off oil exports and would be joined by Iraq and possibly other Arab states. Because of the world tanker shortage, the continued closure of the Suez Canal, and the disruption of the Trans-Arabian Pipeline, a Libyan-Iraqi shutdown would cause significant oil shortages in Europe. The United States no longer had the spare capacity to help compensate for a cutoff of oil supplies to Europe and could free oil for Europe only by rationing oil domestically. Disruptions in oil supply could also damage the international economy, lead to the nationalization of U.S. oil company holdings, impair European security, and increase pressure on the United States to change its policies toward Israel. In addition, the Europeans might try to head off shortages by striking government-to-government deals with Libya, at the expense of U.S. and British companies.[30]

As a result of these concerns, the United States apparently decided not to confront the producing countries. The shah convinced U.S. envoy John Irwin II, who had gone to Tehran to explain the U.S. position, that negotiations with the Persian Gulf producers should be held separately from those with Mediterranean producers (Libya, Algeria, and the portion of Iraqi oil that arrived at the Mediterranean by pipeline). The shah claimed that it would not be possible for Iran and the other Persian Gulf producers to dictate terms to Venezuela or the radical Arab producers, so any attempt by the oil companies to demand that all producers accept the same terms could only be seen as bad faith. Moreover, the Mediterranean producers led by Libya refused to negotiate until after the negotiations with the Persian Gulf producers were concluded. The Tehran negotiations between the companies and the Persian Gulf producers provided for a price increase of 35 cents a barrel, a tax rate of 55 percent, and regular increases in prices during 1971–75 to offset inflation. In April the Tripoli Agreements raised prices an additional 90 cents per barrel for the Mediterranean producers. A month later Nigeria, which had joined OPEC during the Tehran Conference, received terms similar to the Mediterranean producers.[31]

The rise in oil prices in combination with increases in the volume of exports greatly increased the earnings of the major Middle Eastern oil-producing countries. Earnings prospects received further boosts in January 1972 and again in June 1973 when OPEC won an 8.5 percent and then an 11.9 percent revenue adjustment to offset higher than expected inflation and declines in the value of the dollar.[32] Saudi Arabia's oil revenues increased from $1.2 billion in 1970 to $2.7 billion in 1972; Iran's from $1.1 billion in 1970 to $2.4 billion in 1972; Kuwait's from $800 million in 1970 to $1.4 billion in 1972; and Libya's from $1.4 billion in 1970 to $1.6 billion in 1972. Iraq's oil revenues increased from $500 million in 1970 to $800 million in 1971 before falling to $600 million in 1972 as a result of nationalization.[33]

The U.S. Central Intelligence Agency (CIA) calculated that the dramatic increase in oil revenues (around 80 percent of export earnings) would allow Iran "simultaneously to continue its 10 percent growth rate, to expand its military establishment, and generally to increase its foreign reserves."[34] The shah responded by increasing his spending, especially on arms. In talks with the shah in May 1972, President Nixon had agreed to sell Iran laser-guided bombs and sophisticated F-14 and F-15 aircraft. The U.S. military opposed handing over such sophisticated weaponry to Iran, but, in part because of concerns about U.S. arms companies losing sales to other nations, Nixon's promise was soon transformed into a "blank check" for the shah to buy whatever arms he desired. In a July 25 memorandum to the secretaries of state and defense, Kissinger pointed out that "the President has also reiterated that, in general, decisions on the acquisition of military equipment should be left primarily to the government of Iran."[35]

In addition to raising prices and taxes, OPEC countries began to assert themselves on questions of ownership and control. In 1968, OPEC had adopted a resolution declaring that producing countries had the right to participate in the ownership of their oil. Shortly after the Tehran and Tripoli agreements, Venezuela passed its Hydrocarbons Reversion Law, which provided for government takeover of the oil industry when the companies' concessions expired in the 1980s and for an immediate increase in government control of the oil companies. In September 1971, OPEC called for member countries to begin negotiations with the oil companies on the question of participation in the ownership of their oil operations.[36]

In theory, participation was supposed to provide a moderate alternative to nationalization, but the distinction between the two became blurred in practice. Algeria nationalized its oil industry in February 1971, with the government taking a 51 percent share in all companies. In December 1971, Libya nationalized British Petroleum's holdings in retaliation for Britain's supposed role in permitting Iran to seize islands in the Persian Gulf. In June 1972, Iraq nationalized the Iraq Petroleum Company after an eleven-year dispute.[37]

Negotiations between the Arab Gulf states and the oil companies resulted in the General Agreement on Participation of October 1972, which provided for 25 percent government participation to take effect on January 1, 1973, with scheduled increases leading to 51 percent government ownership by the beginning of 1982. The agreements compensated the companies on the basis of updated book value and guaranteed them the right to purchase most of the production of the oil that now belonged to the governments. Iran, which had technically owned its own oil industry since 1951, reached an agreement with the consortium in the spring of 1973 that provided for Iranian management and control of the industry, with the companies continuing to provide operating services and purchasing the bulk of Iranian production at prices equal to those received by the other Gulf states. Not to be outdone, the Libyan government nationalized 51 percent of all the companies operating in Libya in August and September 1973.[38]

There was little the United States, Britain, or the oil companies could do to oppose these actions. As James Akins, head of the State Department Office of Fuels and Energy, pointed out in a widely circulated study in March 1972, the world oil economy had entered a "seller's market." In the short term, allowing prices to rise and gradually ceding control over ownership could avoid a cutoff of oil supplies and the use of oil as a political weapon. Accommodating the producing countries would also buy time until alternative sources of oil could be developed, thus lessening OPEC's leverage. Although higher oil prices could feed inflation and slow economic growth, they could also stimulate increased investment and production, especially in such high-cost areas as Alaska, offshore fields, Canada, Mexico, Venezuela, and the North Sea, making the West less dependent on Middle Eastern oil. Higher oil prices could also encourage conservation and increased efficiency in oil use and increased utilization of alternative sources of energy, in particular coal and nuclear power.[39]

Akins's study prompted the NSC staff to recommend that the government undertake a study of the impact of energy on U.S. national security. Nixon and Kissinger were preoccupied with other matters such as the Vietnam War, the Strategic Arms Limitation talks, and the opening to China and did not approve such a study until March 1973. The resulting study, "National Security and U.S. Energy Policy," completed in August, focused on how increasing oil imports affected U.S. national security. After the United States ended its Mandatory Oil Import Program in April 1973, the import share of U.S. oil consumption rose to 36.3 percent for that year, with about 30 percent of this amount coming from the Middle East. The report foresaw increasing U.S. dependence on imported oil, very large increases in total world demand for oil, and strains on the balance of payments of consuming nations. Middle Eastern producers would meet most of the growth in world demand, and all producers would enjoy sharp increases in revenues. The structure of the world oil market would also change through increased ownership and marketing by producer governments.[40]

The increase in oil imports meant that the United States would become increasingly vulnerable to short-term supply interruptions. The risk was great because a growing share of imports would come from the volatile Middle East, and many of the key producers would have sufficient financial reserves to be able to limit production for a long period without sacrificing their living standards. Although the United States could still meet its military needs for oil, a sharp reduction in oil supplies caused by war in the Middle East or "a politically motivated decision to embargo oil shipments to the U.S." could cause severe economic problems.

The report recognized that growing Arab frustration over lack of progress toward an Arab-Israeli settlement, pressure from radical Arab states, or a radical takeover of Saudi Arabia or other Gulf state could result in the use of oil as a political weapon to force a change in U.S. policies toward Israel. To avoid this, the United States had to "show some movement on the Arab-Israeli problem" and try harder to accommodate the security and economic concerns of the "moderate" Arabs. The "overriding concern" should be to prevent Saudi Arabia from becoming radicalized or "falling under the control of a Qaddafi." There was not much that could be done to lessen U.S. dependence on imports from the Middle East in the short term, but in the longer term the United States should seek to diversify away from Middle Eastern sources and, within the Middle East, away from Saudi Arabia toward Iran, and to build up domestic supply alternatives. The NSC believed that if the United States developed voluntary and mandatory rationing plans, enlarged oil stockpiles, and import sharing plans with its allies it could probably withstand a total cutoff of oil imports well into the 1980s.[41]

Time ran out, however, on October 6, 1973, when Egyptian and Syrian forces attacked Israeli positions in the Sinai and the Golan Heights. The following day, Iraq nationalized Exxon's and Mobil's shares in the Basra Petroleum Company and announced a halt in oil shipments to the United States. Despite warnings from the oil companies that Saudi Arabia would respond to U.S. support of Israel with an oil boycott, the United States began resupplying Israel with arms to replace war losses. On October 17, the Organization of Arab Petroleum Exporting Countries (OAPEC) announced that its members would cut back oil production and exports. Created in 1968 by Saudi Arabia, Kuwait, and Libya, OAPEC in 1973 also included Algeria, Bahrain, Egypt, Iraq, Qatar, Syria, and the United Arab Emirates. Three days later, after the United States announced an increase in aid to Israel, OAPEC imposed an embargo on oil shipments to the United States and the Netherlands. Although it was an OAPEC member, Iraq did not participate in the cutbacks. Instead, it instituted its own boycott of the United States and the Netherlands and other countries targeted by OAPEC and increased its overall oil production.[42]

The oil companies carried out the embargo, but they undercut its political purpose by shifting non-Arab oil to the embargoed countries and distributing the cutbacks so that both embargoed and non-embargoed countries had their oil imports cut by about 15 percent.[43] Although increased production in Canada, Indonesia, Iran, and the Soviet Union partially offset the decline in Arab production, the price of oil skyrocketed, reaching $17 per barrel on the spot market. On December 22, the Persian Gulf producers, acting on behalf of OPEC, raised posted prices to $11.65 per barrel, effective January 1, 1974.[44]

As a result of the price increases, the major oil-exporting countries' revenues skyrocketed. Higher prices benefited the major Middle Eastern oil producers, which, with the exceptions of Libya and Iraq, were conservative pro-Western regimes. From the perspective of the Nixon Doctrine, higher oil prices helped Iran build up its military power and replace Great Britain as "guardian of the gulf" and allowed Saudi Arabia to continue financial support to Egypt to help Sadat reduce his ties with the Soviet Union and reorient Egypt's internal politics away from Arab socialism.[45]

Iran's oil revenues rose from around $2.4 billion in 1972 to $4.4 billion in 1973, $17.8 billion in 1974, and $21.2 billion in 1977.[46] Increased revenues allowed the shah to buy more arms. Between 1972 and 1977, U.S. military sales to Iran amounted to $16.2 billion. Iran's defense expenditures increased 680 percent, from $1.4 billion in 1972 to $9.4 billion in 1977, around 40 percent of Iran's budget. Iran became the largest single purchaser of U.S. military equipment, bolstering its position as the strongest military power in the Persian Gulf region. Non-military trade also increased. Between January 1973 and September 1974, U.S. companies signed contracts and joint ventures with Iran totaling $11.9 billion, and in March 1975 the United States and Iran signed an economic agreement that committed Iran to spend $15 billion on U.S. goods and services over the following five years.[47]

On the other hand, higher oil prices fed inflation and slowed economic growth, thus weakening the West. Although prices stabilized after 1974, and even declined slightly in real terms, some U.S. officials, notably Treasury Secretary William E. Simon, believed that the balance had swung too far toward Iran. Simon found an ally in the Saudis, who believed that excessive price increases could lessen their power by fostering the development of alternatives to Middle Eastern oil. Kissinger, on the other hand, opposed putting pressure on the shah, pointing out that Iran had not participated in the 1973–74 embargo and was an important ally. Iran could not afford lower prices because the shah needed money for arms and development. Eventually, in late 1976, Simon convinced President Gerald Ford to support the Saudis in opposing further price increases.[48]

When the OPEC conference in Doha in December 1976 decided to raise prices by 10 percent in January 1977 and a further 5 percent in July 1977, Saudi Arabia and the United Arab Emirates refused to go along with the decision and raised their prices by only 5 percent. In addition, the Saudis announced that they would hold prices down by increasing production. Instead, OPEC reached a compromise under which the Saudis raised their prices 5 percent in July, bringing them into line with the rest of OPEC, and the other OPEC members agreed to forego the increase scheduled for July.[49]

From the Iranian Revolution to the Carter Doctrine

By 1977, Iran's expenditures were outrunning its revenues, and rapid but uneven economic growth and wrenching social change increased discontent with the shah's regime.[50] Widespread demonstrations against the regime began in early 1978, after a visit during which President Carter praised the shah for making Iran "an island of stability in one of the more troubled areas of the world."[51] The unrest escalated during the year, but Carter and his top officials were busy with other important issues, including Camp David, SALT, negotiations with China, and violence in Nicaragua. By the time they realized how serious the situation was, it was too late to take action to save the shah's regime.[52]

In early December 1978, national security advisor Zbigniew Brzezinski warned President Carter that the United States was confronting the beginning of a major crisis, "in some ways similar to the one in Europe in the late 40's," in an area stretching across the globe in an "arc of crisis" from Bangladesh to Aden. "Fragile social and political institutions in a region of vital importance to us are threatened with fragmentation," and, to make matter worse, pro-Soviet elements could fill the resulting political vacuum. Brzezinski was especially concerned about Iran. At the end of December he warned, "The disintegration of Iran, with Iran repeating the experience of Afghanistan would be the most massive American defeat since the beginning of the Cold War, overshadowing in its *real consequences* the setback in Vietnam."[53]

Unable to convince him to crush the revolt with military force or to organize a military coup to save his regime, the United States reluctantly gave up on the shah, who fled Iran on January 16, 1979.[54] Despite U.S. hopes that the military would rally behind the caretaker government the shah left to run Iran, it soon collapsed, and power passed to a coalition of opposition forces including supporters of the Ayatollah Khomeini, who had returned to Iran from exile at the beginning of February.[55]

The Iranian revolution underlined the West's dependence on Middle Eastern oil. Although the Middle East's share of world oil production declined from 42.2 percent in 1973 to 40.5 percent in 1978, Middle Eastern oil was still vital in the world oil economy. World oil consumption increased from 54.9 million bpd in 1974 to 62.8 million bpd in 1978. Partially shielded from higher oil prices by price controls, U.S. oil consumption increased from 16.6 million bpd in 1974 to around 18.7 million bpd in 1978, with imports accounting for 36.8 percent of consumption in 1974 and 44.7 percent in 1978.[56]

In November 1978, Iranian oil production fell from 5.8 to 1.1 million bpd as oil workers went on strike. Although production soon recovered to 3 million bpd, the largest consortium members, British Petroleum and Shell, cut back sales to other companies, forcing their former customers into the spot market for supplies. In late December, Iran halted oil exports as production dropped below domestic requirements, and this forced more companies into the spot market. Even though increased production in Saudi Arabia and elsewhere largely made up for the loss of Iranian oil, the disruption in trading channels and fears of further unrest in the region drove up spot market prices as companies scrambled for supplies and built up their inventories in anticipation of further price increases. Spot market prices doubled, and OPEC raised its official base price to $14.55 a barrel in March 1979 and permitted individual members to add surcharges if they wished. Prices continued to soar, and at its June meeting OPEC raised the official ceiling to $23.50 a barrel.[57]

The fall of the shah, fears of internal unrest in Saudi Arabia, and soaring oil prices convinced U.S. policymakers that the previous policy of reliance on regional surrogates to guard Western interests in the Middle East was no longer viable. According to Brzezinski, the overthrow of the shah "destroyed the strategic pivot of the U.S.-sponsored shield for the Persian Gulf region." There was also the danger that the Soviets might take advantage of the conditions in Iran to install a pro-Soviet regime.[58]

Concerns about Western vulnerabilities in the Middle East had existed from the onset of the Carter administration. In its first comprehensive assessment of the U.S.-Soviet strategic balance, the administration had identified the Persian Gulf as a vital and vulnerable region and singled out Iran as a potential trouble spot. In addition, Brzezinski believed that Soviet and Cuban involvement in the Horn of Africa, the northeast part of the continent close to the Arabian Peninsula, threatened Western access to the Persian Gulf and its oil and required a strategic response, including forging closer ties with China. He also pointed to Afghanistan, where local communists had seized control in a bloody coup in April 1978. Although there was no indication that the Soviets were behind the coup, the Soviet Union immediately recognized the new government and began sending it military and economic assistance.[59]

Reports about an impending oil shortage in the Soviet Union further fed fears of Soviet designs on the Persian Gulf. In 1977 three studies had predicted that Soviet oil production would peak in 1980 and decline sharply thereafter, forcing the Soviet Union and its Eastern European allies to look outside the Soviet bloc to meet their oil needs. After being mentioned in connection with President Carter's energy plan, some of the studies were made public.[60]

In July 1977, Brzezinski had recommended that the United States improve its capacity to intervene in the Persian Gulf and other non-European areas through the creation of a "highly responsive, global strike force." A month later, President Carter directed the U.S. military to establish a deployment force of light divisions with strategic mobility to deal quickly with crises in such areas as the Persian Gulf. Because of competing priorities and opposition from the military, little was done to implement this directive.[61]

The leaders of Saudi Arabia and other U.S. allies in the Gulf were concerned about the course of events in Iran, fearing that leftist forces, supported by the Soviet Union, could seize control of the revolution. They also feared that Soviet support of leftist governments in Ethiopia and South Yemen and Soviet policy toward the Gulf in general reflected a plan to gain strategic control of the region's oil supplies.[62]

On March 3, 1979, Brzezinski sent a memorandum to the president recommending a "consultative security framework" for the Middle East and a dramatic increase in U.S. military capabilities and presence in the Persian Gulf region. The fall of the shah "added a new and dangerous dimension" to the crisis in the Middle East, and U.S. allies in the region lacked confidence in the willingness of the United States to use its power on behalf of their security. Brzezinski warned that, in addition to the Arab-Israeli conflict and the political radicalism that fed on that conflict and on the vast inequality in wealth in the region, the Soviets were trying to exploit these problems to displace U.S. influence and expand their own "for ideological, strategic, and economic purposes." Instability in the Middle East, he warned, interacted closely with U.S., Western European, and Japanese economic conditions. Brzezinski also recommended that the United States take steps to reduce its dependence on Middle Eastern oil and coordinate action with Western Europe and Japan.[63]

Brzezinski sent Carter an overview of an NSC study on March 30 that argued that the Soviet Union had strengthened its position in the Middle East over the previous two years by gaining new and important bridgeheads in Ethiopia, Afghanistan, and South Yemen while losing a less important one in Somalia. In contrast, the fall of the shah had reduced U.S. influence in the region, with "major implications for the regional balance of power, domestic stability in neighboring areas, the world oil supply, and U.S. intelligence and security interests." To counter Soviet gains, United States should provide technical assistance and political support to strengthen pro-Western regimes in the area; move rapidly to implement previous plans to establish a "quick reaction force" that could be deployed in the area if necessary; increase the U.S. military presence in the Persian Gulf area; and reconstitute CIA covert action capabilities in the area.[64]

At the Vienna summit in June, Carter told Soviet leader Leonid Brezhnev that there were some areas of the world, such as the Persian Gulf and the Arabian Peninsula, where the United States and its allies had "absolutely vital interests." In such areas it was essential that the Soviets exercise restraint in order not to violate U.S. national security interests. Carter also claimed that the United States had not interfered in the internal affairs of Iran and Afghanistan, and it expected similar restraint from the Soviet Union.[65]

Carter's remarks ignored Soviet interests in the region, including concerns about instability in contiguous areas and rising Islamic fundamentalism. Brezhnev told Carter not to blame the Soviet Union for all the changes taking place in the world. He specifically denied that the Soviet Union was trying to gain control of countries in "the so-called arc of crisis" extending from East Africa to South Asia as a way of surrounding the Persian Gulf. The whole idea, he protested, was a "complete fairy tale." Brezhnev also complained about the casual way the United States claimed that remote corners of the world were vital to its interests.[66]

Carter's statement at Vienna reflected the views of his top national security advisors, whom he had instructed to study U.S. interests in the Persian Gulf and Middle East. After looking at the impact of the fall of the shah, Soviet activities in Afghanistan, the Horn of Africa, and Yemen, and increasing U.S. dependence on oil imports, they concluded that "the United States has vital interests in the Middle East area and the Soviet Union, by comparison[,] does not." To guard those interests, the United States should improve its "military surge capabilities" and increase its military presence in the region. The U.S. objective should not be an equal military balance with the Soviet forces in the region but rather "a perceptible military preponderance" in favor of the United States.[67]

Shortly after the Vienna summit, the United States began sending aid to the forces fighting the communist-led government of Afghanistan. Saudi Arabia, Egypt, and Pakistan were already aiding the Afghan antigovernment forces. Brzezinski later boasted that this aid was intended to provoke the Soviets into invading Afghanistan, revealing their "true" nature and getting them bogged down in a Vietnam-like quagmire.[68]

The storming of the U.S. embassy in Tehran by radical supporters of the Ayatollah Khomeini in November 1979 and the taking of more than fifty U.S. diplomats hostage intensified the U.S. search for bases and access rights in the region. In December an interagency intelligence memorandum warned that U.S. influence in the area had declined and that "manifestations of anti-American feeling" had increased. Although the changed circumstances were mainly explained by internal factors such as the Arab-Israeli conflict, higher oil prices, and "the resurgence of a politicized Islam and a rejection of Western culture," the CIA believed that the Soviets were "abetting the growing instability in the region" in order to expand their influence there. The greatest potential for substantial Soviet gains in the near term, the report warned, was in Iran, "where continuing serious instability could give way to a leftist regime more sympathetic to the USSR." Even if Khomeini was able to restore order, which the CIA viewed as unlikely, "it may only be the precursor to more determined efforts to export the revolution."[69]

The United States was seriously considering military action against Iran when Soviet intervention in Afghanistan on Christmas Day 1979 changed the strategic context. The intervention not only seemed to confirm fears that the Soviets would take advantage of regional turmoil to advance their interests; it also led the United States to reconsider moving militarily against Iran. According to Brzezinski, any action against Iran had to be weighed against its impact on the larger goal of containing Soviet ambitions. In particular, Brzezinski believed that to mobilize Islamic opposition to the Soviets the United States had to avoid a military confrontation with Iran.[70]

Brzezinski wrote Carter on December 26 that the United States was "now facing a regional crisis," with both Iran and Afghanistan in turmoil and Pakistan "both unstable internally and extremely apprehensive externally." If the Soviets succeeded in Afghanistan and Pakistan acquiesced, "the age-long dream of Moscow to have direct access to the Indian Ocean will have been fulfilled." The Iranian revolution had led to "the collapse of the balance of power in Southwest Asia, and it could produce Soviet presence right down on the edge of the Arabian and Oman Gulfs."[71]

The CIA, on the other hand, did not believe that the Soviet intervention in Afghanistan constituted the "beginning of a premeditated strategic offensive" but rather a reluctant response to what Soviet leaders feared was the "imminent and otherwise irreversible deterioration" of their position in a country on their border. Though noting that the Soviets would probably try to take advantage of the situation in Iran, the CIA did not believe that the Soviet move in Afghanistan presaged similar action against Iran.[72]

Consciously drawing on his understanding of the Truman Doctrine, Brzezinski wanted to use the Soviet invasion of Afghanistan to push through plans to gain support for more aggressive policies toward the Soviet Union, including sharp increases in military spending and establishing a strategic relationship with China.[73] After NSC discussions of measures the United States could take to respond to the Soviet action, Brzezinski recommended to the president that the United States respond by altering its policy on selling arms to China and "more broadly" by moving "deliberately to fashion a wider security arrangement for the region." Otherwise, Soviet influence could spread rapidly from Afghanistan to Pakistan and Iran, which "would place in direct jeopardy our most vital interests in the Middle East."[74]

Brzezinski explained his reasoning in an interview published in mid-January. The Middle East and Persian Gulf area had joined Western Europe and the Far East as areas of "vital interest," and the fates of these three areas were interdependent, so that a threat to the security and independence of one was a threat to the other two. The Soviet invasion of Afghanistan had altered the strategic situation in Southwest Asia and brought the Soviet Union "within striking distance" of the Indian Ocean and even the Straits of Hormuz. It would enable the Soviets to intimidate Iran and Pakistan, thus putting them on the edge of an area of vital strategic interest and economic significance to the survival of Western Europe, the Far East, and ultimately the United States.[75]

President Carter repeated these themes in his State of the Union address on January 23, 1980, linking the Iranian hostage crisis with the Soviet invasion of Afghanistan as serious challenges to the United States.[76] According to the president, three basic developments had led to the situation the United States faced: the steady growth of Soviet military power and its increased projection abroad; the overwhelming dependence of the Western democracies on Middle Eastern oil; and the turmoil resulting from social, religious, economic, and political change in the developing world, as exemplified by the revolution in Iran.

Carter called the Iranian seizure of American hostages an act of "international terrorism" that violated "the moral and legal standards of the civilized world." He saw the Soviet invasion of Afghanistan as a "broader and more fundamental challenge," the implications of which "could pose the most serious threat to the peace since the Second World War." The Persian Gulf contained more than two-thirds of the world's exportable oil, and Soviet control of Afghanistan would put Soviet military forces within 300 miles of the Indian Ocean and close to the Straits of Hormuz, "a waterway through which most of the world's oil must flow." The Soviets, he concluded, were "attempting to consolidate a strategic position, therefore, that poses a grave threat to the free movement of Middle East oil."

To meet this threat the President announced what became known as the Carter Doctrine: "An attempt by any outside power to gain control of the Persian Gulf region will be regarded as an assault on the vital interests of the United States of America, and such an assault will be repelled by any means necessary, including military force." The long list of steps that the United States would take included increasing military spending, improving U.S. capability to deploy military forces rapidly around the world, working for peace in the Middle East, normalizing relations with China, increasing the U.S. naval presence in the Indian Ocean, making arrangements for facilities that U.S. forces could use in northeast Africa and the Persian Gulf, and providing Pakistan with additional military and economic assistance.

Finally, noting that the crises in Iran and Afghanistan dramatized the fact that "excessive dependence on foreign oil is a clear and present danger to our nation's security," Carter called for a "clear, comprehensive energy policy" that would include a major conservation effort, initiatives to develop solar power, decontrol of oil prices, incentives for the production of coal and other fossil fuels in the United States, and a "massive" investment in the development of synthetic fuels.

Secretary of Defense Harold Brown echoed Carter's concerns in an interview published in mid-February. Concerned that many people did not understand the threat from the Soviet invasion of Afghanistan, Brown warned that the Soviets would be "sorely tempted" to push farther to gain control of the region's oil. "If the industrial democracies are deprived of access to those resources, there would almost certainly be a worldwide economic collapse of the kind that hasn't been seen for almost 50 years, probably worse." "Even an energy self-sufficient America would not be secure in a world in which Western Europe or Japan or Turkey or Brazil could be made energy hostages of a hostile power," Brown noted a few weeks later.[77]

In March, the Defense Department established the Rapid Deployment Joint Task Force at MacDill Air Force Base in Florida. The United States also began negotiations to secure access to facilities in the region and made preparations to preposition equipment on land and on special ships.[78]

In one of its last official acts, the Carter administration elevated the status of the Persian Gulf in terms of U.S. strategic priorities. According to the NSC, Western Europe, especially after NATO's decision in December 1979 to modernize its theater nuclear forces, was more secure, and the development of a strategic partnership with China had improved the strategic balance in East Asia. In contrast, the U.S. ability to defend Western interests in the Persian Gulf region needed to be improved, and this required a greater allocation of U.S. resources to the defense of the Persian Gulf. In a pair of presidential directives signed on January 15, 1981, the Carter administration assigned the Persian Gulf region top priority for resources in the Five Year Defense Plan and second place, after Western Europe, in terms of planning for wartime operations.[79]

Making Connections: Oil and Geopolitics in the 1970s

Looking at Iran and the geopolitics of oil provides a fuller picture of the history of the 1970s. Concerns about access to oil deeply affected U.S. perceptions of the Soviet threat, and cold war concerns strongly influenced the U.S. response to the decade's oil crises.

Essential to both military power and the functioning of modern society, oil fueled American power and prosperity during the twentieth century. Possession of ample domestic oil supplies and control over access to foreign oil helped the United States and its allies win both world wars and played an important role in U.S. efforts to contain communism during the cold war.[80]

In the 1970s, however, domestic oil production began to decline, forcing the United States to import ever-increasing amounts of oil to meet its energy needs. At the same time, war and revolution in the Middle East, which had become the center of world oil production, and the changing dynamics of the cold war raised questions about the ability of the United States to maintain access to the region's oil. When the British decided to pull their military forces out of the Middle East in the midst of the Vietnam War, the Nixon administration turned to Iran to guard Western interests in the Persian Gulf.

Although not directly related to the cold war, the oil crises of the 1970s evoked images of a weakened West, especially since they coincided with the U.S. withdrawal from Vietnam, the Watergate crisis, a wave of revolutions in the third world, and the Soviet Union's achievement of nuclear parity with the United States. In addition to the economic impact, the realization that access to oil at low prices was threatened undermined confidence in Western dominance. The impact was especially sharp in the United States because American popular culture equated the private automobile and personal mobility with individual freedom.

The Soviet Union, in contrast, was an unintended beneficiary of the oil price increases in the 1970s. As new fields in western Siberia entered production, the Soviet Union overtook the United States as the world's leading oil producer in 1974. Although most Soviet oil exports went to Eastern Europe, Cuba, and Vietnam, oil exports to Western Europe and Japan rose sharply during the decade and by 1976 were responsible for half of the Soviet Union's hard currency earnings. The windfall from higher prices allowed the Soviets to import large amounts of Western grain and machinery and may have made it possible for them to increase their involvement in the third world in the 1970s. Reports about an impending oil shortage in the Soviet Union also fed fears of Soviet designs on the Persian Gulf. These concerns, along with the dynamics of the arms race, gave rise to concerns about Soviet "geopolitical momentum" and helped lead to the demise of détente.[81]

Although the sharp increases in oil prices in the 1970s caused severe economic problems for both developed and less developed countries, high oil prices also stimulated the development of higher-cost production outside the Middle East, made alternative sources of energy economically viable, and promoted conservation and efficiency. Higher prices also provided Iran and Saudi Arabia, the two pillars under the Nixon Doctrine, with the money to play their assigned roles in stabilizing the Persian Gulf. Iran and Saudi Arabia bought millions of dollars of military equipment from the United States and other Western countries, and thousands of Western advisors and technicians accompanied the equipment. The result was not increased security, however. The shah's military spending and other efforts to build up Iran were important factors in the demise of his regime.

The Carter administration began considering increasing the U.S. military presence in the Persian Gulf and Indian Ocean region shortly after taking office. The main purpose of building up U.S. capacity to intervene militarily in the region was to be able to deal with indigenous and regional unrest, not to resist an unlikely Soviet attack. The Iranian revolution gave added impetus to these efforts, but it took the Soviet intervention in Afghanistan to overcome divisions within the U.S. government and mobilize public opinion in favor of the United States taking on a direct role in regional security.

Examination of the geopolitics of oil shows how the strategic and economic benefits of controlling world oil significantly shaped U.S. foreign policy in the 1970s. In addition, social and economic patterns that fostered high levels of oil use reinforced U.S. determination to maintain access to foreign oil. Although U.S. officials recognized that part of the solution was to end the U.S. "addiction" to oil, Nixon was not willing to risk political opposition by promoting reduced oil use, and Ford and Carter were unable to convince Congress to take decisive action.[82]

Eventually higher oil prices, along with various conservation policies, led to lower oil use: U.S. per-capita oil consumption peaked at 31.0 barrels a year in 1978 and fell to 27.4 barrels a year in 1980; oil's share of U.S. energy consumption fell from 48.7 percent in 1978 to 45.1 percent in 1980; and U.S. oil imports fell from 47 percent of U.S. supply in 1977 to 37 percent in 1981. Reductions in oil use in Western Europe and Japan were even greater. Most of the decrease in oil use resulted from reduced use of fuel oil for electricity generation and industrial use, however.[83] There was little structural change in the transportation sector, which was the main oil consumer.

The Reagan administration moved away from an emphasis on conservation and emphasized supply-side and military solutions to the problem of dependence on Middle Eastern oil. Supply-side solutions—developing alternative sources of oil and relying more on non-petroleum energy sources—succeeded for a time, but after oil prices collapsed in the mid-1980s, U.S. oil consumption and oil imports resumed their upward path. Supply-side solutions also ignored the environmental impact of increased use of hydrocarbons.

U.S. military power succeeded in thwarting Iraq's attempt to take over Kuwait in 1991, but the sanctions regime put in place after the war to contain Iraq became counterproductive over time. The September 11, 2001, attacks on the United States highlighted the costs of keeping U.S. troops in the Persian Gulf to ensure Western energy security. In addition, although increased oil production outside the Middle East lowered the region's share of world production, these efforts failed collectively to displace Persian Gulf oil from its dominant place in the world oil economy. Moreover, when these policies worked to reduce oil prices, they also reduced incentives to conserve oil and to develop alternatives. As a result, concerns about oil were a factor in the U.S. military intervention in Iraq that began in March 2003, a move the late conservative strategist William E. Odom called "the greatest strategic disaster in American history."[84]

Notes

1. Daniel Yergin, *The Prize: The Epic Quest for Oil, Money, and Power* (New York: Simon and Schuster, 1991), 773; Paul Kennedy, *The Rise and Fall of the Great Powers: Economic Change and Military Conflict from 1500 to 2000* (New York: Random House, 1987); John Lewis Gaddis, *Strategies of Containment: A Critical Appraisal of American National Security Policy during the Cold War*, rev. ed. (New York: Oxford University Press, 2005), 332.

2. Fiona Venn, *Oil Diplomacy in the Twentieth Century* (New York: Macmillan, 1986); Simon Bromley, *American Hegemony and World Oil: The Industry, the State System, and the World Economy* (University Park: Pennsylvania State University Press, 1991); see also David S. Painter, "Oil and World Power," *Diplomatic History* 17 (Winter 1993): 159–70.

3. Henry Kissinger, *Years of Upheaval* (Boston: Little, Brown, 1982), 854.

4. Raymond Garthoff, *Détente and Confrontation: American-Soviet Relations from Nixon to Reagan*, rev. ed. (Washington, D.C.: Brookings Institution, 1994); Odd Arne Westad, *The Global Cold War: Third World Intervention and the Making of Our Times* (Cambridge: Cambridge University Press, 2005); William Bundy, *A Tangled Web: The Making of Foreign Policy in the Nixon Presidency* (New York: Hill and Wang, 1998); Fredrik Logevall and Andrew Preston, eds., *Nixon in the World: American Foreign Relations, 1969–1977* (New York: Oxford University Press, 2008); Melvyn P. Leffler, *For the Soul of Mankind: The United States, the Soviet Union, and the Cold War* (New York: Hill and Wang, 2007; Niall Ferguson, Charles S. Maier, Erez Manela, and Daniel J. Sargent, eds., *The Shock of the Global: The 1970s in Perspective* (Cambridge, Mass.: Harvard University Press, 2010).

5. Yergin, *Prize*; Fiona Venn, *The Oil Crisis* (London: Longman, 2002); Raymond Vernon, ed., *The Oil Crisis* (New York: W. W. Norton, 1976).

6. Unfortunately, the *Foreign Relations of the United States* volumes on the energy crises of the 1970s were not available when this essay was being prepared.

7. *Public Papers of the Presidents of the United States, Richard Nixon, 1969* (Washington, D.C.: Government Printing Office, 1999), 548–54.

8. Henry A. Kissinger, *The White House Years* (Boston: Little, Brown, 1979), 222–25; see also Jeffrey Kimball, "The Nixon Doctrine: A Saga of Misunderstanding," *Presidential Studies Quarterly* 36 (Mar. 2006): 59–74.

9. *Public Papers of the Presidents of the United States, Richard Nixon, 1970* (Washington, D.C.: Government Printing Office, 1999), 9.

10. See the exchanges in U.S. Department of State, *Foreign Relations of the United States* (hereafter FRUS), *1964–68*, 12: *Western Europe*, docs. 288–291. For the background to the British decision, see W. Taylor Fain, *American Ascendance and British Retreat in the Persian Gulf Region* (New York: Palgrave Macmillan, 2008), 141–68; and Stephen G. Galpern, *Money, Oil, and Empire in the Middle East: Sterling and Postwar Imperialism* (Cambridge: Cambridge University Press, 2009), 268–76.

11. Fain, *American Ascendance*, 172.

12. Ibid.; FRUS 1969–76, 24, doc. 82, and E-4, doc. 91; and papers by NSC staff member Chester Crocker, "The United States and the Persian Gulf in the Twilight of the Pax Britannica," attached to Crocker to Kennedy, Oct. 12, 1970, National Security Archive,

George Washington University (hereafter NSA), Presidential Directives, Part II: PR00510; and "Long-Term U.S. Strategy Options in the Persian Gulf, attached to Kennedy to Kissinger, Dec. 30, 1970, NSA, Presidential Directives, Part II, PR00511.

13. Michael MccGwire, *Military Objectives in Soviet Foreign Policy* (Washington, D.C.: Brookings Institution, 1987), 196–203; Robert G. Patman, *The Soviet Union in the Horn of Africa: The Diplomacy of Intervention and Disengagement* (Cambridge: Cambridge University Press, 1990), 81–86. On Diego Garcia, see Michael A. Palmer, *Guardians of the Gulf: A History of America's Expanding Role in the Persian Gulf, 1833–1992* (New York: Free Press, 1992), 95–96.

14. See the documents discussing the Soviet threat in the Indian Ocean and possible U.S. responses in FRUS 1969–76, 24, docs. 36–71.

15. See the statements of Saudi views in FRUS 1969–76, 24, docs. 127, 128, 147, 151, 155, 170; and U.S. analyses of Saudi views in docs. 133, 140, 146, and 169.

16. FRUS 1969–76, 24, doc. 91. For the reasoning behind Memorandum 92, see docs. 2, 82, 83, and 89; and FRUS 1969–76, E-4, docs, 70, 75, and 91.

17. FRUS 1969–1976, E-4, doc. 122.

18. FRUS 1969–76, 24, docs. 86, 93, 133, 135, 140, 146.

19. FRUS 1969–76, E-4, docs. 86, 29, and 164; Gholam Reza Afkhami, *The Life and Times of the Shah* (Berkeley: University of California Press, 2009), 271–72, 301–307; Alvin J. Cottrell, "Iran's Armed Forces under the Pahlavi Dynasty," in *Iran under the Pahlavis*, ed. George Lenczowski (Stanford, Calif.: Hoover Institution Press, 1978), 399–400.

20. Charles Issawi, "The Iranian Economy, 1925–1975: Fifty Years of Economic Development," in Lenczowki, *Iran under the Pahlavis*, 137; Ian Skeet, *OPEC: Twenty-Five Years of Prices and Politics* (Cambridge University Press, 1988), 240; FRUS 1969–76, E-4, docs. 86, 165.

21. FRUS 1969–76, E-4, docs. 29, 59, 89, 108; Robert Stobaugh, "The Evolution of Iranian Oil Policy, 1925–1975," in Lenczowki, *Iran under the Pahlavis*, 228–34.

22. Joel Darmstadter and Hans H. Landsberg, "The Economic Background," in Vernon, *Oil Crisis*, 15–37; see also Steven A. Schneider, *The Oil Price Revolution* (Baltimore: Johns Hopkins University Press, 1983), 49–75.

23. DeGolyer and MacNaughton, *Twentieth Century Petroleum Statistics* (Dallas: DeGolyer and MacNaughton, 2005), 53, 108; Darmstadter and Landsberg, "Economic Background." Contemporary U.S. documents estimated Japanese dependence on Persian Gulf oil at 90 percent.

24. See the analysis prepared by the Interagency Oil Task Force, "The World Oil Situation: NSSM-114," Jan. 24, 1971, NSC Institutional Files, "NSSM-114," Richard Nixon Presidential Library and Museum, Yorba Linda, Calif. (also available in NSA); Frank C. Waddams, *The Libyan Oil Industry* (Baltimore: Johns Hopkins University Press, 1980), 57–59, 191–212.

25. Ian Seymour, *OPEC: Instrument of Change* (London: Macmillan, 1980), 55–62; Waddams, *Libyan Oil Industry*, 267–69.

26. Waddams, *Libyan Oil Industry*, 191–212.

27. U.S. Congress, Committee on Foreign Relations, Subcommittee on Multinational Corporations, *Multinational Oil Corporations and U.S. Foreign Policy* (Washington, D.C.:

Government Printing Office, 1975), 22–25 (hereafter MNC Report); Waddams, *Libyan Oil Industry*, 230–36; Kenneth A. Rodman, *Sanctity versus Sovereignty: The United States and the Nationalization of Natural Resource Investments* (New York: Columbia University Press, 1988), 236–41.

28. MNC Report, 125–30: Yergin, *Prize*, 580–81.

29. C. Fred Bergsten and Harold H. Saunders to Kissinger, "The Developing International Oil Crisis," Jan. 14, 1971, NSC Institutional Files, "NSMM-114," Nixon Library; Kissinger, *Years of Upheaval*, 863.

30. Kissinger, Memorandum for the President, Jan. 18, 1971; and Bergsten and Saunders to Kissinger, "The Developing International Oil Crisis," Jan. 14 ,1971, both in NSC Institutional Files, "NSSM-114," Nixon Library.

31. FRUS 1969–76, E-4, docs. 111, 115; FRUS 1969–76, 24, doc. 94; Stobaugh, "Evolution of Iranian Oil Policy," 238–43. For an overview of the 1971 negotiations, see Yergin, *Prize*, 577–83; Rodman, *Sanctity versus Sovereignty*, 241–45; James Bamberg, *British Petroleum and Global Oil, 1950–1975: The Challenge of Nationalism* (Cambridge: Cambridge University Press, 2000), 447–66; and MNC Report, 130–34.

32. FRUS 1969–76, 24, doc. 94; Stobaugh, "Evolution of Iranian Oil Policy," 243–44.

33. Skeet, *OPEC*, 102–103, 240–42. Outside the Middle East, Venezuela's oil revenues rose from $1.4 billion in 1970 to $1.9 billion in 1972.

34. FRUS 1969–76, E-4, docs. 115, 123, 165; FRUS 1969–76, 24, doc. 94; Stobaugh, "Evolution of Iranian Oil Policy," 238–43.

35. FRUS, 1969–76, E-4, docs. 201, 214; see also documents 204, 205, 210, and 212; James A. Bill, *The Eagle and the Lion: The Tragedy of Iranian-American Relations* (New Haven, Conn.: Yale University Press, 1988), 200–201.

36. Skeet, *OPEC*, 50–57.

37. Rodman, *Sanctity versus Sovereignty*, 245–49, 255–65. On the IPC dispute, see Samir Saul, "Masterly Inactivity as Brinkmanship: The Iraq Petroleum Company's Route to Nationalization, 1958–1972," *International History Review* 29 (Dec. 2007): 746–92.

38. Skeet, *OPEC*, 71–73, 75–78, 80–82; Yergin, *Prize*, 583–85; Rodman, *Sanctity versus Sovereignty*, 249–55, 265–69; Seymour, *OPEC*, 218–30.

39. "The U.S. and the Impending Energy Crisis," November 1972, White House Central Files, Staff and Member Office Files, Subject Files of Charles J. DiBona, Nixon Library. Secretary of State William P. Rogers sent Nixon a draft of the study in Mar. 1972; see Memorandum for the President, Mar. 10, 1972, NSC Institutional Files, "NSSM-174," Nixon Library; see also James E. Akins, "The Oil Crisis: This Time the Wolf Is Here," *Foreign Affairs* 51 (April 1973): 462–90; James E. Akins, "International Cooperative Efforts in Energy Supply," *Annals of the American Academy of Political and Social Science* 410 (Nov. 1973): 75–85.

40. "National Security and U.S. Energy Policy," NSSM-174, August 1973, NSC Institutional Files, "NSSM-174 (Response)"; and Odeen to Kissinger, Aug. 11, 1973, NSC Institutional Files, "SRG Meeting Re: NSSM-174"; both in Nixon Library.

41. "National Security and U.S. Energy Policy," NSSM-174, August 1973; Odeen to Kissinger, Aug. 11, 1973; Odeen to Kissinger, Aug. 15, 1973; and HAK Talking Points, Aug. 16 1973, all in NSC Institutional Files, "SRG Meeting Re: NSSM-174," Nixon Library.

42. Schneider, *Oil Price Revolution*, 222–30; Mary Ann Tétreault, *The Organization of Arab Petroleum Exporting Countries: History, Policies, and Prospects* (Westport, Conn.: Greenwood Press, 1981), 46–50.

43. Robert B. Stobaugh, "The Oil Companies in the Crisis," in Vernon, *Oil Crisis*, 179–202.

44. Yergin, *Prize*, 583–626; Bamberg, *British Petroleum*, 474–89.

45. Schneider, *Oil Price Revolution*, 165; Franz Schurmann, *The Foreign Politics of Richard Nixon: The Grand Design* (Berkeley: University of California Institute of International Studies, 1987), 284–92; Paul Chamberlin, "A World Restored: Religion, Counterrevolution, and the Search for Order in the Middle East, *Diplomatic History* 32 (June 2008): 458–66.

46. Skeet, *OPEC*, 240.

47. Bill, *Eagle and the Lion*, 201–204.

48. Memorandum of Conversation, July 9, 1974; Memorandum of Conversation, Aug. 17, 1974; Memorandum of Conversation, Nov. 10, 1974; Memorandum of Conversation, Feb. 7, 1975; Memorandum of Conversation, Dec. 7, 1976; Memorandum of Conversation, Dec. 14, 1976; all in National Security Adviser, Memoranda of Conversation, 1973–1977, Gerald R. Ford Presidential Library, Ann Arbor, Mich.; Andrew Scott Cooper, "Showdown at Doha: The Secret Oil Deal That Helped Sink the Shah of Iran," *Middle East Journal* 62 (Autumn 2008): 567–91.

49. Skeet, *OPEC*, 134–36.

50. On the relationship between economic growth and the emergence of radical opposition to the shah, see Mark J. Gasiorowski, *U.S. Foreign Policy and the Shah: Building a Client State in Iran* (Ithaca, N.Y.: Cornell University Press, 1991), 142–51, 187–222.

51. Carter praise of shah in Michael H. Hunt, *Crises in U.S. Foreign Policy: An International History Reader* (New Haven, Conn.: Yale University Press, 1996), 400.

52. Zbigniew Brzezinski, *Power and Principle: Memoirs of the National Security Adviser, 1977–1981* (New York: Farrar, Straus, Giroux, 1983), 358; Cyrus Vance, *Hard Choices: Critical Years in America's Foreign Policy* (New York: Simon and Schuster, 1983), 326. For Carter's account, see Jimmy Carter, *Keeping Faith: Memoirs of a President* (New York: Bantam Books, 1982), 438–45. The U.S. ambassador to Iran, William H. Sullivan, chronicles the course of the revolution in *Mission to Iran* (New York, W. W. Norton, 1981). Bill, *Eagle and the Lion*, chap. 7, provides a brief, well-informed account of the U.S. response to the Iranian revolution.

53. Brzezinski to the President, Weekly Report 81, Dec. 2, 1978, Brzezinski Collection, Subject File, Weekly Reports, 71–81; Brzezinski to the President, Weekly Report 83, Dec. 28, 1978, Brzezinski Collection, Subject File, Weekly Reports, 82–90, both in Jimmy Carter Library and Museum, Atlanta, Ga.

54. Special Coordinating Committee (SCC) Meeting, Feb. 11, 1979, NLC-25-37-9-2–3; Brzezinski to the President, Weekly Report 84, Jan. 12, 1979, Brzezinski Collection, Subject File, Weekly Reports 82–90, both in Carter Library; Jimmy Carter, *White House Diary* (New York: Farrah, Straus and Giroux, 2010), 257, 261, 268; Sullivan, *Mission to Iran*, 214–47; Robert E. Huyser, *Mission to Tehran* (New York: Harper and Row, 1986); Afkhami, *Life and Times of the Shah*, 498–510.

55. Brzezinski to the Secretary of State and the Secretary of Defense, Jan. 19, 1979, Brzezinski Collection, Subject File, Alpha Channel (Misc.) 4/78–4/79, Carter Library; Carter, *White House Diary*, 272, 273, 288; Afkhami, *Life and Times of the Shah*, 511–21.

56. BP Statistical Review of World Energy, 2013 (access Statistical Review 1951–2011 via www.BP.com); DeGolyer and MacNaughton, *Twentieth Century Petroleum Statistics*, 53, 108. Percentages for Middle Eastern oil calculated using the total figure for the Middle East plus Algeria, Egypt, and Libya.

57. See Yergin, *Prize*, 684–91, for a succinct analysis.

58. Brzezinski, *Power and Principle*, 193, 356; Henze to Brzezinski, Feb. 13, 1979, (27) National Security Affairs Staff Materials, 2/79, Carter Library.

59. Olav Njølstad, "Shifting Priorities: The Persian Gulf in U.S. Strategic Planning in the Carter Years, *Cold War History* 4 (Apr. 2004): 26, 51, note 19; William E. Odom, "The Cold War Origins of the U.S. Central Command," *Journal of Cold War Studies* 8 (Spring 2005): 57–59; Brzezinski, *Power and Principle*, 177–81, 203–204. On Soviet policy toward Afghanistan, see Westad, *Global Cold War*, 299–306.

60. Brzezinski to the President, NSC Weekly Report 31, Oct. 7, 1977; Military Assistant to Brzezinski, Weekly Report, Sept. 22, 1977; U.S. Central Intelligence Agency, *The International Energy Situation: Outlook to 1985*, ER 77–10240 U, April 1977; CIA, *Prospects for Soviet Oil Production*, ER 77–10270, April 1977; CIA, *Prospects for Soviet Oil Production: A Supplemental Analysis*, July 1977; "Guessing What's There," *Time*, May 9, 1977, 100; Donald L. Bartlett and James B. Steele, "The Oily Americans," *Time*, May 19, 2003, 53–54.

61. Njølstad, "Shifting Priorities," 26–29, 31–32; Brzezinski, *Power and Principle*, 177–78.

62. See the State Department/CIA study, "Regional Implications of Events in Iran," Feb. 20, 1979, NLC-24-102-3-3-6; "Arabian Anxieties: What Soviet Might Do and What U.S. Might Not," *New York Times*, June 16, 1979, 2.

63. Brzezinski to the President, "Consultative Security Framework for the Middle East," Mar. 3, 1979, National Security Affairs Staff Material, Horn/Special, 4/79, Carter Library; Brzezinski, *Power and Principle*, 446.

64. "Comprehensive Net Assessment 1978: Overview," attached to Brzezinski to the President, Mar. 30, 1978, Weekly Report 92, Brzezinski Collection, Subject File, Weekly Reports 91–101, Carter Library; Njølstad, "Shifting Priorities," 32–33.

65. Carter, *Keeping Faith*, 254; Carter, *White House Diary*, 329. For a summary of the meeting, see Memorandum of Conversation, Fourth Plenary Meeting, June 17, 1979, Declassified Documents Reference System (DDRS), CK3100073527.

66. Carter, *Keeping Faith*, 255; Memorandum of Conversation, Fourth Plenary Meeting, June 17, 1979, DDRS CK3100073527; Policy Review Committee Meeting, June 21–22, 1979, NLC-31-11-12-4-8, Carter Library.

67. Policy Review Committee Meetings June 21–22, June 23, 1979, NLC-31-11-12-4-8; Brzezinski to the President, Weekly Report 101, June 22, 1979, Brzezinski Collection, Subject File, Weekly Reports 91–101; Brown, Memorandum for the President, July 11, 1979, Brzezinski Collection, Subject File [Meetings—Vance/Brown/Brzezinski 8/79–9/79]; all in Carter Library; Brzezinski, *Power and Principle*, 447.

68. Robert M. Gates, *From the Shadows: The Ultimate Insider's Story of Five Presidents and How They Won the Cold War* (New York: Simon and Schuster, 1996), 146–48. For Brzezinski's remarks, see David N. Gibbs, "Afghanistan: The Soviet Invasion in Retrospect," *International Politics* 37 (2000), 241–43.

69. CIA, "New Realities in the Middle East," NI IIM 79–10026, Dec. 1979, NLC-6-51-7-7-5, Carter Library.

70. Brzezinski, *Power and Principle*, 482–85; Carter, *White House Diary*, 368, 371, 372, 380.

71. Brzezinski to the President, Dec. 26, 1979, Brzezinski Collection, Geographic File, Southwest Asia/Persian Gulf-Afghanistan [12/26/79–1/4/80], Carter Library.

72. "Soviet Union and Southwest Asia," attached to Turner, Memorandum for the President, Jan. 15, 1980, NLC-133-42-2-2-3, Carter Library.

73. Brzezinski, *Power and Principle*, 431; Melvyn P. Leffler, "From the Truman Doctrine to the Carter Doctrine: Lessons and Dilemmas of the Cold War," *Diplomatic History* 7 (Fall 1983): 245–66.

74. National Security Council Meeting, Jan. 2, 1980, in The *Fall of Détente: Soviet-American Relations during the Carter Years*, ed. Odd Arne Westad (Oslo: Scandinavian University Press, 1997), 332–51; Brzezinski to the President, Jan. 3, 1980, Brzezinski Collection, Geographic File, Southwest Asia/Persian Gulf-Afghanistan [12/26/79–1/4/80], Carter Library.

75. "An Interview with Zbigniew Brzezinski," *Wall Street Journal*, Jan. 15, 1980, 20.

76. Jimmy Carter, *Public Papers of the Presidents of the United States, 1980–1981* (Washington, D.C.: Government Printing Office, 1982), 194–200.

77. "Brown Warns That a Persian Gulf War Could Spread," *New York Times*, Feb. 15, 1980, A3; U.S. Department of State, *American Foreign Policy: Basic Documents, 1977–1980* (Washington, D.C.: Government Printing Office, 1982), 595.

78. Brzezinski, *Power and Principle*, 446, 456; Palmer, *Guardians of the Gulf*, 106–11.

79. Njølstad, "Shifting Priorities," 42–48.

80. See David S. Painter, "Oil and the American Century," *Journal of American History* 99 (June 2012): 24–39.

81. U.S. Central Intelligence Agency, "Soviet Energy Policy toward Eastern Europe: A Research Paper," June 1980, CIA FOIA Electronic Reading Room; Marshall I. Goldman, *The Enigma of Soviet Petroleum: Half-Empty or Half-Full?* (London: George Allen and Unwin, 1980). So far, scholars have found no evidence that Soviet involvement in the so-called arc of crisis was driven by a desire to gain access to the oil resources of the Persian Gulf or to deny the West access to these resources; see Westad, *Global Cold War*, which examines Soviet policy in the third world in the 1970s on the basis of extensive research in Soviet archives as well as secondary sources.

82. See the essays on the Nixon, Ford, and Carter administrations in Craufurd D. Goodwin, ed., *Energy Policy in Perspective: Today's Problems, Yesterday's Solutions* (Washington, D.C.: Brookings Institution, 1981); and Benny Temkin, "State, Ecology, and Interdependence: Policy Responses to the Energy Crisis in the United States," *British Journal of Political Science* 13 (Oct. 1983): 441–62.

83. DeGolyer and MacNaughton, *Twentieth Century Petroleum Statistics*, 53, 108, 110; Edward T. Dowling and Francis G. Hilton, "Oil in the 1980s: An OECD Perspective," in *The Oil Market in the 1980Os: A Decade of Decline*, ed. Siamack Shojai and Bernard S. Katz (New York: Praeger, 1992), 77–80.

84. Michael T. Klare, *Blood and Oil: The Dangers and Consequences of America's Growing Dependency on Imported Petroleum* (New York: Metropolitan Books, 2004); Ian Rutledge, *Addicted to Oil: America's Relentless Drive for Energy Security* (London: I. B. Tauris, 2006); John S. Duffield, "Oil and the Decision to Invade Iraq," in *Why Did the United States Invade Iraq?* ed. Jane K. Cramer and A. Trevor Thrall (London: Routledge, 2011), 145–66. For Odom's remarks at Brown University in April 2006, see www.watsoninstitute.org/events_detail.cfm?id=710; Michael Hammerschlag, "NSA Director Odom Dissects Iraq Blunders," www.hammernews .com/odomspeech.htm.

Chapter 4

The United States, Iran, and the "Oil Weapon"

From Truman to George W. Bush

Steve Marsh

U.S.-Iranian relations have been on a generally downward trend since the 1979 Iranian revolution and the humiliating U.S. embassy hostage crisis. In 2002, President George W. Bush cited Iran as a member of the so-called axis of evil and subsequently accused Teheran repeatedly of destabilizing Iraq and Afghanistan, of supporting terrorism, and of developing a nuclear weapons program. On October 25, 2007, the United States announced a tightening of the sanctions regime upon Iran. Specifically, Executive Order 13382 empowered the authorities to freeze the assets of, and prohibit any U.S. citizen or organization from doing business with, the Iranian Ministry of Defense and the Revolutionary Guard.[1] The obvious hope was that these actions would hurt the Iranian regime directly, deter others from dealing with Iran for fear of American reprisals, and ratchet up the pressure for a stronger line toward Teheran in the United Nations.

Although Iran's alleged support for terrorism and its suspected nuclear weapons program dominated the headlines, oil factors helped to both condition U.S. policy and influence its relative effectiveness. This is nothing new. It is true that Iranian oil per se has never been of great significance in terms of direct American needs and consumption. In 1970 the United

States imported only about 10 percent of its oil, and much of this was from Venezuela. Even during the late 1970s, the United States sourced only approximately 10 percent of its oil imports from its Iranian client, the equivalent of 885,000 barrels per day.[2] By 2001, U.S. oil imports had shot up to 57 percent,[3] but although some sub-Saharan and Persian Gulf States had acquired increased significance, Latin America still provided the bulk of requirements.

Nevertheless, who controls Iranian oil, how it is sold, and for what purposes it is used have long been of U.S. political, strategic, and economic concern. In 1943 the State Department supported an attempt by Socony-Vacuum to establish a concession in Iran, and a consortium deal that resolved the 1950–54 Anglo-Iranian oil crisis finally saw U.S. oil majors enter the former British oil monopoly there. The denial of Iranian oil to the Soviets was also a major concern of successive U.S. administrations throughout the cold war. And the importance of Iranian and wider Middle Eastern oil to Western Europe especially and, by association, to the United States was clearly recognized in American strategic thinking. As early as 1954 the Eisenhower administration estimated that the region would be 90 percent dependent on Middle Eastern oil by 1975 and indefensible in the event of these supplies being lost.[4]

Despite recognizing this dependency concern long ago and becoming the unrivaled superpower upon the end of the cold war, the United States became over time potentially more rather than less vulnerable to Iran's strategic use of oil—both its own oil and its potential to threaten other Middle Eastern supplies or to cooperate with other emergent major energy producers that are not necessarily sympathetic to America. Within this development, processes either begun or accelerated in the 1970s have played an important part, especially the overt disintegration of the oil majors' stranglehold over international oil prices and production and the increasing importance for the United States of market forces in both determining oil prices and, from the second oil crisis onward, responding to energy shocks. At the same time, the failure to develop a long-term energy policy that diversified American energy sources and types sufficiently increased American vulnerability to energy shocks originating from the Middle East and elsewhere. This argument is developed here in three parts. The first analyzes the American response to Iran's first attempt to use its oil as a strategic "weapon" against the United States in the early

1950s. The second projects forward to the George W. Bush era to examine how Iran used its oil weapon to enhance its security and potentially to weaken the United States. The final section focuses on the 1970s to demonstrate how trends developed from, and decisions taken in, this period have contributed not only to a massive U.S. strategic commitment to the Middle East but also to a relative weakening of American ability to counter the Iranian oil weapon.

The Context of the Anglo-Iranian Oil Crisis

American policymakers found the Anglo-Iranian oil crisis of 1950–54 to be a vexing issue that progressively drew them into a mediation role but without sufficient leverage to compel any of the parties to accept a settlement.[5] They initially simply watched to see whether a renegotiation of the Anglo-Iranian Oil Company's (AIOC) 1933 oil concession with Iran would produce a settlement that could have repercussions for American oil interests elsewhere. Beyond this, they regarded the wider situation in Iran with "disinterested optimism." In July 1949 the National Security Council concluded that overt Soviet aggression was unlikely and that the now-outlawed communist Tudeh Party was incapable of overthrowing the regime. The economy was believed backward but "probably about as stable as at any period in recent years." And the government's resources were deemed sufficient to allow it to borrow all the capital it could effectively use from the International Bank for Reconstruction and Development and the Export-Import Bank.[6]

Nevertheless, matters quickly became problematic once control over Iranian oil became a focal point around which disparate sources of domestic discontent and nationalism coalesced. This, aided by a hard-line AIOC approach and complacent British government, caused a supplemental oil agreement that had seemingly resolved the dispute to get stuck in the Iranian Parliament. At the same time that an apparently ordinary commercial renegotiation thus steadily became a political crisis, U.S. perspectives on Iran and the wider Middle East underwent radical change. Shaken by the loss of China and the Soviet atomic test, and viewing the Soviets as contained in Europe after the Berlin blockade and the formation of NATO, U.S. policymakers began to view the Middle East as a natural alternative theater for Soviet ambitions and one that would endanger the West's oil supplies. Iran's oil, shared border with the USSR, political and economic instability, and potential participation in a northern tier for Middle Eastern defense thus quickly promoted Teheran's importance to the United States. This shift in perspective also induced radically more pessimistic assessments of Iran's relative weakness, political leadership, and vulnerability to communism. Within a matter of months the shah of Iran went from being seen as a "dominant, unifying force" to a man possessing monumental indecision, debatable moral fiber, and an attitude perhaps "a little too westernised for an oriental culture."[7] By 1952 the situation in Iran might apparently have been "devised by Karl Marx himself." This was a remarkable transmogrification from April 1948, when Iran had been considered "not an ideal field for the propagation of Communism" on account of Muslim religion, paternalistic tribes, and little apparent inclination in the Iranian people to follow mass movements of any sort.[8]

The Truman administration equated saving Iran from communism with finding both pro-Western Iranian leadership and an oil settlement, not least because oil revenues were Iran's major source of overseas income and critical to restarting a Seven-Year Plan that envisaged £210 million of economic and social development.[9] In addition, an oil settlement would help protect Western European economic recovery and rearmament. It was estimated in January 1951 that a loss of Iranian oil would incur Western Europe an additional dollar oil charge of $700,000,000 and that Britain would be hit especially hard because of its extraordinary dependence on overseas trade.[10]

Neither Britain nor Iran was an easy partner. Traditional Anglo-American sensitivities about oil inclined successive British governments and the AIOC to view initial American advice and subsequent pressure to make concessions to Iran as partial, perhaps to the point that the State Department was "over much influenced by the American Oil Companies, who wish to see our companies driven into an uncompetitive position by constant pressure to raise their royalties and labour conditions."[11] More especially, Washington's transition from viewing the oil dispute as a concession renegotiation to a cold war issue clashed with British determination to prioritize the dispute's commercial character. This was de facto set out in April 1951 by foreign secretary Herbert Morrison in a list of British considerations in the oil dispute: Iran's independence, British government revenue from the AIOC by means of taxation and dividend, the balance-of-payments implications of

Iranian oil, the repercussions of any settlement with Iran on British interests elsewhere, and maintaining the flow of Iranian oil, which was of strategic as well as economic importance.[12] Economic considerations clearly headed Britain's agenda. No settlement could be accepted that jeopardized other British foreign assets. The AIOC must both remain British-registered to safeguard the flow of dividend and taxation payments to the Treasury and retain the distribution responsibilities that earned and saved the dollars that were vital to Britain's balance of payments. Strategic implications of Iranian oil ran a poor second to its economic significance. Britain's postwar over-commitment was principally due to its economic weakness and, unless this was rectified, Britain would be unable in any case to maintain its Middle Eastern presence or to contribute effectively to the containment of communism. Even Iran's independence ranked, in relative terms, low in British priorities, some officials being prepared to consider partitioning it again into British and Russian spheres of interest if that was necessary to ensure that effective control of oil operations in Iran remained with the AIOC.[13]

Sir William Strang, British permanent under-secretary of state for foreign affairs, captured succinctly in April 1951 the fundamental incompatibility of Anglo-American premises that had emerged: "To the Americans, in the fight against Communism in Persia, the Anglo-Iranian Oil Company is expendable. It is not possible for us to start from this premise."[14] Meanwhile, after initially fearing that his apparent xenophobia and ideological neutralism could threaten Iranian security and jeopardize congressional sanction for continuing U.S aid to Teheran,[15] the Truman administration gambled on Iranian prime minister Dr. Mohammad Mossadegh as a bulwark against communism. The hope was clearly that working with a proven nationalist would allow the United States to harness Iranian nationalism and direct it against the Tudeh at the same time that Mossadegh's popularity and established nationalist credentials might enable him to both introduce vitally needed political, social, and economic reforms in Iran and secure nationalist support for a reasonable oil deal.[16] Unfortunately, Britain immediately dismissed Mossadegh as an extreme nationalist who was more likely to deliver Iran to communism than to conclude an oil settlement on acceptable terms. And Mossadegh soon demonstrated that he had little understanding of the international oil business and that his nationalist support base constrained rather than liberated him in oil negotiations. In February 1952, for instance, he demanded that all oil revenues must accrue to Iran and declared that he would "prefer to obtain for the oil 100 percent of 50 cents rather than 50 percent of a dollar."[17] Indeed, Cottam argues that such was Mossadegh's atavistic view of British influence in Iran and the Middle East that he failed to understand that the battle had been won when Britain accepted the nationalization principle.[18]

Furthermore, American policymakers faced additional constraints on their ability to maneuver. Britain's importance to Middle Eastern defense and status as America's premier cold war ally imposed limits upon how far the British government and the AIOC could be pushed to secure an oil settlement. And then there were the potential ramifications of events in Iran for the international oil industry and other overseas commercial interests. A successful expropriation would gravely damage the sanctity of contract, and it was therefore a sine qua non of any settlement that Iran pay the AIOC fair and effective compensation for its nationalized Iranian assets. Also, in 1950 Aramco became the first company to apply in the Middle East the fifty-fifty profit-sharing basis for oil concessions established in Venezuela. If this were to become the accepted standard for the international oil industry, then it was essential that Iran achieve a settlement no better than that secured by Saudi Arabia with Aramco.

Neutralizing the Oil Weapon

Mossadegh's use of the oil weapon came in two forms. The first was the twofold challenge laid down to the sanctity of contract by the act of nationalization without agreed and reasonable compensation and the associated threat to the newly established fifty-fifty profit-sharing benchmark for oil concessions in the Middle East. The second was a much more explicit attempt to use oil for political influence and something of a precursor to the Organization of the Petroleum Exporting Countries (OPEC) era. In early February 1953, Mossadegh threatened to sell Iranian oil at a 50 percent discount to all comers if Britain continued to insist that Iran pay compensation to the AIOC for the loss of future profits. On February 14, he repeated the threat but added that he would fulfill his intent if, in the absence of an agreement with Britain, the U.S. government or American nationals themselves did not undertake to lift substantial quantities of oil.[19] This threatened world oil prices because dumping Iranian oil could potentially drive prices down sharply, which in turn would jeopardize the revenues that were so important in countering communist subversion in other oil-producing states. Also, Mossadegh's implicit threat to supply oil to communist countries would, if realized, oblige the United States to terminate aid to Iran because of U.S. Mutual Security legislation. Moreover, Mossadegh was potentially better placed to fulfill his threat thanks to a greater availability of oil tankers than hitherto.[20]

These threats were, of course, far from mutually exclusive, and in addressing them American policymakers had to avoid sharp fluctuations in world oil prices and supply, prevent Iran from falling to communism, avoid a major split with Britain, and safeguard both the sanctity of contract and the fifty-fifty profit-sharing arrangement. It was vital, too, to prevent Mossadegh from selling oil to communist buyers. Leaving aside safeguarding Iran against communism—a danger the British felt Washington routinely overstated—these considerations placed an emphasis on demonstrating that an expropriation of foreign assets would have unacceptable consequences for a host country and on preventing Iran from dumping oil on the world markets.

The international oil industry and successive British and American governments generally shared these conclusions, and from them came three key responses to neutralize Iran's oil weapon: collective action to alleviate the impact of the loss of Iranian oil supplies, an embargo on Iranian oil for as long as the dispute continued, and then further collective action upon the securing of a settlement to both reactivate Iranian oil production and reintegrate into world oil markets. The former was made necessary by a combination of Mossadegh's nationalization of the AIOC's Iranian assets and Britain's decision to retaliate with a withdrawal of AIOC staff and oil tankers, which effectively closed down Iran's southern oil fields. Crucially, international oil companies worked in conjunction with the British and American governments to minimize quickly the impact on Western Europe especially of the loss of Iranian oil. The AIOC and Aramco immediately met the oil production shortfall by increasing greatly outputs in Kuwait, Iraq, and Saudi Arabia. Remaining potential shortfalls stemming from supply logistics were met not least by the U.S. Petroleum Administration for Defense (PAD) orchestrating Plan of Action No. 1. Under this scheme nineteen American oil companies operating abroad received antitrust protection under the Defense Production Act to form a voluntary arrangement with the objective of supplying 19,429,000 barrels of crude oil and 26,558,000 barrels of refined products to areas in need in the "free world."[21] So effective was this measure that on December 4, 1952, the U.S. government concluded that the loss of Iranian oil had caused no overall shortage of crude petroleum or petroleum products.[22]

The British also imposed an embargo on Iranian oil as part of their war of attrition aimed at economically starving Iran into submission. Would-be purchasers faced a declared British legal intent to sue any party that bought stockpiled or newly produced Iranian oil. The Truman administration reluctantly went along with the embargo. The real key, though, to the embargo's effectiveness was its being de facto underwritten by the oil majors, otherwise known as the Seven Sisters.[23] The American oil majors generally viewed becoming involved in Iran without British approval as "highway robbery" and tantamount to "cutting the industry's own throat" because the precedent set would be fatal for concessionaires elsewhere.[24] And they consequently used their control, direct and indirect, over a large proportion of the free world's tanker fleet, refining capacity, industry contracts, and so forth to deter many independent operators from either buying "illegal" Iranian oil or providing the National Iranian Oil Company with expertise to reactivate Iran's oil fields effectively.

Such was their fear of a bad precedent being set in Iran that the oil majors even served as a bulwark against any weakening of resolve to maintain the embargo by either the British or U.S. government.[25] For example, on October 8, 1952, Secretary of State Acheson laid out preliminary thinking on a three-stage plan to resolve the Iranian oil crisis. The British predictably wanted no part of it, being especially critical of a prospect of a settlement without arbitration. This they regarded as "bad in principle, dangerous in its repercussions and impracticable."[26] Behind the scenes, though, the communist threat to Iran, frustration with the British government and the AIOC, and recognition of Britain's weakening ability to safeguard the Middle East on behalf of the West had all combined to stiffen the Truman administration's willingness to risk greater tension in the special relationship. It consequently put pressure on the oil embargo as part of its efforts to coerce Britain into line. Once initial threats to withdraw American backing of the embargo did not have the desired effect, the State Department announced on December 6 that the decision whether or not to purchase Iranian oil was at the discretion of individual companies and their appraisal of the legal risks involved.[27] This statement was widely interpreted in America as an invitation to handle Iranian oil,[28] but crucially its relative impact depended on the oil majors, upon which the British government and the AIOC were prepared to gamble to nullify the State Department's gambit effectively. As AIOC chairman Sir William Fraser explained, the U.S. position on the oil embargo was not too problematic because the American oil majors were unlikely to break ranks by lifting Iranian oil and only they had the capacity and markets to make a significant difference.[29]

The British were right. The American oil majors felt that any settlement should take into consideration losses beyond the value of aboveground facilities and respect the fifty-fifty principle lest a "Pandora's box" of claims and counterclaims be unleashed that could have devastating repercussions for their own concession arrangements. They were also disappointed that the Truman administration should seek to rescue Mossadegh from his own folly and feared that this would weaken the dissuasion force of the harsh lessons being taught to Iran as a consequence of its abrogation of contract.[30] Furthermore, they refused to participate in the State Department's oil plan without British consent, this despite Acheson telling American oil company representatives that the British should either exploit Iranian resources or allow someone else to do it and that "we were prepared to be very forceful in making this position clear to them and in speaking frankly about the consequences to Anglo-American harmony."[31]

The corollary to the embargo in ensuring stability in oil markets and concessions was the eventual settlement of the Iranian dispute and the reintegration of Iranian supplies into world markets. This was no easy task. In spring 1953, British officials seriously questioned the worth of reaching an oil settlement at that particular time.[32] Oil supplies and tankers were in surplus and the AIOC had successfully diversified to overcome its Iranian difficulties. As J. H. Brook, at the British Ministry of Fuel and Power, explained, "All the calculations we make on offtake and markets suggest that:—it's going to be a hell of a job placing the stuff; it's going to cost us a heap of dollars; there's going to be a cut-back somewhere if we ever get out anything like the quantities the Americans are talking about."[33]

As the Eisenhower administration cast around for an Iranian oil settlement in the aftermath of the coup that it helped sponsor to overthrow Mossadegh in August 1953, it was again obvious that the oil industry itself held the key. Herbert Hoover, Jr., son of a former U.S. president and previously an Iranian government oil consultant, practiced inconclusive shuttle diplomacy for literally months between the interested parties. He concluded that no Iranian government would accept the AIOC back in its original role and that any deal had at least to appear to fulfill the nationalization law of May 1951. He also saw a consortium arrangement as the best way to address overproduction in the world oil markets, previous measures to offset the loss of Iranian oil, and AIOC unpopularity in Iran. Moreover, this consortium had to include the oil majors because only they could provide the massive investment needed, move the large quantities of oil from Iran required, and make the cutbacks necessary elsewhere to accommodate Iranian production without destabilizing the world market. Hoover was, of course, right, and ultimately it was a consortium organized around the oil majors that finally enabled the Anglo-Iranian oil crisis to be settled in 1954. Moreover, though Iran retained acknowledged ownership of the oil fields and refinery, the National Iranian Oil Company did not gain strategic control over production until 1973, and liberation of foreign management of Iranian oil was really achieved only after the 1979 Iranian revolution.[34]

The Resurgence of the Iranian Oil Weapon

The oil embargo on Iran during the Anglo-Iranian oil crisis and the subsequent consortium deal helped to secure Iran as a U.S. cold war client state, demonstrated the severe consequences that would befall any would-be imitator of Mossadegh, and effectively neutralized Teheran's oil weapon. By 1970, Iran and Saudi Arabia were established both as the two leading oil exporters in the world and as America's regional policemen in its twin pillar strategy for Middle Eastern defense. It is thus difficult to dissemble the cold war, regional, and oil components of the shah's influence vis-à-vis the United States during the 1970s. For instance, Iranian oil remained vitally important for Western Europe especially, and Iran's willingness to maintain oil supplies to Israel during the Arab-Israeli war was an undoubted blessing for Washington. Nevertheless, Iran's cumulative importance is obvious. In 1972, President Nixon offered the shah a right to purchase any nonnuclear American weapons, and in 1974 alone the Foreign Military

Sales agreement saw the United States supply him with over $3,950 million of arms. By 1976, Iran had the fifth-largest military force in the world.

Fast-forward thirty years, and Iran's transformation from key ally to leading protagonist was so complete that the George W. Bush administration's 2006 National Security Strategy stated that the United States "may face no greater challenge from a single country than from Iran."[35] The administration foregrounded especially Teheran's nuclear program and suspected roles in the proliferation of weapons of mass destruction, support of terrorism, and destabilization of the already deeply problematic situations in Afghanistan and Iraq. However, the bigger objective was what Vice President Cheney called a "matter of record," namely, Iran's apparent ambitions to establish "hegemonic power" in the Middle East.[36] As the U.S. under-secretary for political affairs, Nicholas Burns, confirmed in the wake of U.N. Security Council approval of a new set of sanctions in March 2006, "Iranians have to understand they don't run the world, they don't have unlimited power, and we are sending a clear message from the rest of the world . . . that we want to contain and block Iranian power, especially on the nuclear side."[37]

This objective, though, proved far more difficult in the context of the Iran of Mahmoud Ahmadinejad than of the Iran of Mohammed Mossadegh. There are several reasons for this, but an important one was Teheran's potential to manipulate its natural resources and threaten those of others as both an offensive and a defensive weapon. Iran's military buildup, especially its acquisition of more advanced weaponry from China and Russia, gave it greater ability to interdict the flow of Persian Gulf oil exports through the Strait of Hormuz than it had in the so-called Tanker War during the conflict with Iraq in the 1980s. Also, Iran declared its intention to set up an international oil bourse on the Persian Gulf island of Kish, which had bloggers and analysts busily speculating about the potential impact on the dollar and, by extension, on U.S. global influence and the world economy.[38] All three extant oil markers were denominated in dollars: North Sea Brent Crude, UAE Dubai Crude, and North America's West Texas Intermediate Crude. If Iran were successful it would establish both a fourth oil marker and a new euro-based pricing mechanism for oil trading. Some commentators even suggested that this possible Iranian challenge increased the likelihood of preventative U.S. military strikes against Teheran.[39]

More certain was that Iran used its oil production and reserves to enhance its economy and its security, especially in terms of limiting America's ability to isolate it politically and economically. First, Iran progressively used its energy to diversify its foreign reserves away from dollars and to thereby minimize the bite of U.S. sanctions. In September 2007, Japan's Nippon Oil became the latest company to agree to buy Iranian oil using yen rather than the traditional U.S. dollars. This meant that 85 percent of Iranian oil exports were then paid for in nondollar currencies (principally euros and yen), and Teheran seemed determined to move the remainder out of dollars too. Second, having learned from Mossadegh's example the importance of having a diversity of markets and collaborators, the National Iranian Oil Company had developed partnerships with Dutch, French, Italian, and Norwegian companies. Although Japan remained Iran's single largest customer for oil, Iran had also established crucial collaborations with Russian and Chinese groups—including a thirty-year, $70 billion contract under which Iran would deliver oil and gas to energy-hungry China. China would also be involved in the development and exploitation of the Yadavaran oil field. Furthermore, China's state-run Zhuhai Zhenrong Corporation became the biggest single buyer of Iranian crude worldwide and agreed to pay for its oil in euros rather than dollars.

All of this reduced the likelihood of the United States being able to orchestrate and enforce either a widespread embargo or sanctions regime on the Iranian energy sector and thus protected Iran's principal means of attracting foreign direct investment and source of vital overseas income; in 2006 oil export revenue accounted for almost half of the government's revenue, estimated at about $46.9 billion.[40] Consider in this light, for example, the U.S. House of Representatives vote in September 2007 of 397–16 in favor of the Iran Counter-Proliferation bill, intended to make sanctions mandatory on energy companies investing more than $20 million in Iran. This bill, were it to become law, had potential at that time to divide the United States from key European partners, with whom the Bush administration wanted to forge a united front against Iran in the United Nations. This division could stem from the implications for European oil and gas interests in Iran—a scenario not that dissimilar to that which encouraged France to oppose U.S. intervention in Iraq in March 2003—or a major wider crisis over third-party sanctions enforcement. Moreover, even if Washington could organize a sanctions regime on the Iranian energy sector by like-minded states, Iran's position was much stronger than during the Mossadegh era because of its ability to sell energy to other powerful states, especially China and Russia, which were resistant to Washington's influence and would be highly unlikely to accept extraterritorial sanctions upon their economic interests.

It is also evident that Teheran used its energy resources to help buy strategic security from Moscow and Beijing especially and to intensify its relations with other regional partners.[41] India signaled interest in a strategic partnership with Iran in the Teheran (2001) and Delhi (2003) Declarations, not least because Iran potentially offered it export markets, much-needed energy sources, land access to markets in Afghanistan and Central Asia, and a key partnership in ensuring the political independence of Afghanistan. It also reportedly declared support for the proposed Iranian international oil bourse. For Russia, Iran was the principal foothold in West Asia and provided a buffer against American encroachment and strategic hemming in from the south. It was also a significant consumer of Russian arms and had strong energy links with Moscow. Indeed, the growing strength of the Russian-Iranian relationship was indicated in February 2005 by Russian president Putin's refusal to bow to U.S. pressure in agreeing to supply fuel to the Bushehr reactor on condition that Iran return spent nuclear fuel rods. As for China, Iran was its only meaningful potential ally in West Asia and a major contributor to meeting China's surging energy needs as the country moved from being almost self-sufficient in oil supplies in 1993 to a predicted import dependence of 50 percent by 2010.[42] It is thus no coincidence that China became Iran's leading trade partner and that the vast majority of arms and technology transfers to Iran came from North Korea and, especially, Russia and China. Neither is it a coincidence that, as the United States sought to develop a hard-line policy through the United Nations toward Teheran, it was China and Russia that were the key Security Council members that sought to moderate successive rounds of sanctions.

In addition there is Iran's potential to disrupt supply to or undermine confidence in world oil markets to consider. In an era of profound complex interdependence and reliance upon hitherto cheap and plentiful oil supplies to lubricate the global economy, this could have serious effects for the U.S. economy and American allies. Moreover, Iran's potential influence tended to be enhanced when energy demand exceeded perceived supply and the U.S. economy was under pressure. Indeed, these two conditions are often related. For instance, research has suggested that oil price rises have preceded most U.S. recessions since 1969 and that "virtually every serious oil price shock was followed by a recession."[43]

During the Bush administration there was little prospect that Iran could muster sufficient influence over even OPEC states, let alone non-OPEC members, to orchestrate an effective multilateral oil embargo of the United States. This does not, however, mean that Iran could not upset world oil markets in other ways. It could, for example, develop differential sales regimes with preferred partners and push within OPEC for reduced production or price hikes. More provocative, it could refuse to sell its oil to selected close allies of the United States, target shipping in the Strait of Hormuz through which almost one-fifth of world oil supplies passed in 2006,[44] and encourage pro-Iranian Shiite groups to disrupt Iraqi production. Arguably its strongest hand, though, was its potential to use its energy resources in response to a preventative U.S. strike against Iranian nuclear sites and defense infrastructure. Such U.S. action would immediately destabilize oil markets that seemingly depended almost as much on confidence as they did on the availability of oil supplies. The battle then would have been for the United States to reestablish confidence and for the Iranians to undermine it. Not only would the Iranians have likely used the options outlined above, but also America's allies and the U.N. Security Council would probably have been deeply divided and Iran would have secured the sympathy of several oil producers. Some of these, notably Libya, Sudan, and Venezuela, might well have embargoed the United States and any fellow travelers and pushed for wider multilateral reprisals. Much could then have depended on a combination of International Energy Agency measures, the duration of the crisis, and the effectiveness of U.S. leadership in dealing with Iran and international opinion.

The 1970s: A Missing Link

The Bush administration kept the option of preventative strikes against Iranian nuclear facilities "on the table" but ultimately played the multilateral U.N. route a little more convincingly than it had when taking military action against Iraq in 2003. How Iran could or would use its oil weapon in such a scenario thus remained the subject of simulations and game plays and a problem bequeathed to the Obama administration. However, this leaves open the question of how the Iranian oil weapon went from impotence in 1953 to significant potency by the time of the Bush administration. Some of the answers are obvious and beyond the scope of this essay—including different geostrategic conditions, new oil markets, deeper economic inter-dependence, ever-increasing demands for oil, and the impact of the global financial crisis from 2008. Equally, though, some of the answers can be found in the crises of the 1970s and American responses to them.

Let us begin by revisiting the Anglo-Iranian oil crisis. The successful neutralization of the Iranian oil weapon in the early 1950s appeared to offer evidence of the West's ability to control host countries and deter rogue oil producers. The embargo had demonstrated the ability and willingness of the oil majors and their home governments to not only sustain solidarity over a considerable period of time but also wreak retribution upon any state that dared challenge their preponderance. British officials maintained that their pressure on Iran was unlikely to cause it irreparable economic damage because Iran was a primitive agricultural community, and a country where 80 percent of the population lives off the land at bare subsistence level "does not 'collapse' economically. It sags."[45] Nevertheless, the embargo did have extremely serious implications for the country. Iran lost its oil markets, oil revenues shrank to a bare trickle, confidence in the economy collapsed, and refining capacity was developed elsewhere. Production in Iran dropped from 242 million barrels in 1950 to 10.6 million in 1952, while elsewhere in the Middle East it soared by 11 percent in 1951 and by a further 8 percent in 1952. Moreover, for all its suffering Iran achieved next to nothing in terms of gaining control over its own natural resources. The consortium that replaced the AIOC in Iran maintained arguably tighter strategic control over pricing and production than the British company had ever done.

Washington certainly carried into the 1970s oil crises assumptions and attitudes based on the successful neutralization of Mossadegh's oil weapon. For instance, American policymakers expressed repeated frustration with the oil majors' unwillingness or inability to confront their host nations. And in the wake of the Libyan nationalizations in 1973 and 1974, President Nixon warned that "oil without a market, as Mr Mossadeq learned many, many years ago, doesn't do a country much good. The inevitable result [of Arab pressure] is that they will lose their markets, and other sources will be developed."[46] Numerous developing countries, for which increased control over their natural resources appeared to be the precondition to development, drew similar conclusions from Mossadegh's fate. Indeed, the dissuasion force of Iran's failed nationalization is demonstrated by the fact that, despite the 1950s and 1960s witnessing rapid decolonization and rising nationalism in the developing world, there were only seven further nationalizations of oil production between 1954 and 1970. Moreover, few of these countries were significant exporters of oil.[47]

Nevertheless, the oil crises of the 1970s revealed that by this time numerous factors had quietly undermined both American and wider Western ascendancy over host nations and their relative ability to contain "rogue oil." Consider each of the key factors that combined to negate Mossadegh's oil threats. First, the oil majors' ability to dictate prices and production to host countries had eroded. This was due in part to changing expectations of host countries of their natural resources, these often being influenced by nationalism. Extracting better terms from oil companies for the exploitation of natural resources was no longer sufficient. Rather, developing countries wanted strategic control, and following the successful examples of the Libyan and Algerian nationalizations this was widely and rapidly achieved during the 1970s. As Kobrin points out, "The pattern of equity ownership of oil-producing concessions by the international oil companies and global vertical integration, which had existed since the nineteenth century, was broken in six short years."[48]

The weakening position of the international oil companies was accelerated by the proliferation of production and weakening solidarity within the industry. The oil majors still enjoyed a dominant position, but a large number of independents had forced their way into international oil. This opened up new areas to exploration and possible production and thereby made it increasingly difficult for the oil majors to orchestrate the type of embargo that they had applied to Iran in the early 1950s—a problem compounded by the inability during the oil shortage of the early 1970s to offset a major loss of production by increasing flows from elsewhere. The emergence of the independents also weakened the cohesion of the oil industry, both through a process of sheer numbers and different operating conditions and through their often having to offer better terms than the oil majors did to secure contracts.[49]

It is also the case that the 1970s oil crises revealed how much the oil majors had lost of their former dominance in terms of technology, distribution facilities, and, frequently, refining capacity. Host states could now purchase or contract on the open market everything from tankers to drilling and engineering services. Moreover, increasing awareness of the potential influence afforded them by their natural resources, coupled with oil nationalism, united both radical regimes and the conservative monarchies of the Arabian Peninsula.[50] In consequence, not only did they accrue technology through foreign direct investment and begin to share it between themselves, they also began to act collectively to counter the dominance of the international oil companies. OPEC, of course, was the principal manifestation of this. All of this meant that the complete isolation of a major oil-producing country was no longer possible.

Finally, Western nations enjoyed fewer options in the 1970s to coerce oil producers into line. Just as Britain considered the use of force to retain the Abadan refinery and southern Iranian oil fields in 1951, so the United States considered military action to secure oil supplies in the early 1970s. However, the likelihood of success was even less then than in 1951. The oil majors could no longer dominate the international oil market to the same degree, the USSR would probably have regarded any such action as highly provocative—especially in an era of détente—and Western military control of the Middle East was arguably weaker than in the early 1950s. Britain had withdrawn from east of Suez and the United States had yet to develop CENTCOM. Neither, judging from the diverse reactions of leading Western states to the 1970s oil crises, would coordinated national responses be as easy to develop to a "rogue" producer, especially in a period of oil shortage. At a time of tight alliance cohesion and oil plenty in the early 1950s, Britain and the United States had had relatively little difficulty in deterring states from either breaking the embargo on Iran or concession jumping. Yet in the 1970s there was an obvious inability or unwillingness to develop a common approach or hold to it. This owed in part to an easing of East-West tension, a shift in the international system from a bipolar to bipolycentric configuration, and the sheer complexity of factors that combined to deliver the oil crises. It also owed to weakened confidence in American leadership, to the varied capacities of states,[51] and to different degrees of dependence upon imported oil. Japan, for instance, had a high degree of foreign policy congruency with the United States. But it also had a high level of dependence on oil imports—around 99 percent of its crude oil requirements, of which approximately 40 percent came from OAPEC members.[52] Ultimately this oil import dependence proved sufficient to push the Japanese government into developing an unusual degree of departure from U.S. policy over the Arab-Israeli conflict.

The increasing inability to isolate a major oil-producing state as a consequence of the emergence of OPEC and the changed pattern of oil ownership and production during the 1970s were not the only adverse trends for the United States that emerged in this period. A related problem was that the balance of power within international oil companies progressively moved away from the United States. Whereas American companies, together with those of other friendly Western nations, once dominated production, price, and supply, the balance of power swung in the 1970s toward organizations such as OPEC. By the time of the George W. Bush administration, OPEC remained a significant influence in the oil industry, as did a resurfacing of resource nationalism and the emergence of a new set of dominant national oil companies, many of which were backed strongly by home states that had poor relations with the United States, including the National Iranian Oil Company. For example, the four remaining members of the original Seven Sisters produced in 2007 approximately 10 percent of the world's oil and gas and held just 3 percent of its reserves. Meanwhile, a new "seven sisters" had emerged, comprising predominantly state-owned companies—Saudi Aramco, Russia's Gazprom, CNPC of China, NIOC of Iran, PDVSA of Venezuela, Petrobras of Brazil, and Petronas of Malaysia. These companies controlled almost one-third of the world's oil and gas production and more than one-third of its oil and gas reserves.[53] Moreover, they cooperated increasingly in the development of each other's reserves to the exclusion of other companies and interests.

Iran's oil weapon also benefited in the long term from the increasing importance to the United States of international oil markets, its embrace of market forces in the late 1970s to help deal with oil shocks, and its failure to develop a robust energy policy that both diversified energy supply and type and addressed the demand side of the equation. Some positive measures were, of course, taken, including concluding burden-sharing agreements among nations and the establishment of the Strategic Petroleum Reserve, the International Energy Agency, and the U.S. Energy Information Administration. Nevertheless, many of the energy-saving trends that emerged from the 1970s disruptions had been captured by 2000,[54] little was done to address demand, and research and development into energy alternatives lost its priority once oil prices fell again in the early 1980s. Even after Congress introduced the 1992 Energy Act amid Gulf War–inspired fears of renewed oil shortages, the outcomes were disappointing. Among the consequences were increasing U.S. energy import dependence, greater direct and indirect vulnerability to events in the Middle East, a hugely expensive and open-ended strategic commitment to that region, an open invitation to other states to free-ride on U.S. commitments there, and too heavy an American policy reliance on key states, especially Saudi Arabia. As for the growing importance of the international oil market and the embrace of market forces rather than government controls to address oil shocks, the former was not necessarily bad, and the latter was not necessarily mistaken. Indeed, in hindsight many commentators argued that excessive regulation and price and allocation controls exacerbated rather than alleviated the impact upon the United States of the 1970s oil crises. Still, though, the importance of the market and market forces foregrounded the variable of confidence, which was something both far more easily shaken than guaranteed and ever less in the hands of the United States to provide for, especially given the shift in the balance of power within international oil and the uncertainties following 9/11.

The second oil shock was in many respects an early warning of the confidence factor because it was triggered not by immediate shortages of oil but by fears of future shortages. The Iranian revolution raised the prospect of a tsunami of Islamic fundamentalism racing across the Arabian Peninsula and engulfing the pro-Western conservative monarchies. It marked the collapse of America's twin pillar strategy for the Middle East and the emergence of an implacable regional opponent of Washington that possessed huge oil production potential and reserves and overlooked directly the crucial Strait of Hormuz. And it resurrected the specter of the first oil shock at a time when confidence in U.S. leadership had already been weakened by factors including the collapse of Bretton Woods, the Vietnam War, and the Watergate scandal.

In addition, not only did the increasing importance of the market, as indicated by the introduction in 1983 of oil futures contracts onto the New York Mercantile Exchange, foreground the confidence variable and embrace energy price volatility as "normal," but also much of the faith placed in this was based on an assumption that large rises in oil prices would be followed by correcting declines rather than there being a longer-term problem.[55] However, the shifting balance of power within the international oil industry coupled with peak oil calculations both challenged this assumption and again potentially strengthened Iran's oil weapon. For example, global demand for oil and gas continued to increase, and proportionately Gulf oil and flows through the Strait of Hormuz became ever more important. New finds of oil and gas were of decreasing frequency and, especially, scale. During the Bush era, the last field found capable of pumping more than 1 million barrels per day was Kazakhstan's Kashagen field in 2000. Also, alternative new supplies were likely to have high resource extraction costs or insecurity factors attached to them. For instance, the International Energy Agency predicted in 2007 that 90 percent of new energy supplies would come from developing countries over the following forty years.[56] This suggested that the United States would face rising competition for energy, upward pressure on prices, and increased price volatility.

In this situation market confidence was already vulnerable and Iran's relative influence had the potential to increase further in the future. On one hand, together with Iraq and Saudi Arabia, Iran accounted for 43 percent of known world oil reserves. And in 2006 the Persian Gulf provided 17 percent of total U.S. net oil imports.[57] On the other, analyses of peak oil suggest that, as oil peaking nears, relatively minor events are likely to exert more pronounced effects on oil prices and future markets[58]—thereby increasing Teheran's potential to hurt the United States through the manipulation of oil markets. Indeed, it was possible that Teheran may have to do nothing more than talk about the oil weapon in one form or another to generate anxiety and destabilize markets. In June 2006, Ayatollah Ali Khamenei, Iran's supreme leader, warned, "You will never be able to protect energy supply in this region. . . . If you make any mistake, definitely shipments of energy from this region will be seriously jeopardised."[59] Writing that same month, prominent Iran expert Ilan Berman observed that "investor jitters over a looming confrontation with Tehran are directly responsible for the recent spike in crude oil prices—and the attendant chorus of voices warning about the dire consequences of seriously bringing Iran to account."[60]

Conclusions

Within sixty years the United States went from being able to neutralize Teheran's oil weapon completely to having to take it increasingly into account when developing policies toward Iran. When Mossadegh first threatened nationalization without agreed compensation and then to sell oil at a 50 percent discount to all comers, the United States helped orchestrate an embargo that avoided a major break with Britain, protected Aramco's fifty-fifty profit-sharing precedent, and maintained confidence in the sanctity of contract. The logic, as Secretary of State Dulles laid out on September 23, 1953, was clear: "Whatever solution is found for [the] Iranian problem will soon be forced on both of us [Britain and America] by all other countries if [it is] in any way more advantageous to them than those now in operation."[61] British and U.S. military power and influence were important in underpinning the embargo. But the real key to its effectiveness was the dominance of the oil majors in everything from price and production to technical expertise and subcontracts. Similarly, government-industry cooperation easily offset the loss of Iranian oil in 1951 and ultimately allowed for its reintegration from 1954 in ways that avoided a supply glut or rapid cuts in oil prices that could destabilize the economies of oil-producing countries especially elsewhere.

According to Henry Kissinger in 2007: "An Iran that practices subversion and seeks regional hegemony—which appears to be the current trend—must be faced with lines it will not be permitted to cross: The industrial nations cannot accept radical forces dominating a region on which their economies depend."[62] The logic was not much different to cold war fear of Soviet penetration of the Middle East, but conditions were much changed for the George W. Bush administration. Its strategies for surgical military strikes on key Iranian facilities and civilian contingency planning for a probable energy shock were necessary measures. Yet had they been used they would likely have at best constituted short-term palliatives for longer-term political, strategic, and energy problems. U.S. energy policy needed to be integrated more effectively into American foreign and security policies and complemented with measures to tackle domestic demand, reduce U.S. import dependence, and promote a liberal global energy market. Yet there is little evidence that the Bush administration attempted this systematically with any more vigor than its counterparts of the 1970s had. Neither was there much apparent appetite in the states or Congress for fundamental reform of the American energy sector, of which the stalled restructuring and deregulation of electric utilities was an example.

Teheran admittedly remained vulnerable to the vagaries of the global economy, American pressure helped restrict its opportunities to develop its energy capacity quickly, and the economic policies especially of the Ahmadinejad regime raised questions about the Iranian government's ability to realize fully the potential political, economic, and strategic benefits that its energy resources offered. Nevertheless, the seriousness of Iran's potential destabilizing threat was reflected in the extensive gaming of an Iran-provoked energy crisis, defensive or offensive, and in the importance attributed therein to U.S. leadership and ability to maintain confidence on the world oil markets.[63] Just why this threat had become so grave for the Bush administration owed in part to the events and consequences of the 1970s. This was the decade above all others that swept away edifices of apparent Western dominance over producer states, set in motion a radical restructuring of patterns of ownership and production, and within oil companies encouraged an eventual shift in the balance of power from the old to the new "seven sisters."

In addition, two key American decisions in response to the oil crises of the 1970s reverberated through to the era of the Bush administration (and beyond). First, successive governments failed to exploit the crises of the 1970s to drive and sustain a robust energy policy that tackled both demand and diversity of energy type and origin. Second, as state controls in response to the first oil shock slowly gave way to greater reliance upon market forces, so confidence in supply and demand and in states' capacity to cope with future shortages and crises became critical. This created a new vulnerability because the United States became progressively less able alone to ensure market confidence. Control over energy resources and the companies that extracted them shifted increasingly toward the developing world and, often, to states that had difficult relations with the United States. And periodic fears of peak oil, sharpening energy resource competition and increasing tension between resource nationalism and resource internationalism encouraged market volatility—and the implications that flowed from this.

All of this meant that, by the time the Bush administration had to deal in earnest with the Iranian threat, Teheran's oil weapon was significantly stronger than when Eisenhower defeated Mossadegh. Teheran had learned lessons from the experience of the Anglo-Iranian oil crisis and had capitalized on changes in the international oil industry and on shifting patterns of supply and demand. The National Iranian Oil Company had diversified its partnerships and become a key member of the new "seven sisters" of overwhelmingly state-owned energy companies. Iran used its energy revenue streams to buy arms and move its foreign exchange into euros. And it used its energy resources to obtain patronage from Russia and China especially, which increased in turn its strategic security and helped to insulate it against intensified U.S. unilateral sanctions and its push for stronger multilateral sanctions. Indeed, it is not unreasonable to suggest that during the Bush administration years these factors, combined with peak oil calculations, increasing long-term global energy demands, and Iran's strategic position, energy reserves, and tense relationship with the United States, afforded Iran greater potential to influence critical energy markets than at any time previously.

Notes

1. BBC News, "Iran Defiant at New US Sanctions," Oct. 26, 2007, http://news.bbc.co.uk/1/hi/world/middle_east/7063188.stm.

2. J. A. Phillips, "The Iranian Oil Crisis," Feb. 28, 1979, www.heritage.org/Research/MiddleEast/bg76.cfm.

3. Paul Rogers, "Oil Security and the Iraq War," Oxford Research Group, May 2006, www.oxfordresearchgroup.org.uk/publications/monthly_briefings/2006/05/oil-security-and-iran.

4. *Foreign Relations of the United States* (hereafter FRUS) 1952–54, vol. 10, NSC Staff study on certain problems relating to Iran, attached to Statement of Policy by the National Security Council, Jan. 2, 1954, 881.

5. S. Marsh, *Anglo-American Relations and Cold War Oil: Crisis in Iran* (Basingstoke: Palgrave, 2003).

6. FRUS 1949, vol. 6, NSC 54, Jan. 27, 1949, 545–52; ibid., 'Analysis and Comparison of US Policies with Regard to Aid to Greece, Turkey, Iran and Afghanistan," u.d. annex, 4.

7. President's Secretary's Files (PSF), Intelligence File, box 257, CIA Reports, ORE 1949, no. 90–100, "Current Situation in Iran," Nov. 9, 1949, 3, Harry S Truman Library, Independence, Mo.; FRUS 1950, vol. 5, State Dept. Paper, "Present Crisis in Iran," u.d., 509–18; ibid., Wiley to Sec. State, Jan. 30, 1950, 463.

8. PSF, General File, box 116, Memos and minutes of Churchill Truman meetings, Pres–PM talks, Jan. 8, 1952, 7; and PSF, Intelligence File, box 258, folder SR Reports no. 4–6, SR 6 "Iran," Apr. 1948, I-11, both in Truman Library.

9. Cost of plan as given by Morrison to the Commons, 491, H.C. Deb., 5s, comprising period 23rd July–4th October, July 30, 1951, c. 966. The plan was developed by Overseas Consultants Inc., an organization of eleven American firms and one British firm of consulting engineers. For the plan's origins and development, see Overseas Consultants Inc., *Report on Seven Year Development Plan for the Plan Organization of the Imperial Government of Iran*, 5 vols. (New York: Overseas Consultants, 1949); L. P. Elwell Sutton, *Persian Oil: A Study in Power Politics* (London: Lawrence and Wishart, 1955), 159–61; M. Elm, *Oil, Power, and Principle: Iran's Oil Nationalization and Its Aftermath* (Syracuse: Syracuse University Press, 1992), 52.

10. Marsh, *Anglo-American Relations*, 30.

11. Cited in G. McGhee, *Envoy to the Middle World: Adventures in Diplomacy* (New York, London: Harper and Row, 1983), 322.

12. K. H. Wieschhoff, "The Salience of Economics in the Formulation of UK Foreign Policy: The Persian Gulf 1945–55," *Kent Papers in Politics and International Relations*, Series 1, no. 13 (1992): 29.

13. For instance, British chancellor of the exchequer Hugh Gaitskell noted regarding partition of Iran between Britain and the Soviets: "This sounds bad but I think it might be the best ultimate solution"; P. M. Williams, ed., *The Diary of Hugh Gaitskell 1945–1956* (London: Cape, 1983), 260.

14. FO 371/91529, Strang to Franks, Apr. 25, 1951, British National Archives (hereafter UKNA).

15. FRUS Iran 1952–54, vol. 10, Sec. State to U.S. Embassy Iran, May 10, 1951, 50–51; LM 73, reel 44, State Dept. to U.S. Embassy Iran, May 7, 1951, U.S. National Archives (hereafter NA).

16. PSF, box 180, Subject file Iran, CIA report "Analysis of Iranian Political Situation," Oct. 12, 1951; and Subject file Iran—W. Averell Harriman, Harriman to Sec. State, July 19, 1951, both in Truman Library; F. Diba, *Mohammad Mossadegh: A Political Biography* (London: Croom Helm, 1986), 96–97. Acheson noted in hindsight that they were slow to perceive that Mossadegh's essential conservatism lessened the prospects of this reform; D. Acheson, *Present at the Creation: My Years in the State Department* (London: Hamilton, 1970), 504.

17. BP 101912, record meeting in Strang's room, Feb. 27, 1952, British Petroleum Archives (hereafter BP); FRUS Iran 1952–54, vol. 10, Berry to Sec. State, Jan. 8, 1952, 306.

18. R. W. Cottam, *Nationalism in Iran* (Pittsburgh: University Pittsburgh Press, 1973), 284.

19. FRUS Iran 1952–54, vol. 10, Henderson to State Dept., Feb. 14, 1953, 665.

20. FRUS Iran 1952–54, vol. 10, Sec. State to U.S. Embassy U.K., Feb. 10, 1953, 663; CAB 128, CC(53) 20th conclusion, Mar. 17, 1953, NA.

21. PSF, report to the National Security Council by Sec. Interior and Petroleum Administrator for Defense, Dec. 8, 1952, 10, Truman Library.

22. LM 73, reel 40, Eakins to Linder, Dec. 4, 1952, NA.

23. This was a group of oil companies comprising Shell, AIOC, Standard (New Jersey), Socony Vacuum, Gulf Corporation, Texas, and Standard of California. Together with their

subsidiaries they controlled, in 1952, 80 percent of oil production outside of the United States.

24. FRUS *1951*, vol. 5, Memorandum of conversation by Richard Funkhouser (Office of Near Eastern Affairs), "Discussion of AIOC Problem with U.S. Oil Companies Operating in Middle East, May 14, 1951, 311.

25. This argument is developed at greater length in S. Marsh, "Crude Diplomacy: Anglo-American Relations and Multinational Oil," *Journal of Contemporary British History* 21, no. 1 (2007): 25–54.

26. FO 371/98702, Ross draft Cabinet paper, U.S. Ideas of a Settlement of the Oil Dispute, u.d., UKNA; FRUS Iran 1952–54, vol. 10, Sec. State to U.S. Embassy U.K., Oct. 12, 1952, 491; FO 371/98702, FO to Washington Embassy, Oct. 25, 1952, UKNA.

The proposed settlement in brief: First, a lump sum settlement would be used to circumvent problems of compensation and counterclaims. This would comprise the purchase of 15 million tons of crude oil and a further 15 million tons of oil-related products. Second, an international oil distribution company would be created to purchase up to 25 million tons of oil and related products per annum from the National Iranian Oil Company for at least a decade. Third, the international oil distribution company would advance $100 million against future oil purchases. Payment would comprise a lump sum of $50 million and $10 million monthly tranches thereafter.

27. Acheson Papers, box 167, Byroade to Acheson, "Recommended Change in U.S. Policy toward Iran," Sept. 16, 1952, Truman Library; BP 101912, 27th meeting of the Cabinet Persia (Official) Committee, Oct. 8, 1952, BP; FO 371/98698, minute Makins, Sept. 26, 1952, UKNA; LM 73, reel 39, Funkhouser to Dorsz, "Resumption of Iranian Oil Movements," Oct. 3, 1951, NA; LM 73, reel 39, Richards memo, Oct. 28, 1952, NA.

28. BP 91032, A. E. C. Drake to Snow, Dec. 9, 1952, BP.

29. BP 91032, extracts minutes Cabinet Persia (Official) Committee 32nd meeting, Oct. 21, 1952, BP.

30. LM 73, reel 39, B. B. Jennings to McGhee, May 17, 1951; and Standard Oil to McGhee, May 18, 1951, NA; FRUS 1951, vol. 5, memo. of conv., Funkhouser, May 14, 1951, 311. Fesharaki has argued that fear of further nationalization was the primary motive for the solidarity of the U.S. companies with the AIOC; F. Fesharaki, *Development of the Iranian Oil Industry: International and Domestic Aspects* (New York: Praeger), 1976, 45.

31. LM 73, reel 34, memo. meeting with oil company representatives, Dec. 4 1952, NA.

32. FO 371/104615, P.E. Ramsbotham, "Persian Oil—Future Policy," Apr. 14, 1953, UKNA.

33. POWE 33 1937, J. H. Brook to J. A. Beckett, u.d., UKNA.

34. F. Venn, *Oil Diplomacy in the Twentieth Century* (Basingstoke: Macmillan, 1986), 117; D. A. Rustow, *Oil and Turmoil in the Middle East* (New York: W. W. Norton, 1982), 99.

35. U.S. National Security Strategy, Mar. 2006, 20, www.whitehouse.gov/nsc/nss/2006/nss2006.pdf.

36. Vice President's Remarks to the Washington Institute for Near East Policy, Oct. 21, 2007, www.whitehouse.gov/news/releases/2007/10/20071021.html.

37. Burns cited in "Russia, China to Iran: Heed U.N.," www.usatoday.com/news/world/2007-03-26-iran_N.htm.

38. See, for example, Niusha Boghrati, "Iran's Oil Bourse: A Threat to the U.S. Economy?" Worldpress.org, Apr. 11, 2006, www.worldpress.org/Mideast/2314.cfm; C. Nunan, "Trading Oil in Euros? Does It Matter?" *Energy Bulletin*, Jan. 29, 2006; K. Petrov, "The Proposed Iranian Oil Bourse," *Energy Bulletin*, Jan. 17, 2006.

39. William Clark, "Petrodollar Warfare: Dollars, Euros and the Upcoming Iranian Oil Bourse," Energy Bulletin, Aug. 2, 2005, www.energybulletin.net/7707.html.

40. D. Sands, "Analysis: Iran Moves to Ditch U.S. Dollar," Sept. 10, 2007, www.upi.com/International_Security/Energy/Analysis/2007/09/10/analysis_iran_moves_to_ditch_us_dollar/6990/.

41. D. Ross, "The Middle East Predicament," *Foreign Affairs* 84, no. 1 (2005): 61–74; S. Kapila, "Iran in the Strategic Matrix of Russia, China and India," South Asia Analysis Group, paper 1284, Mar. 9, 2005; E. Ahrari, "Iran, China and Russia: The Emerging Anti-US Nexus?" *Security Dialogue* 32, no. 4 (2001): 453–66; H. Pant, "The Moscow-Beijing-Delhi 'Strategic Triangle': An Idea Whose Time May Never Come," *Security Dialogue* 35 no. 3 (2004): 311–28.

42. Paul Rogers, "Oil Security and the Iraq War," May 2006, www.oxfordresearchgroup.org.uk.

43. R. L Hirsch, R. Bezdek, and R. Wendling, "Peaking of World Oil Production: Impacts, Mitigation, and Risk Assessment," Feb. 2005, www.netl.doe.gov/publications/others/pdf/oil_peaking_netl.pdf, 28.

44. Persian Gulf region, www.eia.doe.gov/emeu/cabs/Persian_Gulf/Background.html.

45. FO 371/98608, minute Sarell, Jan. 22, 1952, UKNA.

46. Nixon cited in "The Arab's Final Weapon," *Time*, Sept. 17, 1973, www.time.com/time/magazine/article/0,9171,907884-1,00.html.

47. S. Kobrin, "Diffusion as an Explanation of Oil Nationalization," *Journal of Conflict Resolution* 29, no. 1 (1985), 13.

48. Ibid., 17.

49. A. Sampson, *The Seven Sisters: The Great Oil Companies and the World They Shaped* (New York: Viking, 1975).

50. F. Venn, *The Oil Crisis* (Pearson: Harlow, 2002), 37.

51. G. J. Ikenberry, "The Irony of State Strength: Comparative Responses to the Oil Shocks in the 1970s," *International Organization* 40, no. 1 (1986): 105–37.

52. Venn, *Oil Crisis*, 124.

53 Carola Hoyos, "The New Seven Sisters," *Financial Times*, Mar. 11, 2007.

54. Hirsch, Bezdek, and Wendling, "Peaking of World Oil Production," 71.

55. Ibid., 27.

56. R. W. Tillerson (Chairman and CEO, Exxon Mobil Corporation), "Building Bridges of Energy Understanding," Strauss Center for International Security, University of Texas, Oct. 3, 2007, www.exxonmobil.com/Corporate/news_speeches_20071003_rwt.aspx.

57. Persian Gulf Region, www.eia.doe.gov/emeu/cabs/Persian_Gulf/Background.html.

58. Hirsch, Bezdek, and Wendling, "Peaking of World Oil Production," 30.

59. Cited by Alec Russell, "Iran Threatens to Disrupt Gulf Oil Shipments," June 14, 2006, www.telegraph.co.uk/news/main.jhtml?xml=/news/2006/06/05/wiran05.xml.

60. Berman cited in A. Cohen, J. Phillips, and W. Schirano, "Countering Iran's Oil Weapon," *Backgrounder 1982*, Nov. 13, 2006, www.heritage.org/Research/Iran/bg1982.cfm.

61. FRUS 1952–54, vol. 10, Sec. State to U.S. Embassy Iran, Sept. 23, 1953, 802.

62. Henry Kissinger, op-ed in the *Washington Post*, "The Disaster of Hasty Withdrawal," reprinted in full in the *Congressional Record*, Sept. 18, 2007, www.govtrack.us/congress/record .xpd?id=110-s20070918-20.

63. See, for example, J. J. Carafano and A. Cohen, "If Iran Provokes an Energy Crisis: Modelling the Problem in a War Game," July 25, 2007, www.heritage.org/Research/ Energy and Environmnet/CDA07-03.cfm.

Part III

Supply

Chapter 5

Diving into Deep Water
Shell Oil and the Reform of Federal Offshore Oil Leasing

Tyler Priest

T he leasing of offshore territory for oil and gas exploration is one of the most vital but least understood aspects of American energy policy since World War II. For much of this period, offshore leasing was second only to income taxes as a generator of public revenue (usually taking in more from a single offshore lease sale than from all timber sales and onshore mineral leasing for the year combined). More important, it brought forth a huge landscape of industrial development in the ocean. In the Gulf of Mexico in 2012, after more than fifty years of federal leasing, there were nearly three thousand platforms servicing 35,000 wells and more than 30,000 miles of pipeline.

To the extent that historians have revisited the energy crisis of the 1970s, the offshore leasing story has faded amid the discussion about other policy issues, such as price controls and allocations, CAFE standards and other conservation measures, promotion of synthetic fuels, and nuclear power. Many policies introduced did not work as intended or did not endure. On the other hand, policies designed to increase the flexibility of conventional petroleum supplies, such as offshore leasing and the Strategic Petroleum Reserve, did achieve their intentions and remain a key aspect of U.S. energy governance.

Offshore leasing deserves our attention not only for its historical importance but also for its current policy relevance in light of simmering controversies over the management of offshore revenues, proposals to open up new frontier areas (Alaska, Florida, Virginia) to oil and gas exploration, and serious safety and environmental questions following the Macondo/*Deepwater Horizon* oil spill of 2010. The spill focused attention on the past and future of deepwater drilling operations in the Gulf of Mexico. But to assess the risks and challenges of the deepwater Gulf, we must first understand how and why oil companies began exploring there.

The deepwater (1,500 feet or deeper) oil developments in the Gulf of Mexico that began making headlines in the mid-1990s resulted from a major transition in the federal offshore leasing system that dates back to the 1970s. To understand this transition, we must examine how policy changes interacted with evolving technology and oil investment strategies, and how, especially in extractive industries such as offshore oil, environmental factors constrained policy, technology, and strategy.[1] Considering these interactions, in this essay I discuss the evolution of offshore leasing during the 1970s and 1980s from the perspective of Shell Oil, the most aggressive player in the offshore business, the industry's technological leader, and the company most actively involved in shaping federal offshore leasing policy.

Shell Oil's Deepwater Vision

"There's a romance about big, offshore structures," said Pat Dunn, Shell's manager of civil engineering, back in 1989. "There's something about seeing them out there on the frontier."[2] Since the time Shell Oil's New Orleans vice president Bouwe Dykstra teamed up with drilling entrepreneur Doc Laborde to build the submersible drilling vessel *Mr. Charlie*, Shell Oil had carried on a passionate affair with these structures. But as they rapidly evolved, words could hardly describe their mind-boggling size and complexity. In the 1970s, Shell led the industry in dramatically extending the depth threshold for fixed platforms from 350 feet to over 1,000 feet. In the 1980s, Shell's geoscientists and engineers continued to push the offshore frontier in both exploration and production, moving the industry to take another quantum leap, this time off the edge of the continental shelf into truly deep water of 3,000 feet and beyond.

Offshore was more than just a romance for Shell Oil Exploration and Production (Shell E&P). It was its heart and soul, a symbol of long-standing technological leadership, and a main source of income for the entire company. Offshore development was the key component of Shell's multifaceted strategy in the 1970s to expand the quest for energy resources. And the Gulf of Mexico remained the hotbed of activity. There, feverish exploration and platform installation followed the Arab embargo, with Shell's Cognac platform in 1,025 feet of water establishing a benchmark that redefined the concept of deepwater production. Yet, in the midst of the boom, many in the industry believed that the Gulf had begun to play out. Overall production was declining and ultra-deepwater work seemed technologically and economically unfathomable.

In the mid-1970s, Shell and other companies began to shift their long-range sights to other unexplored U.S. offshore provinces, such as the Atlantic basin, California, and Alaska. But political controversies, environmental opposition, and dry holes delayed or limited drilling in most of these areas. Desperate for new reserves, Shell once again staked its future on the Gulf. It embraced advanced seismic technologies, gambled on deepwater leases, and developed new deepwater platform and subsea systems that enabled production beyond the continental shelf. The deepwater play of the 1980s was a tough sell to some of Shell's directors, who were understandably concerned about taking such giant, costly, and speculative steps into the virtual unknown. This might have been the greatest risk Shell Oil in the United States had ever faced. The big question, held out since the early days of offshore after World War II, was revisited: Even if the technology could be developed, would deepwater ever pay? Strong leadership in exploration and production, driven by an abiding confidence in Shell's marine engineering capabilities and faith in the potential of the Gulf to yield large new fields, persuaded the company to take the risk. But to make the risk pay off, the terms of access to offshore territory had to be changed.

The willingness to take on massive technological challenges against conventional wisdom was ingrained in the corporate culture of Shell E&P. Research was closely integrated with operations and engineering, and personnel moved fluidly back and forth between the Bellaire lab and the area offices. Long before the "team concept" or "matrix form" of organization came into vogue, Shell E&P management had encouraged the formation of task forces—collections of people with different skills working on a problem together. This mutual support system emboldened managers to go into big projects with a certain level of technical understanding, confident they would come out with more knowledge than before.

Shell's leadership never wavered in their commitment to offshore. The top two men who led Shell's initial thrust into deeper offshore terrain in the 1970s, president Harry Bridges and his executive vice president, John Redmond, were technically oriented managers who drew faithfully on the talents of the organization. John Bookout, who first replaced the retiring Redmond as leader of Shell E&P and then Bridges at the top of the company, further sharpened Shell Oil's focus on the offshore frontier. Bookout believed in the offshore and was fully conversant with Shell's evolving capabilities in this area. Upon becoming executive vice president of Shell E&P and then Shell's CEO, Bookout emphasized offshore development and campaigned hard to have the U.S. federal government open up the nation's continental shelves to oil exploration. His counterparts in the industry regarded him as one of the brightest and best-informed men among them. In 1981 they elected him chairman of the National Petroleum Council, the industry group that acts as semiofficial consultant to the U.S. secretary of energy. And in 1984 he became chairman of the American Petroleum Institute, the first Shell Oil president since Max Burns in the 1950s to head that prestigious trade association. Bookout's exceptional strength as a leader and wide respect both within and outside the company were instrumental in convincing Shell's board to continue moving deeper offshore.

Top Shell E&P management under Bridges and Bookout was composed of people who had distinguished themselves technically during a period when offshore had taken center stage in the company. Bookout's executive vice president for Shell E&P, Charlie Blackburn, was a petrophysicist and protégé of Gus Archie, the man who invented the field of petrophysics. As vice president of the southern Shell E&P region, Blackburn ran the bidding in the important federal offshore lease sales in the Gulf of Mexico in 1970 and 1972, when Shell first deployed its revolutionary "bright spot" seismic technology, which used advanced digital imaging to pinpoint oil and gas deposits. He also had managed Shell's deft handling of a major platform blowout in Bay Marchand in 1970–1971.[3] Bookout's exploration and production vice presidents, all technically accomplished, contributed in their own way to Shell's offshore vision. On the exploration side, geophysicist Billy Flowers was a prime mover in getting Shell to apply state-of-the-art geophysics offshore. Geologist Bob Nanz, a pillar in the research organization for many years, orchestrated Shell's crusade for greater access to federal offshore lands. Exploration vice presidents Jack Threet and Tom Hart, veterans of exploration in many frontier basins with both Shell Oil and the Royal Dutch/Shell Group, also helped push the company into the deep waters of the Gulf of Mexico. The strong emphasis on technology, so touted by earlier exploration leaders in the company, continued unimpeded under Bookout's exploration program.

Managers on the production side also had demonstrated exceptional technical abilities and offshore experience. For most of Bookout's presidency, Gene Bankston and Don Russell headed the production organization. In the late 1950s, Bankston (vice president for production, 1972–79) had contributed to developing the "big picture" policies on how management decisions should be made in Shell E&P and the economic model that supported the first push into what then was considered deep water (past 200 feet). Russell (vice president for production,1980–86) had been a star researcher at Shell's Bellaire exploration and production research laboratory in the area of reservoir engineering and had helped develop more rigorous quantitative methods for evaluating offshore leases. In the late 1960s, he also had been regional production manager in New Orleans. Under Bankston and Russell, Shell E&P refined and improved its sophisticated methods for preparing economic scenarios for given offshore prospects, using statistical projections of volumes, prices, profitability, drilling costs, and success ratios. As offshore development moved into deeper water, and as competition for leases intensified, production economics became ever more important to formulating bids.

Whereas the 1960s marked the great leap forward in exploration technology, the 1970s witnessed similar progress in offshore production technology. In this area, Shell was well prepared to take the lead. First of all, it was committed to the Gulf of Mexico, which at the beginning of the decade accounted for over 50 percent of the company's domestic crude oil and natural gas liquids production. Shell knew offshore Gulf of Mexico as well as any company in the business. It was the only area in the Shell E&P organization that had kept engineers and technical teams in place continuously over the post–World War II period. In 1971 the marine divisions in New Orleans and Los Angeles were renamed the Offshore East Division and Offshore West Division, respectively, and then in 1979 combined into one large Offshore

Division, the company's largest. In 1972, Shell confirmed its commanding presence in Louisiana by moving 1,500 employees into a giant, new, fifty-one-story skyscraper in downtown New Orleans, called One Shell Square. Housing Shell's Southern Exploration and Production region, it towered above the New Orleans skyline and laid claim to the distinction of the tallest building in the Deep South.[4]

During this period, Shell Oil concentrated on expanding deepwater production capability. In 1972 exploration activity in the Gulf began to taper off. There was still a lot of development in so-called shallow water (out to 300-foot depths), but the industry was not really expanding into deeper water. Lease sales had been postponed because of rising environmental concerns over offshore development and the fallout over several platform disasters. Yet there was another factor that contributed to the lull. The industry was still trying to figure out how to operate at greater depths in the proven oil province of the Gulf of Mexico. Fixed platforms had become standard for waters extending out to 350 feet, but moving deeper—toward 600 feet and beyond—introduced fundamentally new problems. Steel jackets would be more slender and therefore more susceptible to stresses caused by wave dynamics and metal fatigue, which could be safely ignored in shallow water. So Shell continued to explore alternatives for producing at these depths. By the early 1970s the company had elite engineering groups working on a range of different technologies, including subsea wells, fixed platforms, and tension-leg platform designs. Some professional competition existed between the various groups, but it was congenial, for everyone realized that they were striving toward a common goal.

Shell's civil engineering group in the head office and at Bellaire kept finding ways to extend the depth capability of platforms on a more cost-effective basis than could be done with subsea systems. First organized in 1965 and headed by Bob Bea, the Central Engineering Group assumed the task of designing and overseeing fabrication and installation of all of Shell Oil's offshore structures. Previously the operating divisions (e.g., New Orleans, Houston) had performed the engineering. But with the increasing challenges of offshore engineering and Shell Oil's expanding portfolio of leases, a more specialized and concentrated effort was needed. During its first year, the Central Engineering Group designed and managed the construction of thirty-three platforms in water depths ranging from 30 to 300 feet in offshore Louisiana and Texas. This was the most ever designed and constructed in a single year in Shell Oil's history. Dunn declared, "The period 1964–1972 was, in my opinion, the most active in terms of platform technological development in the whole history of the offshore."[5]

Shell stayed at the forefront of innovation, but advances in deepwater production also resulted from the acquisition of knowledge and skills by the industry as a whole. The introduction of the digital computer in the early 1960s had revolutionized design techniques, and bigger launch barges were built to assist the installation of increasingly ponderous platforms. Onshore support industries and communities had sprung up all along the Gulf Coast, in such places as Morgan City and Lafayette, Louisiana, helping to spread and standardize skills among offshore operators. Lessons learned from the destruction of platforms by three devastating hurricanes in the 1960s and several platform disasters in 1969–70, including Union Oil's notorious blowout in the Santa Barbara Channel, accelerated the industry's learning process and helped build technical consensus around all kinds of new design criteria.[6]

The demonstration of successful projects in the tough North Sea environment, furthermore, helped improve practices in the Gulf. In the early 1970s, several North Sea platforms installed in 500 feet of water, under the most inhospitable conditions, provided invaluable knowledge about wave dynamics and metal fatigue. The North Sea also provided an example of how quickly costs rose with increasing depth. Offshore leaders such as Shell Oil knew that safe, reliable platforms could be built in much deeper water—but at a steep price. The big question was, could they afford it?

The Arab oil embargo of 1973 provided the answer. The skyrocketing price of crude oil and an aggressive federal leasing system gave new impetus to offshore expansion. With prices at $10 per barrel instead of $3 per barrel, companies found they could justify much more expensive offshore drilling and development. And the federal government eagerly encouraged them. Under the mandate of Project Independence, the Nixon administration increased the pace of leasing in the Gulf of Mexico and resumed Outer Continental Shelf (OCS) sales off the Atlantic, Pacific, and Alaskan coasts, all of which had been closed to drilling after the Santa Barbara blowout. In 1973, even before the embargo, the government had held sales in the central Gulf, offshore Texas, and in the so-called MAFLA region—offshore Mississippi, Alabama, and Florida. After the embargo, interior secretary Rogers Morton announced that the government aimed in 1975 to lease 10 million acres of offshore property to oil companies, as much as had been handed out in the entire twenty-year history of OCS leasing. Most people outside the government regarded this goal as totally unrealistic. It also raised the hackles of environmentalists, who geared up for confrontation in California and along the Atlantic coast. Nevertheless, the announcement accelerated plans to offer deepwater tracts in the Gulf, where environmental opposition hardly registered.

In anticipation of deepwater sales in the Gulf, Shell E&P's seismic surveys located several attractive features for testing, and some confirmation drilling was completed. One of the most attractive prospects, codenamed "Cognac" by Shell, was on tracts in the Mobile South area. The amplitude reflections, or bright spots, on the seismic records gave a high probability of finding several oil and gas plays on the structure. "The prospect was full of bright spots," said Mike Forrest, the discoverer of the method and geophysical project leader for the Offshore Division at the time. Shell's technical analysts estimated that the field might contain 150 million barrels. Although not large by Middle Eastern standards, this was a potentially major field for the Gulf of Mexico.[7]

Cognac, which was located in 1,000-foot depths, would establish a new offshore frontier, one far deeper than the 200-foot depths that had so concerned some Shell executives in the late 1950s. It would be another giant and risky step, with expensive drilling and facility costs. Shell managers also expected high bids for the tracts, as much as $100 million for a single 5,000-acre tract. This worried president Bridges. He felt that the board of directors needed to be more fully apprized about the methods used for evaluating the bidding before they could consider and approve the company's move into such deep waters. Heretofore, the board had been advised, but not in great detail. It was a security problem. Bids reached into the millions of dollars, and the bidding was competitive. But because of the escalating cost and added risk, Bridges decided to give the board a more detailed strategy presentation.[8]

As the price and competition for offshore leases increased significantly, the process of deciding which tracts to bid on in a lease sale and how much to bid had become an increasingly lengthy and secretive process involving the work of hundreds of people over a period of several years. After the Department of Interior called for nominations on tracts, Shell and other companies would submit a list of tracts to the government, based on the ongoing collection of seismic and geological data and information from previous lease sales. As the sale approached, Shell would undertake intensive seismic work, with a geophysicist and geologist assigned to each prospect—the geophysicist analyzing the seismic data and mapping out the subsurface structure, and the geologist acting as the evaluator, determining which possible oil or gas reservoirs should be pursued and estimating the volumes of oil or gas on each tract. Beginning in the late 1960s the technical team also worked with other parts of Shell E&P—production, economics, platform design, and drilling—to establish a most probable tract value within a range of values, discounting for operational and geological risks. The list of proposed tracts was then culled through a district review and then a division review. A month before the sale, a final division review was held and general bids were attached to the tracts. The head office then reviewed the bids with all the Shell E&P vice presidents, and finally with the executive vice president and president, who placed the final numbers on the sealed bids. The fewer people involved at this stage, the better, since a competitor would need to bid only $1 higher to take a given tract away from Shell.[9]

The meeting with the board on Cognac took place several months before the scheduled March 1974 offshore Louisiana lease sale. Redmond and Nanz made the presentation, outlining Shell's detailed methods of evaluating prospects. They also reviewed the application of Shell Oil's bright spot technology for the very first time. The main objective of the presentation was to show the board that Shell's bids were based on the value of a particular tract to Shell and never at an amount just to be higher than a competitor. After the strategy presentation, Redmond and Nanz discussed the specific tracts they were targeting and the bidding levels. As usual, they did not name the individual tracts or specify their location. But Cognac was one of the prospects reviewed, and the price per tract was around $108 million. The board approved the recommendations and, as Redmond remembered, "we were sure that the strategy presentation had helped in gaining their confidence and support."

Shell E&P had enough confidence and support from the board to bid alone in the approaching March sale. But in this case Shell took on bidding partners. Traditionally the company had not bid with partners, preferring to go it alone and protect its technology. As the prices for leases soared, however, Shell decided to lay off some of the front-end financial risk by taking on partners.[10] This decision also was forced by smaller operators, who were priced out of the picture and had been pressuring the majors to give them some representation in the bidding. But in Shell's case these smaller partners had little input. They did not know what kind of technical work was involved, nor did they even know where the lease was or how much was being spent. Their participation was essentially an investment in Shell's proven record offshore.

Shell Oil bid strong in the March 1974 sale and got most of what it wanted. The U.S. government opened up more than two hundred tracts (940,000 acres) in the central Gulf, including forty-two deepwater tracts (199,000 acres) on the continental slope offered for the first time.[11] Oil companies spent a record $2.16 billion in bonuses in the landmark sale. Only thirteen of the deepwater tracts were bid on, but the eleven bids that were accepted pulled in an impressive $321 million. Exxon was the top spender in the sale ($245 million for six tracts), but the Shell Oil group made the biggest bet on deepwater production—spending $214.3 million in bonuses for three adjoining blocks on the Cognac prospect, officially called Mobile South No. 2, and paying over $112 million for the most prized of the three blocks. Shell did not, however, win all of the blocks on the prospect. A group led by Amoco won the fourth block with a bid of $81 million, beating out Shell by about $10 million.[12]

The Cognac prospect was so far beyond working depths that Shell Oil did not even have a rig that would drill it at the time of the lease sale.[13] The company soon found a semisubmersible, *Pacesetter II*, and reequipped it with added mooring, larger conductors, and other modifications. Not until June 1975 did Shell have the rig ready to drill on the prospect in 1,000 feet of water, more than twice the depth in the Gulf previously plumbed for commercial production. In July the first exploratory well for Shell by *Pacesetter II* struck oil, and the well log showed 140 feet of pay, more than enough to go forward with a platform.[14]

During the next year, eleven more tests were drilled on the four blocks; eight discovered oil and gas. As it turned out, Amoco had obtained the best acreage. From Shell's perspective, the logical way to develop the field was to unitize the operations of the two groups of partners. But Amoco used its reserve estimates as a strong bargaining chip. After two years of difficult negotiations, Shell and Amoco formed a joint venture with Shell as operator (Shell 42.8 percent, Amoco 21 percent). By 1977, when the agreement was signed, Shell was already building the jacket, strengthening the company's bargaining leverage over Amoco, which had not yet figured out how to develop its interests. "They didn't have a design, and we were already building the bottom section of the platform," explained production manager Sam Paine. "So we knew we had them. We traded hard. We didn't back off. And they finally agreed with us."[15]

When Shell bought the leases at the March 1974 sale, however, its engineers had not yet come up with a design concept for producing in 1,000 feet of water. A year earlier, Dunn's civil engineering group had begun to analyze the problems of designing and installing a fixed platform for such a water depth. To withstand the day-to-day waves of deep ocean water as well as the extreme winds and waves of hurricanes, it would have to be mammoth-size and heavily reinforced, dwarfing anything ever built. The base of the structure also had to be sturdy enough to withstand tremendous forces from mudslides. Design, however, was the easy part. Finding a way to install it was the main challenge. Along the Gulf Coast there were no construction yards and launch barges even remotely big enough, or tow-out water depths deep enough, to handle a one-piece, 1,040-foot-tall steel jacket.

The only conceivable solution at the time was to build and install the jacket in sections small enough to be floated and lifted by available equipment, then mate the sections in the water. Such a project would be incredibly complex, requiring untried procedures. At the same time, Exxon was working on installing the "Hondo" platform in 850 feet of water in the Santa Barbara Channel. Exxon engineers decided to launch the jacket in two pieces and then mate them horizontally in protected water. The mated jacket would then be uprighted on the bottom. Shell engineers considered a Hondo-style mating operation but ruled it out because of the risks of this time-consuming procedure in the hurricane-prone Gulf. Instead, they settled on a unique and innovative concept: building and launching the jacket in three pieces, mated vertically, or "stacked," under water. Launching each section would be a separate and relatively quick operation. And the mating would take place deep enough to be protected from strong wave action.

This was easier said than done. "It took many agonizing hours of planning, thinking, re-thinking the problems," said Gordon Sterling, who supervised the detailed engineering design of the structure. "How would the base section be connected safely and securely to the middle and top sections? How would the base be leveled on the Gulf floor? There were many other questions. But the answers started coming." Shell's project managers assigned an elite team of specialists to engineer various parts: engineering and pile hammer design; base section launch; electronic instrumentation package for guiding and monitoring each step of the installation; hydraulically actuated mud mats for leveling the base section; and ballasting, pile cementing, and flooding systems for installation. Said Dan Godfrey, the engineer in charge of fabricating the base, "Whether a man was working on the base, the middle or the top section of this construction effort, no one—even old timers in the yard—had ever been a part of something so special."[16]

Shell awarded the contract for building and installing the Cognac structure to J. Ray McDermott, one of the leading offshore construction firms in the Gulf. In April 1975, even before exploration drilling had started, steel was ordered, and in December fabrication of the base section began at McDermott's Bayou Boeuf yard in Morgan City. Slowly, tons and tons of steel filtered into the yard. Joint by joint, brace by brace, the base began to take form like a giant jigsaw puzzle. It grew even larger than initially planned. The discovery of huge reserves and the exploding U.S. demand for oil in the mid-1970s compelled Shell to speed up development. Consequently, the original jacket, designed for a forty-well, single-rig platform, was enlarged to accommodate two drilling rigs and sixty-two wells. In turn, the total weight of the jacket increased from 19,000 tons to 49,000 tons, with an attendant escalation in costs.

On a fast-track schedule, a project of such unprecedented magnitude and complexity was bound to run into problems and setbacks. The general design for the jacket as a single piece was essentially a deepwater application of the basic American Petroleum Institute (API) drilling-production platform used in shallower waters. But translating the design into metal called for exceptional accuracy in fabrication to ensure that the pieces would fit when mated. All measurements had to be temperature-calibrated to take into account the expansion and contraction of the steel, from the hot Louisiana sun to the cold depths of the Gulf of Mexico. Braces even millimeters out of alignment, for example, had to be replaced. "In terms of construction tolerances, there's absolutely no comparison with any other job," Godfrey noted. "If the sections don't mate, you can write off the whole thing."[17]

Shell and McDermott employed space-age technology to ensure that the sections would fit. A survey by Boeing Aerospace predicted how well the sections would match under varying offshore conditions. An on-site construction survey used infrared devices to check Boeing's figures. The fit of jacket members was checked with photogrammetry, a computerized method of correlating photographed targets developed by the U.S. Army for high-altitude mapping. Even the most sophisticated measuring techniques, however, could not completely eliminate uncertainty.[18]

These doubts would never be completely laid to rest until the sections were mated in the water. Installation was even trickier and more worrisome than fabrication. There could be virtually no margin for error. Over half of the three hundred man-years of engineering logged on the entire project dealt with installation procedures.[19] In the summer and fall of 1977 the massive base section was installed, and the following summer the middle and top sections were fastened together with the base. For the numerous engineers who had labored for years over the installation, the successful mating of the sections brought relief and jubilation. In September 1979, more than five years after the leases were bought, the Cognac platform began producing oil and gas. By the summer of 1981 all the wells had been drilled, permanent production facilities had been installed, and the world's deepest platform-to-shore oil pipeline had been laid. At the end of 1982, according to one Shell Oil brochure, Cognac was producing 72,000 barrels of oil and 100 million cubic feet of gas per day.

Cognac was the most sophisticated fixed platform installation ever completed; at $240 million, the platform was also the most costly. From start to finish, the overall project cost Shell and its co-owners nearly $800 million. Other companies built subsequent platforms in similar depths with less steel and launched them in one piece from larger barges for much less money. But these projects could not have happened without the deepwater precedent established by Cognac. It marked an unparalleled advance in the technology of offshore structures, setting records for the deepest water, largest number of wells, and heaviest steel platform, among numerous other innovations. In 1980 the American Society of Civil Engineers honored Cognac with it annual Outstanding Civil Engineering Achievement award, the first ever received by an oil company. Along with Exxon's Hondo and developments in the North Sea, Cognac opened a new era for truly enormous, offshore engineering/construction projects. It introduced the "team" or "project line" concept to the industry, marrying disciplines such as naval architecture, structural engineering, and mechanical engineering. Company engineers also worked closer with the fabrication and installation contractors than ever before, taking project management to a whole new level.

Controversies, Delays, and Disappointments

As the energy crisis of the mid-1970s intensified, and as onshore prospects in the United States declined, U.S. oil companies looked increasingly offshore to expand domestic reserves. But even with Cognac paving the way deeper into the Gulf of Mexico, many oilmen, including Shell Oil's own, believed that after twenty-five years of development only lean prospects remained in the Gulf. The best hope for increasing national reserves, they insisted, was to open up the unexplored sedimentary basins off the east and west coasts and off Alaska.[20]

The industry's drive to explore these areas, however, collided with opposition from environmentalists and coastal communities. Of the 19,000 wells drilled in U.S. waters up to 1975, only four had caused major oil spills. But those four had been relatively recent and spectacular. As the industry moved into deeper, rougher waters, environmentalists feared that the likelihood of spills increased—with potentially ruinous consequences for marine ecology and recreational beaches along places like Long Island and Southern California. New England fishermen, furthermore, did not want oil companies invading their territory. Governors and politicians from coastal states, unprepared to cope with the onshore consequences of an aggressive leasing program, objected to providing costly services and facilities for offshore development. They wanted to be consulted about the federal leasing program, which they increasingly argued would be inconsistent with the requirements of state coastal and marine management programs. "People seem to want new oil sources developed, but they don't want it where they live," complained Bookout. "We have been far less willing to open up our continental shelves than most countries."[21]

Bookout emerged as a vocal and articulate spokesman for expanded access to "frontier" areas. "Offshore represents the major domestic potential yet to be explored," he repeatedly emphasized. Other Shell executives also spoke out. Already sensitized to environmental concerns and convinced of the need to establish a more open relationship with the public, Shell Oil sent its exploration and production managers out to plead the case to government officials and coastal communities. Exploration vice president Nanz spearheaded the effort, organizing and presenting detailed information before numerous groups on what he called the "Offshore Imperative." Nanz and other Shell representatives participated in industry efforts organized by the API and coordinated with the National Ocean Industries Association (NOIA) to help overcome local and government resistance to offshore development. "We did a lot of work with fishing groups in different areas, because they were one of our primary opponents," remembered O. J. Shirley, Shell's Southern E&P Region safety and environmental conservation manager, who was active in these efforts. "We worked with the governor of Massachusetts in trying to get access to Georges Bank. We worked with New Jersey people for access to the mid-Atlantic. It was easy to identify who our adversaries were, and we tried to get an opportunity to speak to them."[22]

It was a tough battle. Adversaries were not easily converted. Oil company representatives struggled to convince people of the industry's renewed commitment to safety and environmental protection. Shirley had been a founder of the Clean Gulf Associates (CGA), an industry organization formed in 1972 to upgrade oil-spill-handling capabilities in the Gulf. As lease sales were scheduled in the mid-Atlantic, some of the same companies organized a new group, called the Clean Atlantic Associates (CAA), with Shirley as its first chairman. The CAA compiled an oil spill contingency manual, identified areas of particular sensitivity to oil spills, and planned to stockpile oil spill equipment for the North Atlantic, Mid-Atlantic, and South Atlantic regions.[23] The CAA sought to puncture the stereotype of offshore oilmen as insensitive to the environment and demonstrated the industry's willingness to abide by rigorous environmental protection standards. "Through strong personal contact, one-on-one discussions, and actual friendships, we formed relationships with the environmental community," said Shirley.

These efforts helped break down public resistance, but obtaining leases and permits to drill still entailed protracted legal struggles. "It looked like, sometimes, that we were never going to get there," said Shirley, "but, looking back, we gained access to almost every area that we wanted to drill offshore." One promising area was the Baltimore Canyon trench off the coasts of Delaware and New Jersey. In a 1976 federal sale, Shell and partners obtained twelve tracts in relatively shallow waters of the Baltimore Canyon. The sale was contested in court, and not until March 1978, when the Supreme Court refused to hear an appeal of a lower court decision validating the sale, was drilling allowed to proceed.[24] A string of dry holes from the 1976 sale, however, including several by Shell Oil, dampened enthusiasm for a second sale held in 1979. Shell had been hoping for a bonanza, "one or more giant fields the size of Mexico's Golden Lane," said Jack Threet.

There had been geological reason to hope for such fields. "We knew we had reservoirs and we were almost certain we had traps," he explained. "But we think there was probably not enough oil generated in the Atlantic Basin to migrate into those traps."[25]

As companies began to write off the Baltimore Canyon, attention shifted to another promising area—the Georges Bank trough southeast of Cape Cod, Massachusetts. But drilling there encountered even greater opposition. In 1976 the Conservation Law Foundation and the State of Massachusetts filed suit to block sales in Georges Bank. After two years of legal wrangling, the Supreme Court refused to grant a final request to cancel the Georges Bank sale, which was finally held in December 1979. Shell and its bidding partners won three tracts for a price of $86 million. Obtaining permits to drill, however, dragged on for many months. In 1978, Congress passed the Outer Continental Shelf Land Act Amendments (OCSLAA), which opened up the offshore leasing process to wider public participation, involving more government agencies, with the intention of building public confidence in this activity. At least in New England, however, this act further delayed drilling. The permits issued by the U.S. Geological Survey and Environmental Protection Agency and approved by state agencies in Connecticut, Massachusetts, Rhode Island, and Maine—pertaining to mud discharge, spill equipment, and protection of fisheries—were among the most stringent ever applied to offshore drilling.[26]

In 1981, once all the appropriate permits had been obtained, Shell finally drilled its first exploratory well in the Georges Bank. But, alas, this and subsequent wells turned up dry. It was a good gamble against long odds, because even with high costs the rewards looked rich enough to justify the search. But after years of fighting the modern-day "Battle of the Atlantic" for access to the eastern continental shelf, the industry found little tangible reward, except for a better geological understanding of this offshore basin and a better appreciation of the political dimensions of offshore development outside the Gulf of Mexico.

Californians put up even fiercer resistance to offshore drilling than easterners. Offshore development was not new to California, but it had proceeded along a different and stranger trajectory than in the Gulf. Beginning in the 1930s, drilling platforms built from piers had been erected from Santa Barbara down the coast to Long Beach. Because the ocean floor of the Pacific sloped off sharply from the shore, companies could not move deeper gradually as they could in the Gulf. Large structures that would have been placed far beyond view in the Gulf, therefore, were clearly visible from California beaches. In the late 1950s, to appease residents who did not want their scenic ocean view spoiled by drilling rigs, artificial islands made of sand and rock were introduced to house and beautify them. In the 1960s, the THUMS Group—Texaco, Humble, Union, Mobil, and Shell—extended this artificial island concept by building four 10-acre islands off Long Beach. Each had elaborate façades to camouflage rigs and equipment and give the impression of real estate developments rather than offshore facilities.[27] Leasing off California came to sudden halt, nevertheless, after the 1969 Santa Barbara oil spill, which galvanized local groups statewide to agitate for restrictions on offshore development.

Despite early setbacks, the movement gained political strength. In 1974, after the moratorium on drilling was lifted, the State of California unsuccessfully tried to block the first federal lease sale, maintaining that it did not meet the requirements of the National Environmental Policy Act. In the December 1975 sale, held in Los Angeles, Shell Oil and its partners spent $123 million, most of this for two 5,700-acre leases on a prospect called Beta, in water ranging from 220 to 1,000 feet in San Pedro Bay off Long Beach. The sale bolstered anti-industry forces, however, creating enough pressure to cancel the two federal sales proposed for 1976 and 1978. A suit brought by the County of Santa Barbara postponed the next sale, originally scheduled for 1977, until 1979. Meanwhile, the California Coastal Commission (CCC), backed by Governor Edmund G. Brown, issued ever more stringent requirements for federal leasing to ensure that it was consistent with the state's federally authorized coastal management program. Subsequent lease sales became so embroiled in lawsuits and subject to the withdrawal of the most attractive tracts due to environmental concerns that development of offshore California screeched to a halt. Beginning in 1982, Congress inserted prohibitions into the Department of Interior's appropriation that effectively shut down leasing on the OCS of both the east and west coasts.[28]

Within this antagonistic political climate, Shell Oil pressed forward with the development of its Beta prospect. Of all the tracts leased in the 1975 sale, Beta yielded the only commercial discovery, in July and August 1976. Exploratory drilling revealed an estimated 150 million-barrel field, and Shell badly needed this oil to supply its West Coast refineries, which had been forced to purchase increasing amounts of crude from other companies. But bringing the field into production would prove to be neither simple nor inexpensive. Platform designs had to account for the possible impact of shock waves generated from earthquakes. Although the advent of powerful computers had improved the seismic analysis of offshore structures, knowledge of earthquake design was still not that developed, even by the early 1970s. Ensuring that a platform had enough structural resilience to absorb the energy of severe earth tremors, therefore, required conservative and thus costly designs. Development strategy also had to take into consideration the fact that the reservoir contained heavy oil and low natural pressures. Water injection and downhole electrical pumps would be needed to produce the oil. Shell used sophisticated computer simulation techniques to predict reservoir performance, studied various alternate development plans, and eventually decided to build two offshore structures instead of a combined drilling/production platform. The two-platform complex allowed for the most efficient development of the Beta field and provided the large amounts of space needed to support the processing equipment.

Political and regulatory obstacles, driven by growing opposition to offshore oil in California, hindered the project more than design considerations. But Shell was determined to see the project through by meeting or exceeding all state and federal safety requirements and environmental standards. Early on, Shell teams spelled out detailed development plans in face-to-face meetings with numerous community and civic groups, as well as with the appropriate local, state, and federal officials. They covered all the major impacts of the Beta project, including safety, air and water quality, marine traffic, oil spill prevention, and onshore activities. "The path that we adopted was to be completely open with them," said Phil Carroll, division production manager for Shell Western E&P at the time. "No surprises or attempts to sneak something by. We did everything we could to accommodate them."[29]

Still, the permitting process dragged on for two years. Of the eleven different local, state, and federal agencies from which Shell had to obtain permits, the California Air Resources Board (CARB) threw up the most difficult roadblocks. Shell Western E&P managers took a calculated risk, ordering fabrication of the components just as they began applying for permits. Brown & Root constructed the two platform jackets for Shell in Labuan, East Malaysia; the deck sections, pilings, and conductors were made in Japan. "I was frequently asked," remembered Carroll, "'My God, why don't you stop building those things until you are sure you can get the permits?'"

But because the field required two major platforms in 260 feet of water, Shell compressed the construction schedule, contracting for components from multiple international contractors to speed up fabrication. In late 1979 the jacket for the drilling platform called Ellen was literally being towed by barge across the Pacific Ocean before Shell had obtained all the permits. Carroll planned to tow the jackets right out to location in San Pedro Bay and invite television crews out to see a major new source of energy desperately needed by the nation, but which was being held up by regulatory red tape. Fortunately, the permits came through in time to avoid a showdown.

Gaining permission to develop the Beta field was an impressive feat. Shell's frank and open discussions with government officials and community leaders cleared up many misconceptions about the impact of the project and paved the way through the permitting process. In early 1980, Shell installed the production platform Elly, linked by a 200-foot bridge to its sister drilling platform Ellen. Four years later, as the development drilling program on Ellen drew to a close, Shell installed a mammoth 700-foot drilling platform called Eureka to develop the much deeper southern portion of the field. Built by Kaiser Steel at Vallejo, near San Francisco, Eureka was the largest single-piece jacket installed up to that point on the West Coast and the sixth-largest overall in the world. All told, the Beta project cost $700 million and touched nearly ever organization in the company over the course of a decade. By the late 1980s, Beta had hit peak production of about 20,000 barrels per day, the industry's only commercial success from the 1975 lease sale in Los Angeles.[30]

In Alaska, the last frontier area off U.S. coasts, Shell was not so fortunate. The first stumble came in the Cook Inlet, where Shell had enjoyed previous success in the Middle Ground Shoal Field. In a December 1973 state lease sale, the company tried to expand on that success by acquiring five tracts in the Kachemak Bay area of Cook Inlet. As Shell prepared to develop the leases, however, the coastal communities rose up against offshore operations in the bay, a pristine, picturesque setting. In June 1976, after a protracted series of hearings, the state imposed a one-year moratorium on drilling in the bay. A year later, state legislation authorizing condemnation of leases in the Kachemak Bay forced Shell to sell the leases back to the state.

Undaunted, Shell remained faithful to Alaska's oil potential and optimistic about the industry's chances at getting access to it. In the mid-1970s, Shell and other oil companies believed that federal territory in the Gulf of Alaska might have the same kind of big, concentrated oil deposits that were found at

Prudhoe Bay. Sales of Gulf of Alaska leases by the federal government were supposed to follow the state sales at Prudhoe Bay, but the Santa Barbara blowout incurred the wrath of environmentalists and held up sales for years as research was done on the hazards of drilling there.[31] Finally, in April 1976, the federal government put the acreage up for lease, after failed attempts by the State of Alaska to block it. This sale, *Business Week* announced at the time, "may very well hold the last hope for an oilfield big enough to reverse the nation's four-year decline in oil production."[32] The sale also offered Shell exploration managers a chance to redeem the company in Alaska after their failure at Prudhoe Bay.

The Gulf of Alaska was Shell's top candidate among the seventeen potential OCS oil and gas provinces listed by the Bureau of Land Management (BLM) in 1974. It was also a forbidding frontier region, one of the most hostile in the world. Its fierce, chilling winds drove waves cresting at 100 feet. Fog often made helicopter transport impossible. Moreover, it was a seismically active area that would require earthquake-resistant platforms. "The Gulf of Alaska," said John Swearingen, chairman of Standard Oil of Indiana (Amoco), "will make the North Sea look like a kiddie pool."[33] Shell estimated that a production platform in 300 feet of water in the Gulf of Alaska would cost as much as one in 1,000 feet of water in the Gulf of Mexico. Unfortunately, the techniques Shell had laboriously developed for evaluating leases in the Gulf of Mexico were not applicable there. "There was no information other than seismic," remembered Marlan Downey, exploration manager for the Alaska division. "There wasn't a history of production. There wasn't anything that told you whether or not there would really be commercial oil there."[34] Nonetheless, Shell was anxious to find out. In preparing for the sale, its Alaska division geophysicists identified several major structures. Although they did not find any verifiable bright spots on the seismic data, they saw hints of an unusual type of undersaturated oil that did not have gas. So they decided to bid aggressively, taking on ARCO as a partner, though, to spread the risk. The Shell-ARCO partners were the high bidders in the sale, together spending $276 million (Shell's share being $148 million) out of an industry total of $572 million, according to Shell's 1976 annual report. They won twenty-nine tracts totaling 165,000 acres (nine of eleven prospects on which Shell bid).

And they drilled nothing but dry holes. There was no source rock. It appeared that temperatures never got high enough in the formation to cook up the oil. "Everything looked good and the structures were there," said Nanz. "Except oil was not generated in the particular ones we sampled."[35] These dry holes were also expensive. Stormy weather and high formation pressures made drilling from semisubmersible rigs difficult, resulting in drilling costs from $10 million to $23 million per well, according to the company's 1977 annual report. Shell's Gulf of Alaska venture was a complete failure, a miserable disappointment. When a second lease sale in the eastern Gulf of Alaska came up a few years later, Nanz resisted any temptation to place another bet. "I feel like that monkey they put on the sled down there at NASA in the acceleration chamber," Nanz told his geologists. "He did not want to get back on that sled again and that is how I feel about this sale."

Shell went to the sale but acquired only five tracts for $1.4 million. It was saving its money for other sales in Alaska's western and northern waters. Despite a string of controversies, delays, and failures in other frontier areas, Shell's exploration leaders still believed in the potential of offshore Alaska. In 1978 the company announced that it expected Alaska to provide 58 percent of the country's future crude and condensate discoveries. There were some very large structures off the Alaska shore. If oil and gas had migrated out there, these structures could be "company makers." As one executive described Shell's thinking, "There was a huge, world-class field up there onshore. So there just had to be something, right, in the offshore?"[36]

New Urgency

During the 1970s, offshore oil in the United States became the subject of rising political controversy. Environmental opposition and the "not-in-my-backyard" syndrome thwarted the industry's efforts to explore many frontier areas of the OCS. As the oil industry also came under intense scrutiny for alleged profiteering after the Arab embargo, questions about the competitiveness of offshore leasing increasingly entered the discussion. Critics charged that the bidding system based on cash bonuses with fixed royalties did not always give the federal government a "fair value" on leases and that joint bidding by the major oil companies kept the smaller independents from operating in deeper waters.

Oilmen scoffed at the suggestion that lease sales were not competitive. They argued that, even though smaller companies did not have the capital to develop leases on their own, many of them were often included in successful offers. Oil companies emphasized that skyrocketing lease prices were ample evidence that the system was highly competitive. In 1977, Nanz pointed out that, of the average winning bids in the previous twenty OCS sales, 45 percent of the bonus was "left on the table"—it was not needed to get the lease. "It's been more than competitive," he commented. "More like frantic."[37]

With OCSLAA, Congress attempted to reform the bidding process to make it even more competitive. The amendments required the Department of Interior, during a five-year experimental period beginning in September 1978, to try new bidding systems that reduced the amount of front-end money needed to obtain leases and thereby, in theory, enable more companies to purchase leases. The traditional format consisted of a cash bonus bid for a given tract with a fixed percentage royalty on what was produced, whereas the alternative systems included those that derived income for the federal government largely through variable royalties bids or net profit sharing rather than through cash bonus bids. Shell, like other companies, did not like rising cash bonuses but still favored the traditional system over most of the alternatives, which company officials argued would only encourage speculation, impose new administrative burdens, and delay exploration.[38]

With the deepening of the energy crisis in the United States, the last thing the Department of Interior wanted to do was delay or impede domestic exploration. The December 1978 overthrow of the shah of Iran by Shiite Muslim revolutionaries cut off petroleum exports from Iran, lifting world crude oil prices from $13 per barrel to $34 per barrel and precipitating a full-blown panic at the pump. In March 1979, U.S. Secretary of the Interior Cecil Andrus, as directed by the OCS amendments, announced a five-year offshore leasing schedule aimed at expediting exploration and development. The program would average five sales a year with emphasis on the Gulf of Mexico and Alaska. Faced with new urgency to develop domestic oil deposits, over the next several years the Department of the Interior continued to rely on the tried-and-true system of cash-bonus leasing and experimented with the different systems only in a limited way. After studying the comparisons, Interior found that these systems produced no statistically meaningful differences in industry competition, a view that the Supreme Court upheld in 1981.[39]

As Interior expanded its leasing program, Shell Oil geared up for the biggest push the company had ever made into the offshore United States. Company officials had often criticized Interior's leasing timetable in the past and thus were exhilarated by the promise of new areas being opened for exploration. Over the years, Shell had placed bigger and bigger bets on offshore development. Now, Bookout and his lieutenants were prepared to stake the whole company's future on it. In their minds there was really no alternative for a company whose central realm of business was in the United States. They could not see any more major finds onshore. Nanz estimated that nearly 60 percent of the oil yet to be found in the United States was offshore, most of it under federal control.[40] The risks of pushing into the offshore frontier were staggering—huge bonuses, expensive drilling, and if all went well up to that point the monumental costs of development. But they had to be taken for Shell to have a future as a major oil producer. The exploration department was looking for large-scope projects; these would involve higher risks, but if they came about they would remake the company. "We worked so hard," remembered Mike Forrest. "Shell needed to find 200 million barrels of oil a year just to stay even, to replace production."[41]

By the mid-1980s, roughly 60 percent of Shell's exploration dollars went to the offshore effort in the United States. "Exploration has been called a poker game," Jack Threet mused in 1984. "But there's more to it than that. In this game, we don't have chips or coins or dollar bills that can change hands over and over again. We're dealing with a declining resource base, and every barrel we find is never going to be found again."[42] Two places Shell believed in were Alaska and deepwater Gulf of Mexico. Environmental opposition had basically shut down leasing off California and Florida. Drilling in the North Atlantic and eastern Gulf of Mexico (the MAFLA region) had found little. There were really no other virgin areas in the United States to explore for large oil accumulations. Shell believed that large oil fields would be discovered in Alaska and included the risked reserves there in the company's ten-year long-term plan in the late 1970s and early 1980s. Shell's exploration leaders still held the Gulf of Mexico in high regard, but the economics of so-called deep water was still controversial (see below), so the deepwater Gulf of Mexico did not really make it into Shell's long-term plan until the mid-1980s.

Despite the Gulf of Alaska bust, Interior and oil company officials considered other parts of offshore Alaska to have the highest resource potential anywhere in the United States. It was big-structure country. For years, Nanz had led the charge in lobbying the Interior Department to accelerate leasing in Alaskan waters—particularly the Bering Sea and Beaufort Sea basins. After the second oil shock, his words finally appeared to carry more weight. In June 1979, Secretary Andrus revised the leasing schedule announced in March to give earlier consideration to the Alaskan sales. Although not entirely satisfied with the proposed pace of leasing, Nanz was encouraged by the announcement. He asserted that the technology was available for exploring most Alaskan offshore basins. But extreme weather would make it difficult. Ice prevented seismic boats from even getting into Alaska's northern waters, except for maybe one year out of every five. Drilling crews would have to cope with minus-60 degree temperatures and 24-hour darkness in the winter. Furthermore, there was no clear-cut method for producing oil from such an ice-ridden environment. Yet Shell's credo held that, if the fields could be found and the economic conditions were favorable, the technology would arrive to bring them into production. In 1979, with the price of crude soaring near $40 per barrel and the phasing out of price controls in the United States, almost any project seemed possible.

Shell Oil believed as fervently as anyone that Alaska might be the savior of the U.S. oil industry. Shell Western E&P performed exhaustive geophysical work on all of Alaska's offshore basins and, with Amoco as a key bidding partner, forked out millions of dollars in a succession of lease sales held between 1979 and 1985. In 1979, Shell spent $69 million in partnership with Amoco on leases in the first sale in the Diapir basin of the Beaufort Sea, north of Prudhoe Bay. In October 1982 the company joined Amoco, Union Oil, and Koch Oil in purchasing leases in another part of the Diapir basin, mostly on parcels that covered a huge structure called Mukluk. In April 1983, Shell Oil spent $78 million in a joint venture with Amoco and Marathon to acquire leases in the St. George and Norton basins of the Bering Sea. A year later a Shell-Amoco combine dominated a sale of tracts in the Navarin basin, with Shell putting up $175 million of the winning bids. The last major area was the Chukchi Sea, for which, in lease sales held in 1985 and 1988, Shell outspent the competition for large tracts. In the final analysis, Shell spent more money and acquired more acreage than any other company in offshore Alaska lease sales.

All areas held tremendous promise. The Beaufort Sea possessed giant structures, Mukluk in particular. It looked much like the neighboring

Prudhoe Bay field, with the same reservoir rock, source rock, and geological history. Even though Mukluk was only a 1–2 billion barrel prospect, the industry—led by British Petroleum and its U.S. affiliate Sohio—had high hopes for it, spending nearly $1.5 billion on Mukluk leases. Most of the tracts were in 40–100 feet of water covered with ice as thick as ten feet for eight months of the year. Shell and other companies turned to building artificial islands out of gravel to drill their exploration wells. Tragically, though, Mukluk turned out to be the most expensive dry hole in history.[43] Oil stains in the rocks indicated that it had once been a giant oil field. But some time in geological history the structure had been breeched, allowing oil to leak to the surface, or regional tilting had caused the oil to migrate elsewhere. "We drilled in the right place," said Richard Bray, the president of Sohio's production company. "We were simply 30 million years too late." Although Shell geologists had not assigned as high a probability of finding oil at Mukluk as some other companies, and thus did not bet as heavily on it (the company spent $162 million on leases), Shell Oil shared in the costly disappointment.[44]

Shell and the industry did not fare any better in the other basins off Alaska. Either they found no source rocks or the deposits they did find were not large enough to be commercially viable. The company collected massive amounts of data on every prospect, drilled in every basin, and came up empty. The last gasp was in the remote, hostile waters of the Chukchi Sea. Shell had obtained acreage on several sizeable structures and, after struggling to satisfy environmental concerns in gaining a federal drilling permit, discovered oil. The federal drilling permit was approved none too soon, on March 23, 1989, literally one day before the *Exxon Valdez* oil tanker rammed into a reef in Alaska's Prince William Sound and spilled 240,000 barrels of petroleum into those pristine waters. Even then, Shell had to jump through many hoops to prove it had the capability to drill in the tempestuous Arctic waters, building a $15 million oil spill barge with state-of-the-art cleanup equipment.[45]

The Chukchi deposits were too expensive to develop. The technological challenges were supreme, even for Shell. Because enormous sheets of floating ice would demolish conventional drilling and production platforms, the company looked at installing big ice-breaker platforms and pipelines that could resist ice scouring. Even if the technology could have been found, however, the falling price of oil by the late 1980s made the development of the Chukchi deposits out of the question. "It may have been a blessing in disguise that we didn't find commercial quantities," admitted Jack Little, the head of Shell Western E&P at the time. "We probably would have found the technological problems to be almost insurmountable."[46] During the 1980s, Shell spent an estimated $2 billion on leases and drilling offshore Alaska and came away with nothing to show for it.[47] So ended, for the time being, Shell's arduous, thirty-year quest to find bonanza reserves in Alaska.

Deepwater Vistas

As the failures followed one upon another in Alaska and other frontier areas, Shell started to shift the exploration spotlight back on the Gulf of Mexico, a proven oil province that in the late 1970s showed renewed signs of life with rising oil and gas prices. During 1975–77, Shell had actually deemphasized the "Cenozoic play" in the Gulf in favor of exploration elsewhere. In 1970–74, Shell bid on 64 percent of the volumes discovered by the industry in the Gulf but on only 22 percent during the next three years. The company focused on geopressured natural gas prospects in the ultra-deep producing horizons of the Texas Miocene. Discoveries in 1975 at Prospects Manifold (Eugene Island 136) and Calcite (East Cameron 57) encouraged this search, and Shell subsequently dominated the Corsair sandstone trend with discoveries at Picaroon (Brazos A19, A20) and Doubloon (Brazos A23) in the June 1977 and May 1978 sales.[48]

The June 1977 Gulf of Mexico lease sale surprised industry observers by taking in $1.17 billion in high bids. Shell placed second in the bidding, with $100 million in winning bonuses. Anticipated higher natural gas prices from the staged decontrol of gas, observers presumed, spurred on the bidding. Indeed, most of the discoveries on these leases—including significant ones by Shell on the Brazos and Matagora Island tracts—were made in gas-rich areas stretching from the mouth of the Mississippi River westward to the Mustang Island area near Corpus Christi, Texas. Over the next several years, lease acreage in the Gulf continued to draw spirited bidding. In the December 1978 sale, Shell outspent all others, laying out $184 million for ten tracts, again in natural gas–producing areas. Two years later, Shell and its leasing partners announced a $1.2 billion program for developing fourteen central and western Gulf of Mexico gas fields discovered on these leases.[49]

Despite success discovering natural gas in the West Louisiana and Texas Miocene, Shell managers felt they could have done better. The company still led the industry in Gulf of Mexico discoveries in the mid-1970s with an ultimate estimated volume (in 1987) of 349 million barrels, compared to 229 million for its closest competitor, Gulf Oil. But these results did not meet the extremely high standards that Shell explorationists set for themselves in the Gulf. As the 1987 lookback study concluded, Shell had "lost a good opportunity to add volumes mostly by Bright Spot discoveries," and "a lot of smaller companies did well on the 78 percent of the volume SOI [Shell Offshore, Inc.] did not bid."[50]

Beginning in 1979, motivated by the sense of missed opportunity in previous years and dimmed prospects in other offshore areas, Shell Oil expanded exploration in the Gulf. Meanwhile, BLM accelerated its lease sales. In 1981 there were a record seven offshore sales held in the United States. "We had a lot of lease sales. We went through a lot of lease sale reviews," remembered Charlie Blackburn. Competition for leases in the Gulf became fiercer than ever. The oil price shock of 1979 and the perception that offshore prospects were declining created a feeding frenzy for what was left. Bonus bids skyrocketed in the Gulf, shattering all previous records. "The bidding just got ridiculous," said Blackburn. "The whole business got ridiculous!" The September 1980 sale in New Orleans brought in $2.8 billion; Shell Oil purchased sixteen tracts for a whopping $316 million, second-highest in the sale. "I got a three-letter description: W-O-W!" said John Rankin, manager of the BLM's New Orleans offshore office, after the sale.[51]

Shell's exploration managers became increasingly dissatisfied with the direction of BLM's leasing program in the Gulf. First, there was the question of steeply increasing costs, as Blackburn indicated. Bonus bids, even those by Shell Oil, the most accurate and cost-efficient explorer in the industry, were too high for the potential volume available. During 1979–82 the company's bonus per barrel of oil discovered soared to $3.94, from well under $1 for the previous eight years, while the ratio for the top companies in the industry increased by a factor of at least four or five. Shell tried to maintain its advantage by bidding on deeper, more subtle traps rather than compete only on the few bright spots nominated, yet most discoveries were on bright spot prospects such as Roberto, Hornet, Cougar, Boxer, Glenda, Wasp, Peccary, Hobbit, and Cheetah. And the company made only two geopressured discoveries at Onyx and Persian.

Although all nice discoveries, these were still predominantly gas deposits containing lower average volumes than those discovered by Shell Oil in the preceding years.[52]

In Shell's view, the second problem, which contributed to escalating costs, was the federal government's method of rationing leases through the nominating process. The relatively small amount of nominated acreage actually offered in sales was creating an artificial shortage of exploration opportunities. "Tract selection," as the BLM method was called, offered tracts or blocks in a piecemeal fashion, which hindered more efficient exploration strategies involving basin-wide assessments or the pursuit of structural trends that transcended tract boundaries. The Department of Interior's policy of stipulating a two-year time limit before the release of well logs compounded the problem. Often, when a company had a discovery on a given tract, it would fail to get a promising offset tract nominated before having to surrender its well logs on the discovery. This policy both increased the cost and inhibited the development of prospects that spanned multiple blocks. Billy Flowers remembered Picaroon and Cougar as two important discoveries with open offset tracts that were not being followed up in 1980.[53]

Cougar was particularly important in that it held clues to finding petroleum in deep water—depths beyond the record 1,000 feet set by Cognac. At Cougar, Shell had found hydrocarbons in the "turbidite sands" associated with deepwater geology. The company had been focused on so-called deep water since the late 1950s. But the definition of the concept had changed over time—first deeper than 60 feet, then deeper than 200 feet, deeper than 600 feet, deeper than 1,000 feet. The only constant definition of deep water over time has been "the depth of the water just past the deepest platform." The modern concept, in use since about the early 1980s, refers to depths deeper than 1,000–1,500 feet, the maximum depth for a conventional six-leg platform, although every company has had their own definition.

The deepwater realm was still largely uncharted territory in 1980. The soaring price of bonuses, the small amount of acreage offered in the sales, and the short time horizon of leases stipulated by BLM prevented pioneering moves into these depths. But Shell geologists believed such depths held interesting possibilities. Combining information from deep-water cores drilled by the *Eureka* drillship in the mid-1960s with a 1977 regional seismic survey that probed the edge of the continental shelf and down the slope, Shell geoscientists detected some huge structures, salt pillars that were different from the conventional Gulf Coast salt dome. These pillars had squeezed up from the mother layer of salt, called the Louann sheet, 165 million years ago when cycles of seawater had rushed in and evaporated as the Gulf of Mexico was slowly forming.

Geologists speculated that ancient "turbidity currents"—underwater rivers formed by suspended sediment—might have carried significant amounts of sand out into deeper water, forming reservoirs to trap oil against the salt pillars. Whereas reservoirs on the shelf were highly faulted and required numerous wells to develop, deepwater reservoirs, if they were there beyond the edge of the shelf, might be large and continuous. In the 1979 and 1980 Gulf lease sales, Billy Flowers and Bill Broman, exploration manager for the Offshore Division, nominated some of these prospects, which ranged out beyond 1,300-foot depths. But because the industry as a whole was not yet concerned about those depths, the Department of Interior would not put them up for sale. Large areas had no calls for nominations, and some were not even blocked out yet.

Shell exploration managers decided that they needed more wide-open lease sales with longer lease terms to bring those areas into play. They put together a traveling road show of talks and presentations to high-level Department of Interior and U.S. Geological Survey officials to persuade them to open up deepwater areas for leasing. Instead of maximizing bonus bids in small sales, they argued, the government could take in more aggregate revenue in the form of royalties through larger, broad-area sales. But lease terms would have to be revised to provide incentive to the companies. They told the officials that the standard five-year leases and one-sixth royalty would not promote deepwater development. Something on the order of ten-year leases and one-eighth royalty would provide more incentive. They also pointed out the need for a safe supply for the country and the effect it would have on the U.S. balance-of-payments situation. "And we did something we had never done before," remembered Flowers. "We showed them prospects." Flowers and Broman were careful not to give away crucial information or overstate the potential, but they wanted to let government officials know that there *was* potential out there. They presented seismic data on some of these deepwater structures that made shallower water tracts, which had been put up for sale, pale by comparison.

One prospect in particular, codenamed "Bullwinkle," showed three likely oil pays.[54]

Lobbying by Shell and other companies planted a seed with Interior officials that grew after the 1980 election of Ronald Reagan as president. Shell officials found a much more receptive audience in the new administration. Reagan's secretary of the interior, James Watt, believed fervently in letting the market determine energy outcomes and in releasing federal lands for exploration. Executives from other oil firms also lobbied for reforms to the leasing program, but according to J. Robin West, assistant secretary of the interior for policy, budget, and administration under Watt, none were as effective or forthright as representatives from Shell Oil. "Charlie Blackburn was the one who tried to really work with us and help us understand what were the pros and cons, what was reasonable, what was not reasonable," remembered West. "Some of the other guys . . . would come in like potentates with vast entourages and they would lecture us about what they wanted and leave." Lloyd Otteman remembered making a presentation with Flowers to the undersecretary of the interior, Don Hodel, laying out their proposal for broad-area leasing with a slew of maps and view graphs. In the middle of the meeting, Hodel received a call from Secretary Watt. "He said he needed to go see Jim and he said he needed what we got," said Otteman. "And he just gathered everything up and went off! Later, he came back, and it wasn't too long after that they came out with 'area-wide leasing.'"[55]

Area-wide leasing, which was part of a new five-year leasing program announced by Watt in May 1981, opened up the bidding on any unleased tracts in an entire planning area (e.g., the western, central, or eastern Gulf of Mexico). Millions of acres would be placed on the auction block at one time. For tracts in waters deeper than 900 meters (about 2,950 feet), the program also offered ten-year leases and one-eighth royalty. Watt's area-wide leasing plan aimed to allow oil companies to explore areas they believed to be most favorable rather than areas selected by the government through the nominating process. Area-wide leasing promised to reduce some of the competition and thus lower the costs of bidding; companies with independent data could submit smaller bids on deepwater tracts because the probability of another bid on a given tract was relatively low. It was the most effective way, on the other hand, of accelerating the pace of exploration in federal offshore waters. After years of vocally advocating such a leasing program, Shell Oil could take some credit for helping bring about this major policy change.

The new policy and outspoken and confrontational style of Secretary Watt were not, however, universally popular. They drew protest from small oil firms and renewed political opposition at both the state and federal levels. Critics complained that the new system would give the majors, who had superior capital and technological capabilities for plying deepwater environments, a substantial edge over the independents. Environmentalists worried about a new wave of environmentally risky offshore development. The Pacific coast states, Florida, and several environmental groups went to court to block Watt's program. While the process was under litigation, Watt combined all offshore leasing, regulation, and royalty management functions in the new Minerals Management Service (MMS) within the Department of Interior, streamlining the leasing process and concentrating the growing pressure against the OCS leasing program in one agency. Legal and legislative challenges to the program failed in the U.S. Court of Appeals, and in May 1983 the MMS held the first big area-wide sale in the Gulf of Mexico, opening up over 37 million acres to bids, more than ten times what had normally been offered previously.[56]

To prepare the company for the new, big deepwater play, Shell had already embarked on a program to establish the viability and safety of deepwater drilling. Up to that point, nobody had drilled deeper than 1,500 feet in the Gulf of Mexico, and there were only a handful of wells in the world deeper than 3,000 feet, none of them in the United States. In 1981, after Watt's announcement of the new leasing program and as Shell exploration managers geared up for the play, Bookout had gathered top management together and recalls telling them: "I cannot in good conscience fund and launch this kind of program unless we can develop it. You've got to give me confidence you can get to 3,000 feet, and I want something on the drawing board saying you can get to 6,000 feet." The head office then assigned Carl Wickizer, manager of Production Operations Research, to conduct a feasibility study of "ultra deepwater" drilling and development in water depths beyond 6,000 feet in the Atlantic Ocean. After earlier exploration failures in the shallow waters of the Baltimore Canyon, Shell decided to see what the different geology of the deeper water in that area held. In the December 1981 lease sale, Shell obtained tracts in water extending to 7,500-foot depths in the Baltimore Canyon and Wilmington Canyon areas.[57]

Many critics of deepwater offshore leasing claimed that a technology barrier existed at 6,000 feet. Shell was determined to prove them wrong. In 1982 the company contracted with SONAT for the dynamically positioned drillship *Discoverer Seven Seas*, one of four vessels in the world rated for 6,000 feet of water. Shell then spent over $40 million extending the ship's depth capability to over 7,500 feet, adding a new large marine riser, a new long baseline dynamic positioning system with enhanced software and hardware, a new remote-operated vehicle designed for greater depths, and other modifications. Before the *Seven Seas* could begin drilling, however, Shell had to disprove the previous conclusion of the U.S. Geological Survey that the ocean floor in the area was too unstable for safe drilling.[58]

The company did this in 1981–82 by deploying its proprietary "deep-tow" technology. Deep-tow was a fish containing side-scan sonar that produced high-resolution images of the ocean floor and accurately revealed geological or man-made hazards. The deep-tow survey produced a new perspective on the seafloor geology of the area, showing a generally stable bottom topography and thus paving the way for deepwater drilling on Shell's leases. In late 1983, one hundred miles southeast of Atlantic City, New Jersey, the *Seven Seas* drilled an exploratory well in a world record water depth of 6,448 feet in the Wilmington Canyon. Although the drilling program in the Atlantic, which included two other deepwater wells, did not discover oil, the successful demonstration of drilling at such extreme depths established the industry's capability to drill in water depths beyond 6,000 feet. Just as important for Shell, it inspired confidence in the company's senior management about exploring in any deepwater frontier.[59]

Although the *Seven Seas* did not drill the first ultra-deepwater well until late 1983, Shell Oil was confident enough in the early feasibility study to bid aggressively in the first area-wide lease auction in May 1983. Gulf of Mexico Sale 72 as it was called, shattered all records. The industry leased 656 tracts for $3.47 billion. Under the leadership of Flowers, offshore vice president, and Doug Beckmann, exploration general manager, and with the enthusiastic senior management support from Blackburn and Bookout, Shell Oil put together an ambitious bidding strategy, spending $270 million for sixty blocks.[60] Several of the prospects it bought—Bullwinkle, Tahoe, Popeye— were in 1,300–3,000 feet. In October, Shell made a promising discovery on Bullwinkle, in 1,350-foot waters of the Green Canyon area. Producing from this depth would be a daunting challenge, but Shell's civil engineering department believed that the fixed-platform concept could be stretched to that limit. As engineers began to design such a structure, exploration managers were already thinking about venturing farther out. In 1984 the *Seven Seas* moved into the Gulf to drill on the deeper leases obtained in the 1983 sale.

As with each historic step into deeper water, production lagged behind exploration. Fixed-platform technology could not be extended much beyond the depth of Bullwinkle. Either subsea wellheads or some kind of floating or compliant structure would be required. Concepts for all kinds of deepwater producing systems were beginning to come on stream. In 1981, Conoco had installed in the Hutton field in the North Sea a tension-leg platform, which was an innovative concept using large steel tendons to tether a floating platform to the seafloor. But the costs of all these concepts presented serious questions. Subsea completions were still expensive and not yet perfected. And Conoco's Hutton platform had experienced giant cost overruns. Shell Oil had to count on significant technological development and favorable economic scenarios to produce oil from 1,500–2,000 feet of water, let alone in anything much deeper.

One of the leading methods Shell's production department considered for those depths was subsea wellheads linked by pipeline back to a fixed platform in shallower water. But offshore pipelining faced distance limitations and economic constraints. Shell's production managers had a rule: the company could explore no farther than fifteen miles past 600 feet of water, the practical depth limit and distance for installing marine pipelines at the time. Because of these concerns, Shell Offshore's exploration managers made only a few bids in the April 1984 sale, the second major area-wide sale in the Gulf. They were caught off guard, however, when other companies, notably Exxon and Placid Oil, acquired acreage in water deeper than Shell had been prepared to go.[61]

The results from this sale prompted a flurry of meetings and discussions at Shell about what its deepwater strategy should be. Exploration managers wanted to eliminate the fifteen-mile rule and probe the extreme depths. Upon transferring from the Alaska Exploration Division in 1984 to become general manager of exploration for Shell Offshore, Mike Forrest told the production managers in New Orleans: "We just spent millions of dollars on prospects in the Bering Sea where there is no infrastructure and there is no proven oil source rock. And yet, in the Gulf of Mexico, we are not willing to take risks going out into deeper water? This is a proven oil province!"

Shell's problem in the other U.S. offshore basins was unfavorable geology. The problem in the Gulf was water depth. The geology problem could not be solved. But the water-depth problem could, as Shell had proved again and again.

Shortly after the April lease sale, Flowers obtained meetings with Threet, Blackburn, and Bookout to make the case for pushing farther into the Gulf. All three appreciated the urgency, given the competition, and they resolved that Shell would drop the fifteen-mile rule and begin gathering seismic data from ultra-deep water, using a $45 million, state-of-the-art seismic vessel, the *Shell America*, which had just been launched and outfitted. "We decided to drop everything we were doing on the shelf and put the *Shell America* to work in deep water," recalled Flowers. As long as a football field and 60 feet wide, the *Shell America* was one of the biggest, fastest, and most sophisticated seismic ships ever built. It housed a massive array of equipment and could deploy eight floats, all by a computerized launching mechanism. It not only could gather more specific data but did so with more speed and precision than ever imagined. It also had the space to accommodate large processing systems, giving Shell the capability to do much faster processing of its offshore seismic data. The *Shell America* virtually revolutionized marine data acquisition.[62]

Time was short before the next Gulf of Mexico area-wide lease sale in July. On the auction block were large tracts in the western Gulf. The *Shell America* immediately set out to gather as much data as possible. Because of the time constraint, however, the vessel had to focus on specific locations. Tom Velleca, general manager of geophysics in Houston, urged the Offshore Division to organize a team to search quickly for prospects in the Garden Banks area. Located in waters ranging from 1,000 to 4,000 feet, the prospects they worked on were considered very speculative. The geophysicists did not have as much seismic coverage as they would have liked. The *Shell America* had time to shoot only one seismic line across some of them. In fact, the areas they prepared bids for were more accurately classified as "leads" rather than "prospects." Previously, Shell had only bid on prospects, for which the company had good data. But the exploration managers thought that now was the time to take calculated risks. The introduction of area-wide sales had opened up huge swaths of virgin territory that could be purchased very cheaply. The lease sale team in New Orleans—Flowers, Beckman, and Don Frederick—poured over the surveys in preparation for the July sale.

Shell bid on ten leads in the sale and won seven of them, and for minuscule bonus prices compared to what the company had been paying prior to the introduction of area-wide sales.[63]

Two blocks acquired in the sale covered a prospect called "Auger," located in 2,900 feet of water. Further exploration identified it as the prospect with the most potential. The outline of the bright spots extended west into two open blocks. In those blocks, Shell geophysicist Mike Dunn mapped an amplitude anomaly at a subsurface depth of 19,000 feet, well beyond the horizon of conventional thinking about bright spots. Dunn and other geophysicists were convinced, however, that the amplitude effects were real. In the next area-wide sale, held in May 1985, Shell leased the two open blocks. "We were so afraid that other companies would go after the blocks," said Forrest, "that we bid $5 million on one and $2 million on the other. It turned out we didn't have any competition at all."[64]

In the May 1985 sale, Shell expanded its deepwater play into water depths of 5,000–6,000 feet. Critical to this play was the need for thick, continuous oil sands that could yield large fields and large reserves per well. Because turbidity currents off the continental shelf dumped such immense quantities of sands in one place, geologists had reason to believe that reservoirs there would tend to be far larger than shelf reservoirs. According to Broman, some of Shell's earlier geological research predicted that, unlike in the deltaic setting, where oil pays were found on the crests of salt dome structures, the turbidite sands deposited beyond the shelf would have largely avoided such crests. The seismic probes, therefore, were shot across the flanks of these structures, "down dip" from the crests. Shell's geoscience team mapped the salt ridges and the regional synclines where turbidite sands might have funneled into deep water. Meantime, Flowers and Forrest pressed the production managers on what size oil fields in water depths between 3,000 and 6,000 feet, using "to be designed" technology, would make deepwater producing operations economic. Gene Voiland and Carl Wickizer, production department managers, finally stated that, if the exploration group discovered fields of at least 100 million barrels, the engineers would find a way to make the discoveries pay.[65]

While these discussions were taking place, Shell drilled an exploration test well in 3,000 feet of water on Prospect Powell that had been leased in the April 1984 sale. Drillers located the well to penetrate a very strong, shallow bright spot anomaly plus a deeper, poor-quality bright spot. Drilling

indicated that the shallow anomaly was not associated with oil or gas. However, Frederick excitedly reported the discovery of a 40-foot thick oil pay at
the deep level. Further drilling and seismic surveys showed that the trap was
entirely stratigraphic, likely to contain huge amounts of oil, certainly enough
to meet the economic criteria set by the production department.[66]

Armed with this bit of intelligence, Shell Oil dominated the May 1985
sale. With partners or alone, the company was the high bidder on eighty-six
of 108 blocks for which it submitted a bid, in a variety of areas. Its share in
the high bids totaled more than $200 million. While most other deepwater
lessees did not show interest in acquiring additional deepwater acreage, Shell
took a giant plunge. It obtained tracts in the Green Canyon area ranging out
to 7,500 feet.[67] It acquired prospects code-named Mensa and Ursa, among
others. Combined with the tracts leased in the 1983 and 1984 area-wide
sales, Shell now had huge areas of deepwater acreage in the Gulf of Mexico.
Although no one at the time knew the extent of what this acreage held,
Shell's deepwater play would pioneer the most spectacular new offshore
frontier ever encountered.

. . . .

According to conventional media accounts, the origin of the deepwater era
in the Gulf of Mexico dates to the mid-1990s, when first production at Shell
Oil's Auger prospect was achieved.[68] But the path to deep water has a longer
history, starting with federal leasing reform dating back to the late 1970s
that allowed Shell Oil to implement an aggressive deepwater exploration
and production strategy. The area-wide leasing system introduced in 1983,
in response to years of lobbying by Shell and other companies, gave oil companies easier and cheaper access to offshore territory, thus helping uncover
valuable new domestic sources of oil and gas. The acceleration of deepwater
oil development came at a price, however, in the form of declining public
revenues from offshore leasing.[69] The 2010 Macondo/*Deepwater Horizon* oil
spill also convinced many Americans that the more permissive regulatory
approach to deepwater oil development introduced intolerably high safety
and environmental risks.

In the three years since the disaster, a host of reforms have been implemented to improve regulatory oversight and ensure safe operating practices.[70]
The long-range effects of these reforms remain to be seen. Meanwhile, the

leasing system in the Gulf of Mexico created in the early 1980s remains relatively unchanged, and deepwater exploration and development have returned at a vigorous pace. Efforts to open up new offshore territory, such as in the Arctic and along other parts of the U.S. coast, have resumed, leading to a replay of many of the debates from the 1970s and earlier.[71] The story of offshore leasing and development in the Gulf of Mexico in the 1970s and 1980s provides valuable perspective on current debates in the United States over offshore oil, as well as for making international comparisons. The government policies that have opened up the deepwater Gulf may provide a template for reforms to government leasing policies elsewhere. The globalization of oil markets has forced nations and governments to become internationally competitive in attracting oil investments. Further research into the terms of access offered offshore might provide new insight into how different governments value energy security and how different companies calculate the risks of deepwater development.

Before we see a transition to some kind of alternative energy regime, we will experience a relatively lengthy period in which an ever-growing amount of petroleum supply comes from marine environments. For the United States, this transition began in the 1970s. As the U.S. experience during the past thirty years demonstrates, the "end of easy oil" does not mean an abrupt transition away from conventional oil but an initial transition to difficult oil, like that produced from 5,000–10,000 feet of water. Deepwater is an important gauge with which we can measure the ability of global oil supply to keep pace with galloping demand. It is also a good indicator of how government and business negotiate the delicate balance between "energy security" and protecting the health of the marine and coastal environment.

Notes

This essay is adapted from Tyler Priest, *The Offshore Imperative: Shell Oil's Search for Petroleum in Postwar America* (College Station: Texas A&M University Press, 2007), 180–226.

1. For elaboration, see Tyler Priest, "Extraction Not Creation: The History of Offshore Petroleum in the Gulf of Mexico," *Enterprise and Society* 8, no. 2 (2007): 227–65.

2. "Rising above the Crowd," *Shell News* 6 (1989): 28.

3. "The Offshore: Heading for Deep Water," *Shell News* 4 (1981): 3.

4. "The Significance of One Shell Square," *Times-Picayune Dixie Roto Sunday Magazine*, special issue, Sept. 3, 1972.

5. Pat Dunn, "Deepwater Production: 1950–2000," OTC Paper 7627, presented at the 26th Annual Offshore Technology Conference, Houston, Tex., May 2–5, 1994.

6. "Offshore Technology Meeting Spotlights Progress, Problems," *Offshore*, June 5, 1969, 23–24.

7. Mike Forrest, "'Toast' Was on the Breakfast Menu," *AAPG Explorer*, June 2000, www .aapg.org.

8. Here and in the following discussion, John Redmond information is from notes on the Cognac project and interview with author, Nov. 12, 1999, Houston, Tex. Most of the oral history interviews cited in this essay are archived at the M. D. Anderson Library Special Collections, University of Houston, Houston, Tex.

9. "On the Block: One Million Acres," *Shell News* 1 (1982): 6.

10. The Shell group consisted of Shell (41.67 percent), Conoco (33.33 percent), Sonat Exploration (10.42 percent), Drillamex (4.17 percent), Barber Oil (4.17 percent), Florida Gas Exploration (4.16 percent), and Offshore Co. (2.08 percent); "Bidders Snub Most Deepwater Tracts," *Oil and Gas Journal*, Apr. 8, 1974, 36–40.

11. "Deepwater Tracts Will Be Offered on Continental Slope for the First Time," *Offshore*, Oct. 1973, 52–53.

12. "Bidders Snub Most Deepwater Tracts," 36–40.

13. Lloyd Otteman interview with author, May 17, 2001, Houston, Tex.

14. Forrest, "'Toast' Was on the Breakfast Menu."

15. Sam Paine interview with author, June 8, 1999, Houston, Tex.

16. Sterling and Godfrey in "Cognac Goes Down Smoothly," *Shell News* 6 (1977), 2–3.

17. Dunn, "Deepwater Production"; "Cognac Rises into a Class of Its Own off Louisiana Coast," *Offshore Engineer*, July 1978, 32.

18. Paine interview.

19. "Tallest Offshore Oil Platform Sets Water Installation Records," *Engineering-News Record*, Aug. 24, 1978, 20.

20. "Why Oilmen Are So Cool to Offshore Leases," *Business Week*, Apr. 26, 1976, 72–73.

21. "Shell's New President Must Find More Oil," *Business Week*, Dec. 22, 1975, 20.

22. "Bookout Says OCS Search Is Urgent, Industry Prepared," *Oil and Gas Journal*, May 19, 1975, 153; R. H. Nanz, "The Offshore Imperative: The Need for and Potential of Offshore Exploration," presented at Colloquium on Conventional Energy Sources and the Environment, University of Delaware, Newark, Apr. 30, 1975. Recollections of O. J. Shirley from interview with Tom Stewart, June 8, 1999, Houston, Tex.

23. "'Be Prepared' Is the Motto of Clean Atlantic Associates," *Shell News* 4 (1978): 10–13.

24. "Spuddin' the Atlantic," *Shell News* 4 (1978): 6.

25. "Prospects Darken for Baltimore Canyon," *Oil and Gas Journal*, Mar. 5, 1979, 72; Threet in "The Time to Start Looking Is Now," *Shell News* 6 (1984): 15, and Threet interview with Bruce Beauboeuf, Dec. 5, 1997, Houston, Tex.

26. "Georges Bank: First Step in the North Atlantic," *Shell News* 6 (1981): 2–4.

27. "Long Beach Builds Four Treasure Islands," *Shell News* 35, no. 1 (1967): 24–27.

28. Robert Gramling, *Oil on the Edge: Offshore Development, Conflict, Gridlock* (Albany: State University of New York Press, 1996), 118–26.

29. Phil Carroll interview with author, Joseph Pratt, and Sam Morton, June 3, 1998, Houston, Tex.

30. "Archimedean in Scope," *Shell News* 5 (1984): 31–33.

31. "Oilmen Turn Cool in Alaska," *Business Week*, Sept. 23, 1972, 43.

32. "Why Oilmen Are So Cool to Offshore Leases," 72.

33. Ibid., 73.

34. Marlon Downey interview with author, Sept. 24, 1999, Dallas, Tex.

35. Bob Nanz interview with author, Sept. 15, 1998, Houston, Tex.

36. "Shell: Alaska Holds 58% of Future U.S. Oil Finds," *Oil and Gas Journal*, Nov. 20, 1978, 214; Jack Little interview author and Sam Morton, Feb. 14, 2000, Houston, Tex.

37. R. H. Nanz, Shell Oil Company, "'What We Need' to Increase Domestic Oil and Gas Supplies" (February 1977), opinion piece provided to author by Nanz; "Shell Backs Offshore Cash-Bonus System," *Oil and Gas Journal*, Apr. 29, 1974, 18.

38. "Caution Urged in Using New Offshore Bid System," *Oil and Gas Journal*, Feb. 25, 1980, 32.

39. "Interior's OCS Leasing Plan Advances," *Oil and Gas Journal*, Feb. 6, 1982, 70.

40. Nanz, "What We Need."

41. Mike Forrest interview with author, June 29, 1999, Houston, Tex.

42. "Time to Start Looking," 16.

43. "High OCS Sale 71 Bids Top $2 Billion," *Oil and Gas Journal*, Oct. 18, 1982, 48–50; "Diapir Basin High Bids Hit $877 Million," *Oil and Gas Journal*, Aug. 27, 1984, 38.

44. Bray quoted in Daniel Yergin, *The Prize: The Epic Quest for Oil, Money, and Power* (New York: Simon and Schuster, 1991), 733; "The Great Arctic Energy Rush," *International Business Week*, Jan. 24, 1983, 70–74.

45. Little interview.

46. Ibid.

47. Forrest interview.

48. D. A. Holmes, "1970–1986 Lookback of Offshore Lease Sales Gulf of Mexico Cenozoic," Interoffice Memorandum, Shell Offshore Inc. (Aug. 24, 1987). Copy provided to author by Mr. Holmes.

49. "Gulf Lease Sale Ranks Seventh off U.S.," *Oil and Gas Journal*, July 4, 1977, 33–34; "Success Ratio High on June '77 Leases in Gulf," *Oil and Gas Journal*, Oct. 30, 1978, 27–31; "Development Slated for 14 Gulf of Mexico Fields," *Oil and Gas Journal*, Apr. 7, 1980, 46.

50. Holmes, "1970–1986 Lookback."

51. Charlie Blackburn interview with author, Sept. 23, 1999, Dallas, Tex. Rankin quoted in "Gulf Lease Sale Shatters Two Records," *Oil and Gas Journal*, Oct. 6, 1980, 34.

52. Holmes, "1970–1986 Lookback."

53. "At Issue: Land Access," *Shell News* 5 (1981): 18–19; Charles Frederick Lester, "The Search for Dialogue in the Administrative State: The Politics, Policy, and Law of Offshore Development" (Ph.D. dissertation, University of California, Berkeley, 1992), 91–93; Billy Flowers interview with author, June 18, 1999, Tyler, Tex.

54. Bill Broman interview with author, Dec. 15, 1999, The Woodlands, Tex.

55. J. Robinson West interview with author, Nov. 18, 2002, Washington, D.C.; Otteman interview.

56. "Interior's OCS Leasing Plan Advances," *Oil and Gas Journal*, Feb. 8, 1982, 70–71.

57. "Ocean Drilling Over a Mile Down: A Subject Deep but Not Dark," *Shell News* 2 (1983): 1–6; Carl Wickizer communication to author, June 27, 2001.

58. Carl Wickizer, "Out on the Horizon: Creating Means to Drill and Produce beyond Current Water Depth Limitations," presentation to Offshore Northern Seas Conference, Stavanger, Norway, Aug. 23, 1988.

59. "Revealing Secrets of the Deep," *Venture* 2 (1987): 1–3.

60. "Gulf of Mexico Exposure Totals $4.5 Billion," *Oil and Gas Journal*, June 6, 1983, 48.

61. "Gulf Deepwater Tracts Spark Bidding," *Oil and Gas Journal*, Apr. 30, 1984, 36.

62. "*Shell America:* A Sophisticated Ship, She Is," *Shell News* 3 (1984): 1–9.

63. Helen Thorpe, "Oil and Water," *Texas Monthly*, Feb. 1996, 143.

64. Ibid.

65. Mike Forrest, "'Bright' Investments Paid Off," *AAPG Explorer*, July 2000, www.aapg .org.

66. Ibid.

67. "Deep Water, Mobile Bay Tracts Spark Sale 98," *Oil and Gas Journal*, May 27, 1985, 46–47.

68. See, for example, Agis Salpukas, "2,860 Feet under the Sea, a Record-Breaking Well," *New York Times*, Apr. 24, 1994, 9.

69. See the critique of the reformed offshore leasing system in Juan Carlos Boué, *A Question of Rigs, of Rules, or of Rigging the Rules? Understanding the Profitability and Prospects of Upstream Oil Activities in the Gulf of Mexico* (New York: Oxford University Press, 2007).

70. For recent developments, see Jonathan L. Ramseur and Curry L. Hagerty, "*Deepwater Horizon* Oil Spill: Recent Activities and Ongoing Developments," Congressional Research Service Report 42942 (Jan. 13, 2013).

71. See, for example, "Should the U.S. Expand Offshore Drilling," *Wall Street Journal*, Apr. 12, 2013, http://online.wsj.com/article/SB10001424127887324020505045783986108510 42612.html.

Chapter 6

The U.S. Strategic Petroleum Reserve and Energy Security Lessons of the 1970s

Bruce Beaubouef

The decade of the 1970s saw a transformation in the American political economy, especially as it applied to petroleum-based energy. For much of the twentieth century, the United States had surplus oil-producing capacity. The American petroleum industry controlled the bulk of those natural reserves, and the government deferred to industry, trusting it to provide extra supplies whenever they were needed. In the two world wars, the industry delivered the petroleum products the American military needed.

By 1970, however, depletion of American petroleum reserves deprived the United States of its surplus oil-producing capacity. That development, along with rising energy demand, meant growing dependence on foreign oil. This made the United States vulnerable to blackmail via the "oil weapon." These vulnerabilities were laid bare by the Arab oil embargo of 1973–74, and subsequently by the Iranian revolution of 1979, both of which brought major oil supply disruptions.

No longer able to deliver affordable petroleum products consistently to protect American energy security fully, the industry lost some of its political power. Subsequent shortages, price increases, and environmental accidents further eroded its political clout. These developments, and an outcry from frustrated energy consumers, emboldened the federal government to take a more active role in energy policy.

In the wake of the shortages, the government stepped into the marketplace with a wide array of energy policy tools to help protect national economic security. These included the Strategic Petroleum Reserve (SPR), which provided man-made, publicly owned, surplus capacity. By the early 1980s, the U.S. government had determined that the SPR was the best, least intrusive means of protecting energy consumption. In an energy supply shortfall, the reserve could combat the two things that threatened energy consumption—shortages and price increases.

The United States had long had the world's largest oil appetite. By 1970 it represented 5 percent of the world's population but consumed about one-third of so-called free-world oil. By 1973, American per-capita energy consumption was six times the rate of the rest of the world, and oil was supplying about half of U.S. energy needs. Thus, by the early 1970s, three developments, or trends, would profoundly undermine American energy autonomy: growing demand, declining domestic production, and growing dependence upon Middle Eastern oil.

With demand chasing supply in a tight market, a huge wave of general inflation spread throughout the nation, driven by rising oil and energy prices. By 1970 the consumer price index was nearly 6 percent. The ensuing public outcry over high prices and energy shortages brought forth a level of government intervention into the economy not seen since the World War II. The vast regulatory program that resulted would be as complex and confusing as it was controversial.

The initial federal response came in August 1971, when President Richard M. Nixon announced an unprecedented across-the-board series of wage and price controls. Importantly, they included controls on crude oil and natural gas prices. Although most controls were lifted by 1974, continuing public disenchantment with energy shortages and price increases meant that crude oil price controls would be continued, in various forms, until 1981. However, the controls had ironic and perverse effects. By holding down domestic prices, they reduced incentives for domestic production. At the same time, they encouraged consumption, furthering American demand for and dependence on imported oil. Notably, these price controls may have been the first example of the government's attempt to protect energy consumers on a large scale.

As demand continued to rise, other regulatory restrictions on supply—instituted in an age of abundance—were lifted. The American oil-producing states ended prorationing in 1971 and 1972, and the federal government abandoned import restrictions in 1973. After this, net imports quickly ramped up to 6.2 million barrels per day (bpd), a 180 percent increase from the 2.2 million bpd of 1967. As a share of total oil consumption over that time, imported oil increased from 19 to 36 percent. By 1977 imported oil would account for nearly half (48 percent) of total U.S. petroleum consumption.

The First Oil Shock

American dependency upon foreign oil, combined with the loss of spare production capacity, made the United States more vulnerable in 1973 than it had been during previous Middle Eastern crises. Although the continental United States still provided nearly 75 percent of American crude oil demand in 1973, 25 percent came from foreign sources. Of that 25 percent, more than 8 percent came from the Middle East.[1]

Responding to U.S. aid to Israel, on October 16, 1973, ministers from OPEC's "Gulf Six"—Saudi Arabia, Kuwait, Iran, Iraq, Qatar, and the United Arab Emirates—decided to raise prices by 70 percent, from $2.90 per barrel to $5.11 per barrel. OPEC had already become a trendsetter in world oil prices; by 1973, OPEC producers accounted for nearly 54 percent of the world's crude oil production. The Yom Kippur War, and American aid to Israel, merely presented the opportunity for the Arab member nations to strengthen their positions vis-à-vis the oil-consuming nations of the West. But unlike the world market of 1967, when the oil weapon was last used, now U.S. and Western dependency on Middle Eastern oil gave OPEC crucial leverage. In December the Gulf Six raised their crude oil prices again, this time to $11.65 per barrel. With the increase, OPEC had quadrupled its prices from its pre-October 1973 levels.[2]

However, members of OAPEC, the Organization of Arab Petroleum Exporting Countries, felt that the OPEC price increases represented a fair and necessary readjustment of their economic position, with regard to their own oil reserves. Something else was needed to send another, even stronger political statement. On October 17, 1973, the day after OPEC had instituted its price hikes, the OAPEC members met and decided to wield the oil weapon once again. They would embargo oil from Israel's allies, primarily the United States, the Netherlands, Portugal, Rhodesia, and South Africa. Unlike the 1967 attempt, in 1973 the Arab oil-producing states cut production in addition to halting exports. The OAPEC production rate declined from 20.8 million bpd in September 1973, just before the embargo, to 15.8 million bpd in November 1973, nearly a 25 percent decrease. In the following embargo months OAPEC production did increase, but only slightly. From November 1973 to February 1974, the OAPEC production rate was still 19 percent lower than it had been in September 1973. On March 17, 1974, the OAPEC ministers resumed production and export operations. But they had established new and higher prices and greater control over their own hydrocarbon resources. For the entire five-month embargo period, the average disruption had been 2.6 million bpd.[3]

Although Western oil companies tried to offset the loss by increasing their efforts in other areas, world oil production fell by 5 percent. U.S. oil supply, already tight, became even tighter. From September 1973 to January 1974, petroleum imports into the United States fell by 2.7 million bpd. With prices high and supply tight, American petroleum consumption was dealt a serious blow. On a daily annual average, consumption fell from 17.31 million bpd in 1973 to 16.7 million bpd in 1974, a 3.5 percent decline. The effect of that decrease, however, was much larger than that seemingly small number implies.[4]

With the OAPEC embargo and OPEC price hikes, the United States entered what could arguably be described as its first significant national energy crisis. The domestic oil market had already been tight, and panic purchases sent prices further upward, from the new OPEC price of $5.11 per barrel to more than $16 per barrel in some markets. In 1973 the nation's power utilities depended upon petroleum for 22 percent of their boiler fuel needs. Many of them continued to operate at full capacity, with limited generating resources. Americans formed long lines to purchase gasoline at retail stations across the country. Tempers often flared as Americans struggled against each other for a decreasing slice of the nation's energy pie.[5]

Project Independence

In a nationwide radio and television address on November 7, 1973, President Nixon announced that the U.S. economy would fall 10 percent short of its petroleum needs in the upcoming winter. To meet the immediate crisis, Nixon called for authority to allocate crude oil and petroleum products; increased production from the Elk Hills, California, and Alaska National Petroleum Reserve sites; reestablished daylight savings time; and announced several mandatory and voluntary conservation measures, including lower thermostat settings for businesses and individuals and restrictions on gasoline purchases.

To address the long-term crisis, Nixon announced Project Independence, a series of proposals whose lofty goal was energy self-sufficiency by 1980. These proposals included deregulation of natural gas prices; lower air-quality standards for factories and the automobile industry; acceleration of the building and licensing of nuclear power plants; suspension of electric utilities' efforts to switch from coal to oil for power generation; incentives to increase coal and lignite production; and increased offshore drilling. The president also asked Congress to pass enabling legislation mandating that no further environmental hindrances be placed upon the Alaskan pipeline project. Other proposals that became law included lowering the highway speed limit to 55 miles per hour and tax breaks for home and office insulation. There was, however, no proposal for a strategic petroleum reserve.[6]

In preparing its *Project Independence Report*, the Federal Energy Administration (FEA) had to study the feasibility of the Project Independence proposals. The need for top-level energy planning came in the very midst of the Watergate crisis, the presidential transition from Nixon to Gerald R. Ford, and the first oil shock. In this tumultuous context, the FEA, using state-of-the-art computer analysis, went about the business of brainstorming for ways to reduce American dependence on foreign oil.

The two-part *Project Independence Report* was published in November 1974. In it, the FEA concluded that reducing American vulnerability to oil supply disruptions was a more realistic (and hence more desirable) national goal than achieving energy self-sufficiency. The United States could dramatically cut back its oil imports, but only at a severe socioeconomic cost. It would be much better, the report argued, to reduce the impact of a future disruption. The FEA abstained from making any direct recommendations (and thus political controversy), choosing instead to explore the feasibility and effectiveness of various energy policy alternatives. The agency did conclude, however, that options to reduce vulnerability fell into three categories: accelerating domestic production, energy conservation and demand management, and strategic oil stockpiling.[7]

Though the FEA ostensibly eschewed recommendations, the report strongly supported oil stockpiling. "Emergency storage," the FEA concluded, "is cost-effective in reducing the impact of an embargo." The report found that strategic stockpiles, though not inexpensive, were cheaper than most other energy security options. It considered a scenario in which the United States faced a year-long disruption of 1 million bpd. Under this scenario, the FEA concluded, a 500 million barrel reserve should be the target goal, with the possibility of going to 1 billion barrels. The agency estimated that the cost of developing a 500 million barrel reserve would be $6.3 billion—far less than the estimated $30 to $40 billion loss in GNP that would result from the disruption scenario considered in the report. The FEA also noted that oil purchases might have some impact on international spot oil prices but concluded that the effect would be largely contained to the first year of acquisition. Gradually, the market would get accustomed to U.S. government's presence and role as a large-scale oil purchaser.[8]

In many ways, the FEA's *Project Independence Report* echoed the National Petroleum Council's (NPC) final *Emergency Preparedness* report, which had been published two months earlier—perhaps because it borrowed heavily from it. As had the NPC, the FEA chose underground salt caverns over in situ reserves and steel tanks as the best means of storing oil. Government-owned salt dome storage, the FEA concluded, would be more secure and better protected than would be the case with private, in situ reserves. Salt dome storage also appeared to be the least expensive alternative and would be significantly cheaper than steel tank storage: $1.50 per barrel in salt domes versus $12 per barrel in steel tanks. Being 2,000 feet underground, it would also be more secure against natural disasters and sabotage.

Salt dome storage would also be cheaper and more secure than other underground storage alternatives, such as mined caverns, salt beds, abandoned mines, and depleted reservoirs. The rock salt that characterized the domes was generally impervious to liquid petroleum and gas. Studies showed that it moved like plastic to seal fractures, even earthquake damage (which was almost unheard of in the Gulf Coast region), and had the compressive strength of concrete. Since oil and salt (like water) did not mix, the oil in the caverns would tend to improve in quality over time as impurities settled out. Storing oil underground also avoided the need for aboveground tank farms, considered an eyesore by some.

As for the location, the FEA found the Texas-Louisiana Gulf Coast to be the most desirable region, for many reasons. The area had numerous salt dome formations capable of accommodating several hundred million barrels, at depths of 2,000 feet or more. Even a relatively small number of salt domes could hold considerable amounts of petroleum. Moreover, with large-scale storage possible, one could achieve high delivery rates, up to several million barrels per day.

Storing crude oil, the FEA concluded, offered the most flexibility. It could be refined into whatever mix of petroleum products were needed at the time. In contrast, stored refined petroleum products could degrade or change quality over time. Any such problems with crude could be remedied through refining. Finally, crude oil and residual oil (the other fuel the FEA recommended stockpiling) were the least covered by private inventories, and they were also the least costly to store.

The best refining and transportation facilities were also in the area. The Texas-Louisiana Gulf Coast region was home to what was arguably the world's foremost petroleum refining and transportation complex, as it had been for most of the twentieth century. The FEA also noted that by 1978 there would be expanded deepwater terminals and seaports along the Gulf Coast, which would enhance shipping access to East and West Coast markets; and additional pipelines, which would expand transportation facilities to Rocky Mountain and Midwest refineries. From the Gulf Coast, crude oil and refined products could be sent anywhere in the country.

Yet even with considerable existing storage capacity along the Gulf Coast, several hundred million barrels of salt dome storage would need to be created through leaching. During this process, oil could be pumped into the cavity, displacing some brine and floating on the remainder. It was thus an additional benefit to using salt dome storage that oil could be stored at the same time as storage space was being created. In the event of a drawdown, brine would be pumped into the salt dome cavity, forcing up the oil. This leaching process would take about six years, with completion in 1979—about the same time that the extra Gulf Coast port terminals and pipeline facilities would be completed. It was thus feasible, the FEA concluded, to have 500 million barrels of crude oil stored in Gulf Coast salt domes by the early 1980s.[9]

The New Politics of Oil

In the wake of the first oil shock, the theory that the major oil companies had contrived the crisis by holding oil supplies off the market spread like wildfire through the news media and the public consciousness. Nor did public cynicism limit itself to the industry. Both the federal government and the Arab oil-exporting nations were also blamed for the energy crisis. Reports in *Time, Newsweek, New Republic, Washington Post,* and other journals and newspapers throughout the nation suggested that the majors had urged Saudi Arabia to raise prices, had held production in check, had oil tankers sitting off the East Coast until prices increased, or had otherwise manufactured the crisis in some fashion. In the same muckraking, investigative vein as Ida Tarbell's *History of Standard Oil,* books such as Anthony Sampson's *The Seven Sisters* and Robert Engler's *Brotherhood of Oil* began to appear, claiming that the energy shortages had been the result of collusive machinations by the international majors, the OPEC nations, and the U.S. government.[10]

This perception was abetted by the fact that the majors reported vastly increased profits during the embargo. In early 1974 some of the largest oil companies released their 1973 earnings. They were up dramatically. The leading thirty oil companies averaged a 71 percent increase in 1973 over 1972. Exxon had an 80 percent increase; Gulf Oil enjoyed a 90 percent increase. Even worse, from a public image perspective, was that in 1974 many of the top industry executives received advances in their salaries. All of this tended to heighten public suspicion of the industry.

While public suspicion of the industry remained, so would stringent government regulation. It was in this context that the federal government took on a much larger role in the energy arena—not just price and supply controls but also new programs designed to secure supplies of traditional energy and support development of alternative energy. As for oil stockpiling, public distrust of the industry would preclude its involvement in a program whose very purpose was to protect energy consumers from the vagaries of the oil market. For its part, the industry would be only too glad not to be burdened with the mandate of building and funding the reserve.[11]

The political economy of oil changed dramatically after the first oil shock. Until the 1970s, oil companies, especially producers, had held the upper hand. As long as they produced the oil that a modern society needed, at a reasonable price, government had largely acceded to their wishes. This could be seen with the oil import quotas for the independents; the foreign tax credit (the "golden gimmick") for the international majors; and the depletion allowance and prorationing for domestic producers.

By the early 1970s, however, Lower 48 producers had lost their surge production capacity. This eroded the industry's ability to deliver petroleum products to the American public consistently and affordably. Once the industry could no longer step up production in an emergency, the decades-old deference melted away. Political power then shifted to American energy consumers and purchasers, including individual end users, manufacturing businesses, utility companies, and certain downstream segments of the oil industry such as independent refiners, marketers, and transporters. After the first oil shock, energy consumers gained power—and for a time, ascendancy—in the American political economy of oil. The disruption underscored the important role that energy consumption played in the economy. Public officials, recognizing this, sought to protect it through legislative and executive action. Later, consumer organizations and groups would develop and actively defend their interests. But in the early 1970s, government officials went out of their way to protect energy consumption patterns. Much of the 1973 Emergency Petroleum Allocation Act (EPAA) had been written to protect energy consumers, and the congressional protections for consumers and end users continued throughout the 1970s. Even in the wake of the embargo, only with reluctance did policymakers even think about demand restraint. "It's not our job to force the country to change consumption patterns," said Jared G. Carter, deputy undersecretary of the interior, in October 1974. Nor was that sentiment limited to the Nixon-Ford administration. After the first oil shock, as an article in the journal *Environment* put it, the consumer was now "in charge."[12]

The Ford Energy Independence Proposal
and Congressional Response

When Gerald R. Ford assumed the presidency in August 1974, he also inherited the energy crisis. In contrast to his predecessor, Ford desired a more pro-market response to the crisis, based largely upon price decontrol as a means of spurring domestic production. Many of the senior officials in the Ford administration felt that the *Project Independence Report* had called for an excessively large government role in the marketplace. Besides lifting price controls, strategic oil stockpiling programs were also part of the Ford "Energy Independence" legislative package.

In early 1975, Robert L. Davies, a systems analyst for the Department of Defense (DOD), led a group of FEA officials in developing the parts of the bill pertaining to a civilian strategic oil reserve. Davies came to the job with plenty of relevant experience. Under the DOD he had advocated oil stockpiling as an alternative to naval requests for more ships to protect the growing number of supertanker fleets of the world. During the Arab oil embargo, he joined the Federal Energy Office, predecessor to the FEA, and participated in the writing of the *Project Independence Report*. Under Davies, the FEA group undertook several studies to consider the costs and benefits of salt dome storage, man-made cavern storage, aboveground steel tank storage, and their environmental impact. It is hard to imagine that they did not also examine the NPC studies, which had been the authoritative source on these issues for years. For the Energy Independence bill, the FEA recommended a 500 million barrel crude oil reserve for a three-month (90-day) disruption scenario, and a 1 billion barrel reserve for a six-month disruption scenario.[13]

Ford's Energy Independence Act was proposed to Congress in January 1975. In the bill, the president proposed the creation of a domestic civilian oil reserve of up to 1 billion barrels and a military oil reserve of up to 300 million barrels. He also called for increased development of the naval reserves. Much of the rest of Ford's energy program was dedicated to removing the government from the marketplace. He proposed an import fee of up to $3 per barrel (a tariff being less bureaucratic than a quota); decontrol of domestic crude oil and natural gas prices; increased coal production and increased use of coal for power generation; and relaxation of environmental laws. The bill also proposed a synthetic fuels program. In accord with the

Project Independence Report, Ford called for a windfall profits tax, to be tied to price decontrol.

Yet the president's desire for a more market-oriented approach came at a politically inopportune time. A significant portion of the American public had become suspicious of the U.S. oil industry. In this political climate, it would be hard for Congress to pass anything that might be interpreted as a "giveaway" to the industry, and a publicly owned strategic oil reserve seemed like a useful insurance policy.[14]

The congressional response to the president's proposal was the Energy Policy and Conservation Act (EPCA) [PL 94–163] of 1975. It bore little resemblance to the market-oriented nature of the Ford legislative package. In the midst of rising oil prices, inflation, and recession, the political pressure for continued price controls and supply allocation was overwhelming. Deregulation would be a bitter tonic and would be especially harsh on consumers and end users. A Congress controlled by Democrats, therefore, was less than receptive to the pro-market aspects of the Ford proposals. When Congress received the Ford omnibus bill in January 1975, it began gutting the entire package.

Ultimately, the House-Senate conference bill on the energy package was reported out on December 9, approved that day, and sent to President Ford on December 17. Though a few of the Ford proposals survived, the ensuing legislation bore little resemblance to his proposed Energy Independence Act. The EPCA protected consumers and end users and had a decidedly stronger government approach than the Ford administration desired. It continued the controversial oil and gas price controls, and the allocation authorities were extended as well. The FEA was once again given the authority to implement and enforce the new regulations, and as with their efforts under the EPAA the attempt to control prices and allocate oil and gas proved to be just as complex, controversial, and difficult.[15]

The EPCA also mandated higher fuel efficiency standards for a host of products, including automobiles (which later became the Corporate Automobile Fuel Efficiency, or CAFE, standards) and electrical appliances. It required manufacturers of electrical appliances to label their products with information on their energy efficiency. The EPCA also ratified U.S. participation in the International Energy Program, whereby Great Britain, Japan, West Germany, and the United States would all develop emergency petroleum reserve systems that they could coordinate and share in the event of another supply disruption.

The EPCA did give the president some of the increased powers he had asked for: the president could require power plants to use coal rather than oil, order the development of new coal mines, and allocate and appropriate domestic oil and gas reserves. The president could also order mandatory conservation measures and oil and natural gas rationing. Ford's proposal for increased authority to impose tariffs, however, was rejected, as were his proposals for stepping up production on the naval reserves and for the creation of a new military reserve. The only success Ford achieved on the energy reserves issue was the establishment of the SPR. Unlike most of the other measures, the oil stockpiling concept enjoyed support from Congress and the White House.[16]

The SPR Established

The SPR was created under Section 151(a) of the EPCA, which declared that "the storage of substantial quantities of petroleum products will diminish the vulnerability of the United States to the effects of a severe energy supply disruption." With the EPCA, Congress declared it to be the policy of the United States to provide for the creation of a strategic petroleum reserve for the storage of up to 1 billion barrels, and not less than 150 million barrels of "petroleum products" (defined widely), to be in place three years from the date of passage of the EPCA.[17]

The legislation in fact contained several important storage targets for the SPR, to be reached in a sequential and chronological manner, to the "maximum extent practicable and except to the extent that any change in the storage schedule is justified." Of these, perhaps most important were three: the 150 million barrel figure for the Early Storage Reserve (ESR) program, which represented the FEA's estimate of existing storage capacity that could be secured fairly soon; the 500 million barrel figure, representing both NPC and FEA estimates of the amount of oil needed to withstand a 90-day (three-month) disruption in the supply of foreign oil; and the 1 billion barrel total figure, the amount the FEA estimated would be needed to withstand a six-month disruption. In the EPCA, Congress accepted the 500 million barrel target as the primary, long-term stockpiling goal, a position supported in the NPC's *Emergency Preparedness* and *Petroleum Storage for National Security* reports and the FEA's own *Project Independence Report*. It did so in Section 154(c)(1)(A) and (2), with a convoluted formula that involved taking the highest average monthly import level in a three-month period, within the 24-month period preceding enactment of the legislation. That amount, calculated to be 495 million barrels, but usually increased to 500 million barrels for convenience, was to be in place by the end of 1982. The EPCA also established the Strategic Petroleum Reserve Office within the FEA to oversee the establishment, management, and maintenance of the reserve. Ultimately, six storage sites were selected in Louisiana and Texas, a marine terminal was built on the Mississippi River, and a field office established in New Orleans.

The very limited role for private industry in the SPR program made the American stockpiling strategy unique. Among reserves in the International Energy Agency (IEA) nations, the U.S. strategic reserve would be the only such program that was wholly government owned and operated, and without obligation on the domestic oil industry. In the United States this was the natural result of strong public distrust of the industry and overwhelming industry opposition to inventory requirements. It was perhaps ironic that, for different reasons, conservatives and liberals both came to agree that the SPR should be totally owned, operated, and funded by the federal government.

The EPCA authorized $1.1 billion for the development and implementation of the SPR, including planning, administration, and acquisition of storage and related facilities. The legislation pointedly excluded oil acquisition from the authorization, most likely because of recent price instability. The bill did provide that funds for oil acquisition could come from future appropriations. Though President Ford did not like much of the EPCA—notably the continuance of price controls and supply allocation and the defeat of his oil import tariff—he knew that it was likely to be the only major energy bill presented for his signature. Anxious to show that he was active on the energy issue, the president signed the EPCA into law on December 22, 1975.[18]

The legislative establishment of the SPR represented the first time in American history that an emergency petroleum reserve system had been created purely for the civilian economy. The Naval Petroleum Reserve had been created for the exclusive use of the navy. Ickes's Petroleum Reserves

Corporation had been alternatively designed for military and civilian purposes but had been short-lived. By the 1970s, the political context had changed. Protection of energy consumers was the new regulatory imperative.

Among all the various proposals and programs for enhancing American energy security, the SPR enjoyed the most support because it was the least contentious. Unlike price controls, supply reallocation, rationing, or relaxing environmental standards, the SPR ran afoul of no one's vested interest—on paper, at least. Moreover, the SPR concept—sans private inventory requirements—had been actively promoted by the NPC and supported by industry in general. But having created the SPR in law, the challenge of actually building it lay ahead. There would be no shortage of delays, challenges, and obstacles. These were physical, technological, political, and bureaucratic in nature and revealed the inherent contradictions and messiness of American democracy. Even with the imperative of national security, it was not until the mid-1980s that the reserve came to be a credible energy policy tool and operationally ready to help enhance American energy security.

The Lessons of the 1970s

The history of the SPR program points to the virtue of passive, supply-side energy policy tools. As the U.S. and IEA response to energy emergencies matured and evolved after the first oil shock, both government and industry officials concluded that passive, supply-side measures were preferable to the intrusive, micro-managing regulatory approach taken in the 1970s. In the energy arena, strategic oil reserves were the best examples of such tools; they dictated neither price nor distribution. Policymakers determined that in a supply crisis it was better to make extra oil available and let the market determine the distribution, allocation, and to a large extent price of oil as well.

Coercive measures, such as mandatory price and allocation controls, demand restraint, and rationing more often than not brought perverse and undesirable results. Price controls kept a lid on domestic prices but also reduced incentives for domestic production and stoked demand for imported oil, thereby placing upward pressure on world oil prices. Nor were they politically popular. Demand restraint proved infeasible to implement. In a crisis, especially, demand restraint runs directly counter to the consumer's primal urge for energy security.

Not surprisingly, IEA governments have abandoned a major reliance on restraint as a tool of emergency energy policy. In the wake of the first oil shock, some European governments had contended that they would respond to a future crisis with a heavy dose of demand restraint. The disruption of 1979 undermined those claims. As they did in the United States, European consumers hoarded in response to the crisis. The result was predictable: oil and gasoline tanks were filled, worsening the panic. After the second oil shock, IEA officials accepted that an agency cannot force a country to exercise demand restraint. They recognized that, in a supply crisis, consumer desire for energy security would render calls for demand restraint ineffectual.

Perhaps the most damaging argument against a strong reliance on demand restraint—even more than the extreme difficulty in effecting it— is that such measures do not alleviate or reduce the economic damage of a disruption. After 1979, U.S. and IEA officials recognized that supply-side measures, in the form of stock drawdowns, should form the bulwark of any collective disruption response. They recognized that in disruption scenarios supply-side measures will be more effective than demand-side measures in mitigating shortages and price increases. Unlike policies that sought to manage shortage through coercive measures, oil supplied from strategic reserves would actually alleviate shortage, lessening pressure on prices while allowing private market mechanisms to work and keeping government intervention to a minimum. In short, unlike the other energy emergency measures, strategic stockpile drawdowns would help reduce prices and calm markets without the unpleasant side effects.

In the wake of the second oil shock, the United States abandoned reliance on coercive, micro-managing tools and embraced the SPR as its emergency energy policy tool of choice. From the regulatory debacle of the 1970s, the government had learned not only the futility but also the danger of trying to micro-manage the world's largest energy economy. Finding a market-oriented strategy became the sine qua non of emergency energy policy, arguably of energy policy in general. Reagan administration officials, in particular, found that the SPR fit that bill. In a supply crisis, the reserve would help alleviate shortages, thereby reducing panic and related political pressures for more controlling regulatory measures. Thus, the government's adoption of the SPR as its primary energy emergency response was part of a larger move toward a market-oriented regulatory approach.

By 1990, U.S. and IEA officials recognized the necessity of accommo-dating the first panic surge of demand in a crisis with extra supply. When energy consumers are nervous about supply, a heavy reliance on demand restraint is not only futile but also counterproductive. Indeed, when one looks at the makeup of the IEA's emergency response program adopted on January 11, 1991, one sees that it relied predominantly on stock draw (79.6 percent) and had a much less reliance on demand restraint (17.2 percent). And apparently it will continue to be U.S. policy to use SPR drawdowns as a substitute for demand restraint measures in any future IEA emergency response program.[19]

Successive U.S. presidential administrations have supported the SPR because it allows the market to work. The January 1991 drawdown is a good example of this passive, supply-side approach: it was a move to make extra oil available to the market, to mitigate potential shortfalls and avert panic, but not control the price or flow of oil. The government placed a significant amount of oil up for sale and then let the market decide how much it wanted. With the program's price indexing system, the market affected the price of that oil as well. SPR drawdowns thus blunted panic-related oil price spikes, without micro-management of the price or distribution of oil. SPR drawdowns, if sold at competitive auc-tion, conformed with the market. To be sure, the government still retains no small amount of power: in its decision on whether to drawdown the reserve; in its decision of how much oil to offer; and in its right to refuse to sell SPR oil for any offer deemed too low. And even when not being drawn down, the SPR played even more of a passive, but still valuable, role. The potential for a drawdown presents an ever-present possibility that extra oil might be placed on the market.

A policy commitment to a pro–energy consumer, petroleum-based economy ruled out many of the most powerful conservation approaches, especially that of demand restraint. Efforts by the Nixon, Ford, and Carter administrations to tax gas guzzler automobiles, or to implement an energy consumption tax (usually in the form of a gasoline tax, or a tariff excise tax) all ran afoul of consumers. These tools restrained energy demand by driving up energy costs, by either reducing supply or taxing consumption. They therefore had little White House support, or met stiff resistance in Congress, and were either defeated outright or rendered innocuous in law. Oil import tariffs, in particular, were opposed by consumers, refiners, importers, and much of the rest of the petroleum industry. Consumers opposed tariffs and fees because they restricted the availability and affordability of petroleum products. Industry segments that opposed them took up the defense of the consumer. For these reasons, oil import tariffs and fees became politically untenable. They were opposed, seemingly, by all except domestic producers, who were in the 1990s drowning in a flood of cheap oil, as they had been in the 1950s and 1970s. No longer able to wholly satisfy the ravenous appetite of the American energy consumer, they had fallen considerably in the pecking order of the American political economy. After 1973 their cries for protection went unheeded.

Decontrol of oil prices had only a brief demand restraint effect, since world oil prices began a prolonged decline starting in 1982. Indeed, the generally low-to-moderate oil prices of the 1980s and 1990s heightened and stoked demand. Even the restraining effects of the CAFE standards waned, as Americans increasingly opted to drive less fuel-efficient sport utility vehicles. And for markets dependent on petroleum—transportation, shipping, manufacturing, and heating in the Northeast—renewable and synthetic energies made at best marginal gains. Other measures to reduce demand for petroleum-based energy have probably realized their maximum effectiveness. Given the parameters of the American political economy, there was, and is, little that can stand in the way of rising consumer demand. Fuel-efficient technologies can slow demand but have to be attractive to the end user market. They have not always been. As a result, American energy demand continues to reach new heights. If the government's answers to the energy crisis were to endure and be effective, they had to conform to the petroleum-based energy economy and not attempt to curtail demand or shift or alter existing energy consumption patterns.

Conclusion

The SPR has become the most functionable energy security policy tool that the federal government has. Whatever energy policy tool the government hoped to wield over the long run had to please, or at least not anger, energy consumers, in the first place, and the petroleum industry, in the second place. To endure, emergency energy policy tools had to conform with the structure of a petroleum-based energy economy. With the internal combustion engine the primary example, the economic, physical, and transportation infrastructure of American society has been committed to a petroleum-based economy for nearly a century. In fact, the SPR, by affirming the petroleum-based economy, helps to define "energy crisis" as a potentially recurring but ultimately short-term phenomenon in which the status quo will soon prevail once more. Such a definition is a blow to those who see in an energy crisis a rationale for the development of alternative energies.

Notes

This essay borrows from my book *The Strategic Petroleum Reserve: U.S. Energy Security and Oil Politics, 1975–2005* (College Station: Texas A&M University Press, 2007).

1. U.S. Energy Information Administration, Department of Energy, *Annual Energy Review: 1994* (Washington, D.C.: Government Printing Office, July 1995), 139, Table 5.1; 293, Table 11.5; 303, Table 11.10.

2. A detailed yet concise review of the Arab oil embargo and OPEC price increases can be found in "A Review of the Energy Situation during the Arab Oil Embargo," Subject File Strategic Petroleum Reserve [3] [OA 7239], Bush Presidential Records, Judy Smith Files, Press Office/White House Collection, Box 18, George Bush Presidential Library, College Station, Tex. In the secondary literature, see Daniel Yergin, *The Prize: The Epic Quest for Oil, Money, and Power* (New York: Simon and Schuster, 1992), 614; Martin V. Melosi, *Coping with Abundance: Energy and Environment in Industrial America* (New York: Knopf, 1985), 280; DeGoyler and MacNaughton, *Twentieth Century Petroleum Statistics* (Dallas: DeGolyer and MacNaughton, 2005), 3, 13; *Petroleum Intelligence Weekly*, Oct. 20, 1980, 11–12.

3. Memo, "Office of Energy and Strategic Resource Policy Comments on Strategic Petroleum Reserve Papers Forwarded under Cover of General Scowcroft's Memorandum of October 22, 1976," Nov. 12, 1976, folder Convenience F6-Economic-Energy-Institutional (4), box A5, Henry A. Kissinger and Brent Scowcroft: Temporary Parallel File, 4, Gerald R. Ford Presidential Library and Museum, Ann Arbor, Mich.; "A Review of the Energy Situation during the Arab Oil Embargo," Subject File Strategic Petroleum Reserve [3] [OA 7239], Bush Presidential Records, Judy Smith Files, Press Office/White House Collection, Box 18, Bush Library.

4. Office of International Energy Affairs, Federal Energy Administration, *U.S. Oil Companies and the Arab Oil Embargo: The International Allocation of Constricted Supplies*, prepared for the Subcommittee on Multinational Corporations of the Committee on Foreign Relations, U.S. Senate, Jan. 27, 1975 (Washington, D.C.: Government Printing Office, 1975), 7–8, 13–18.

5. In 1973, American power utilities depended upon petroleum products—residual fuel oil, distillate, fuel oil, jet fuel, and petroleum coke—for 22 percent of their boiler fuel needs. By 1997 that figure had fallen to 3.8 percent; U.S Energy Information Administration (EIA), Department of Energy, *Annual Energy Review, 1997* (Washington, D.C: EIA/DOE, 1998), 223.

6. "Address to the Nation about Policies to Deal with Energy Shortages, November 7, 1973," in *Public Papers of the Presidents of the United States: Richard Nixon, 1973* (Washington, D.C.: Government Printing Office, 1975), 916–22.

7. Federal Energy Administration (FEA), *Project Independence Report* (Washington, D.C.: Government Printing Office, Nov. 1974), Part 18–11, 19; Part 2, 377–91. For an abridged version, see FEA, *Project Independence: A Summary* (Washington, D.C.: Government Printing Office, Nov. 1974); Joel Havemann and James G. Phillips, "Energy Report/Independence Blueprint Weighs Various Options," *National Journal Reports* 6, no. 44 (1974): 1636.

8. FEA, *Project Independence Report*, Part 1, 10.

9. On the experience with petroleum storage in salt domes, Bill Spanke, in "Salting Away Oil," *Popular Science* 212 (Jan. 1978), 77, stated that "natural gas and petroleum products have been stored safely in salt caverns in several other countries for 20 years." The FEA's SPR plan of 1976 stated that "the feasibility of using leached caverns in salt for storage has been demonstrated in France and Germany, where significant amounts of hydrocarbons are currently being stored." "In the United States," the report went on, "natural gas and petroleum-related products have been stored in salt caverns for 20 years. In North America, over 900 solution-mined caverns, with a total capacity exceeding 300 million barrels, have been developed in salt deposits and converted for storage"; SPR Office/FEA, *Strategic Petroleum Reserve Plan*, Dec. 15, 1976 (Washington, D.C.: Government Printing Office, 1977), 72, 92–93; see also "Energy: Salt Domes," *Newsweek*, Aug. 22, 1977, 55.

10. Anthony Sampson, *The Seven Sisters: The Great Oil Companies and the World They Made* (New York: Viking Press, 1975); Robert Engler, *The Brotherhood of Oil: Energy Policy and the Public Interest* (Chicago: University of Chicago Press, 1977); also in this vein, see Robert Sherrill, *The Oil Follies of 1970–1980: How the Petroleum Industry Stole the Show (and Much More Besides)* (New York: Anchor Press, 1983).

11. U.S. Congress, Senate, Committee on Insular and Interior Affairs, Special Subcommittee on Integrated Oil Operations, *Market Performance and Competition in the Petroleum Industry, Part 3*, 93d Cong., 1st Sess., Dec. 12 and 13, 1973 (Washington, D.C.: Government Printing Office, 1973), 829–69, 1032–53, 1873–76; Louis M. Kohlmeier, "Antitrust Legislation Becomes Fashionable," *National Journal Reports*, Nov. 2, 1974, 1659; U.S. Federal Trade Commission, Staff Report, *Concentration Levels and Trends in the Energy Sector of the U.S. Economy* (Washington, D.C.: Government Printing Office, 1974), 67–95; U.S. Senate, Committee on the Judiciary, Subcommittee on Antitrust and Monopoly, *Interfuel Competition*, 94th Cong., 1st Sess., June 17, 18, 19, July 14, Oct. 21, and 22,

1975 (Washington, D.C.: Government Printing Office, 1975); Senate, Committee on the Judiciary, Subcommittee on Antitrust and Monopoly, *The Petroleum Industry, Part 3*, 94th Cong., 1st Sess., Jan. 21–Feb. 18, 1976 (Washington, D.C.: Government Printing Office, 1976), 1852–80; Senate, Committee on Insular and Interior Affairs, Special Subcommittee on Integrated Oil Operations, *The Structure of the U.S. Petroleum Industry: A Summary of Survey Data*, 94th Cong., 2nd Sess. (Washington, D.C.: Government Printing Office, 1976); Edward M. Kennedy, "Big Oil's Ominous Energy Monopoly," *Business and Society Review*, Summer 1978, 16–20; Chase Manhattan Bank, *Annual Financial Analysis of a Group of Petroleum Companies, 1975* (New York: Chase Manhattan Bank, Sept. 1976), and *Annual Financial Analysis of a Group of Petroleum Companies, 1978* (New York: Chase Manhattan Bank, Dec. 1979); American Petroleum Institute, *Witnesses for Oil: The Case against Dismemberment* (Washington, D.C.: API, 1976), 35–190.

12. "OCS Oil: Mammoth Lease Plan Encounters Heavy Opposition," *Science*, Nov. 15, 1974, 615 ("not our job"); Michael J. P. Boland, "The Consumer Is in Charge," *Environment* 25, no. 2 (1983): 10–15, 35–37.

13. FEA, *Project Independence Report*, 9; Havemann and Phillips, "Energy Report/Independence Blueprint," 1653; Thomas H. Tietenberg, *Energy Planning and Policy: The Political Economy of Project Independence* (Lexington, Mass.: Lexington Books, 1976), 88–92.

14. The FEA's *Project Independence Report* had explored the possibilities of thermal standards for homes and offices and efficiency standards for automobiles and electrical products; Treasury Department officials balked at being "in the business of deciding for John Q. Public how warm his bedroom should be and how bright his dining room should be." Other executive branch departments and agencies had other criticisms. See FEA, *Project Independence Report*, 9; Havemann and Phillips, "Energy Report/Independence Blueprint," 1653; de Marchi, "Ford Administration," 482–87; Tietenberg, *Energy Planning and Policy*, 88–92; author's interview with Frank Zarb, (former) FEA administrator, Jan. 28, 1997. Ford's proposed Energy Independence Act of 1975 was introduced in the House as H.R. 2633 and 2650, in the Senate as S. 594.

15. In 1976 a task force appointed by President Ford concluded that the costs of FEA's regulatory program—particularly the entitlements program—outweighed its benefits; see Paul W. MacAvoy, ed., *Federal Energy Administration Regulation: A Report of the Presidential Task Force* (Washington, D.C.: American Enterprise Institute for Public Policy Research, 1977), 139–76.

16. U.S. Congress, Energy Policy and Conservation Act of 1975 (hereafter EPCA), PL 94-163, *Statutes At Large*, Vol. 89, Dec. 22, 1975, 89 Stat. 871–969 (Washington, D.C.: Government Printing Office, 1977); "Senate Considers Strategic Energy Reserves," *Congressional Quarterly Weekly Report*, July 5, 1975, 1437–38; "Senate Action: Energy Reserves," *Congressional Quarterly Weekly Report*, July 12, 1975, 1503; "House Passes Energy Policy Bill, 255–148," *Congressional Quarterly Weekly Report*, Sept. 27, 1975, 2043; "Final Energy Bill Faces Uncertain Fate," *Congressional Quarterly Weekly Report*, Dec. 13, 1975, 2689–93; *Cong. Rec.*, 94th Cong., 1st Sess., May 7, 1975, Vol. 121, Part 11 (Washington, D.C: Government Printing Office, 1975), 13490–91; *Cong. Rec.*, 94th Cong., 1st Sess., June 20, 1975, Vol. 121, Part 15 (Washington, D.C: Government Printing Office, 1975), 20001–

20002. See also Letter, Donald H. Rumsfeld, Secretary of Defense, to Frank G. Zarb, FEA Administrator, Nov. 3, 1976, folder Convenience File—Economics—Energy—Institutional 4, box A5, Henry A. Kissinger and Brent Scowcroft: Temporary Parallel File, Ford Library.

17. The Arab embargo had involved the raw material crude oil, not refined petroleum products such as automobile gasoline and jet fuel. But the 1975 EPCA legislation (Sec. 3[3]) included crude oil under an umbrella definition of petroleum products.

18. EPCA, Sec. 166(2), 89 Stat. 890 (SPR appropriation). In a Dec. 8, 1975, letter from Jim Cannon, President Ford's domestic advisor, to FEA administrator Frank G. Zarb on the desirability of the EPCA legislation, Cannon wrote:

> On its merits, the legislation seems right on the margin of whether it is good enough to sign, or so bad it has to be vetoed. From the standpoint of the President's policy decision to reduce the Federal government, the bill is bad because it would increase Federal intervention. However, I believe there is a larger question throughout the country: "Will Washington ever get together on an energy program?" . . . Consequently, I recommend that the President sign this imperfect bill with a candid message pointing to the good and the bad in the bill, and stating that amendments will be sent to Congress to correct these faults. (Letter, Cannon to Zarb, Dec. 8, 1975, folder 11/25/75 to 12/12/75, Frank G. Zarb Papers, box 2, Ford Library)

19. Fuel switching and increased indigenous production measures played even less of a role, at 2.57 percent, and 0.64 percent, respectively. In a prepared statement to the Senate Committee on Energy and Natural Resources in May 1997, John Pierce Ferriter, IEA deputy executive director, commenting on the recent budget-related SPR drawdowns, said that "the U.S. government intends to use the SPR . . . [as] permitted by the IEP Agreement—as a substitute for full implementation of the treaty's demand restraint obligations"; U.S. Senate, *Miscellaneous Energy Policy and Conservation Act Bills,* May 13, 1997, 50.

Chapter 7

The Development and Demise of the Agrifuels Ethanol Plant, 1978–1988

A Case Study in U.S. Energy Policy

Jason P. Theriot

I n the 1970s, as the nation grappled with the economic roller coaster created by the energy crisis, it became evident to some Americans that an abundant, cheap supply of petroleum no longer existed. Jimmy Carter, the thirty-ninth president of the United States, entered the White House in 1977 with a country swirling in economic disarray. One of his solutions to America's growing energy crisis was the creation of a domestic alcohol fuels industry—a massive, federally backed crash program to produce ethanol from crops, thereby reducing the nation's dependence on foreign oil imports. The Agrifuels Refining Corporation in south Louisiana became one of several firms across the country that evolved from this era of alternative energy solutions and high oil prices. A history of the Agrifuels plant represents a microcosm of the problems policymakers and private investors faced in developing a new alcohol fuels industry in the 1970s and 1980s.

Looking back to the mid-1970s, the production of alcohol fuels from biomass was certainly technologically feasible. In fact, the commercialization of ethanol as a motor fuel used in this country dates back to before World War I.[1] After that time, however, the low price of petroleum made ethanol uncompetitive. But with the dramatic and seemingly irreversible rise in global oil prices beginning in 1973 and extending into the early 1980s, ethanol, as an alternative fuel source with a seemingly abundant supply of domestic raw materials, suddenly became economically feasible.[2] Two main factors made the ethanol industry a reality: private initiative and government support. For nearly a decade, beginning in 1978, government incentive programs enacted through legislation created a profitable business opportunity that attracted corporations, entrepreneurs, and farmers to invest in a new, unproven, yet lucrative, industry. In this essay I examine the rapid creation and subsequent collapse of the U.S. ethanol industry and evaluate the reasons behind alternative energy policy failure.

The case study of the Agrifuels Refining Corporation serves as a lens through which to view change in domestic energy policy from 1978 to 1988. It examines the inadequacies of government-backed crash programs and the apparent inability of the U.S. political system to negotiate successfully through long-term shifts and cycles in economic conditions. The success and failure of alternative energy sources are almost always contingent on oil prices and political ideology. The collision of divergent political ideals, coupled with a debilitating economic recession and the shocking collapse of oil prices in the mid-1980s, resulted in the demise of ethanol and other biomass fuels, leaving behind the rusted remains of dozens of foreclosed ethanol refineries scattered across the nation's landscape at the expense of the U.S. taxpayers and private investors.

Answers to the Energy Crisis

The energy crisis of the 1970s posed enormous challenges for political leaders in making and managing effective energy policies. For the decades after World War II, energy policies operated under an umbrella of energy surplus. However, beginning with the Arab embargo and the first oil price shock in 1973–74, and with continued high prices throughout the decade, peaking again in 1979 after the Iranian hostage crisis, American political and business leaders scrambled to readjust their strategies and policies to deal with the energy shortage. As new policies sought ways to curtail America's dependence on imported oil, to reduce real energy prices, and to lessen the burden of high prices on American consumers, consumption continued to rise.[3]

Throughout the 1970s, American policymakers were faced with an array of tough choices to solve the energy problem. Most of the policy decisions centered on price controls and the supply side of the equation, with the goal of improving domestic supplies through alternatives to imported energy, new technologies, and economies of scale. These choices also included, among other things, the expansion of nuclear power, the development of synthetic fuels, and the opening of more domestic oil and gas reserves to exploration and production. Some argued for Americans to take drastic conservation measures—"making sacrifices"—in changing their standard of living to ride out the energy crisis and to secure future energy independence.[4] Others argued for more "productive conservation" measures through more efficient use of energy in homes, businesses, and factories.[5] By the late 1970s, however, energy policymakers could not come to a general consensus on any of these strategies and instead offered a little something for everybody, including environmental groups.

President Carter responded to the nation's energy crisis by implementing emergency conservation measures, beginning with natural gas deregulation, and enacting legislation to subsidize the multibillion dollar development of alternative energy sources.[6] In 1977 his administration supported the creation of the Department of Energy (DOE) to administer a series of new policy changes and programs. The Energy Tax Act of 1978, an alternative fuel component to Carter's national energy package, called for federal subsidies in the form of tax incentives for the development and use of biomass-based fuels such as ethanol. The act made alcohol fuel, known then as "gasohol," exempt from the 4 cent federal excise tax on gasoline up to 1984.[7] Two years later the Biomass Energy and Alcohol Fuels Act under the Energy Security Act of 1980 established the Federal Gasohol Plan, a joint effort between the DOE and the Department of Agriculture (USDA) to meet the goals set forth by the Carter administration. The centerpiece of the act included $1.2 billion for loans and loan guarantees to finance the construction of ethanol plants.[8] In addition, the controversial Crude Oil Windfall Profits Tax, enacted the same year, extended the federal excise tax exemption for gasohol from 1984 to 1992. It also established an additional incentive, the 10 percent Energy Investment Tax Credit, for plants and refineries that used fuel sources other than petroleum to make biofuels.[9]

Federal government action had planted the seeds for developing an industry to meet a national crisis, applying a technological and political solution to an economic problem. The states soon followed suit. By 1980, twenty-five states had adopted some form of gasohol subsidy to encourage ethanol development and use as a motor fuel. Five states, including Louisiana, exempted gasohol from at least 8 cents per gallon of the respective state's motor fuel tax.[10]

With the new industry's profit potential, albeit contingent on federal and state subsidies, startup plants began to churn out ethanol, and the gasohol blend began to appear in the market. In the summer of 1979, Amoco began marketing the gasoline-alcohol mixture in thirteen midwestern states. That year more than a thousand service stations across the country began selling gasohol. Texaco soon entered the market and led the pack, with 1,700 stations selling the blend. Alcohol fuel sales jumped from 85 million gallons in 1981 to 234 million gallons in 1982. During this time, nine thousand gas stations nationwide sold the gasohol product.[11] The gasohol craze indeed provided American energy consumers with an affordable alternative at the gas pump.

Throughout the late 1970s, numerous independent and government-funded studies highlighted the potential benefits of alcohol fuels and the impact of incentive programs on both the new industry and the economy. In February 1981, the U.S. National Alcohol Fuels Commission—a federal oversight commission established in 1979—released its much anticipated eighteen-month study on the potential near-term use of alcohol as a motor fuel extender. The final report, *Fuel Alcohol: An Energy Alternative for the 1980s*, recommended continuing the government subsidies and expanding ethanol development.[12] Ethanol, it appeared, had a promising future in America.

Agrifuels Refining Corporation

Louisiana sugar cane is vital to the state's economy and has been for nearly two centuries. Since World War II, the industry has been dependent on federal programs to keep the domestic price of sugar up. In the late 1970s depressed sugar prices, foreign competition, energy shortages, and high energy prices weighed heavy on Louisiana farmers during what one report called their "darkest hour."[13] In 1978, five sugar mills in south Louisiana went out of business, and the crisis threatened several more closures.[14]

In the wake of these mounting problems and in response to the promising uses of energy crops for fuel, Carlos Toca, a plant manager at the Cajun Sugar Cooperative in New Iberia, Louisiana, and Dailey Berard, a local businessman, found the answer with a product they called "agrifuels," using sugar cane and sweet sorghum to produce ethanol. In April 1979, Toca, Berard, and two other local investors formed the Louisiana Agri-Fuels Corporation, later renamed Agrifuels Refining Corporation, in what appeared to be a profitable and personally rewarding homegrown business venture.[15]

At that moment, support for the new industry had reached the state capitol in Baton Rouge. Pro-ethanol and farming lobbyists, such as the newly formed Louisiana Gasohol Association, led the way. In the regular spring 1979 session the Louisiana legislature passed a handful of laws and resolutions that provided incentives for the development of a statewide ethanol, or gasohol, program. Act no. 793 (House Bill no. 571) eliminated the state's excise motor fuel tax of 8 cents a gallon for gasohol, providing the underpinnings for expanding biomass-based alcohol fuel production in Louisiana. Coupled with the federal exemption of 4 cents from the gasoline tax (Energy Tax Act 1978), the new state incentives created an investment opportunity for entrepreneurs and potential stakeholders in the lucrative new ethanol industry.[16]

Gasohol quickly became big business in Louisiana. Nine firms announced plans to build ethanol plants across the state. The ten-year exemption of gasohol from the state gasoline tax, signed into law by Governor Edwin Edwards in July 1979, attracted the attention of ethanol's long-time rival, the oil industry. By decade's end south Louisiana had merged its refining and farming expertise to the benefit of both industries and "to such an extent that the gasohol industry could become a significant part of the Louisiana economy," one report boasted.[17]

In June 1980, Oasis Petroleum Corporation, an independent oil producer from California with self-service gasoline stations on the West Coast, agreed to purchase 100 percent of stock in Agrifuels Refining Corporation.[18] The original partners remained on the board of directors and received additional financial considerations. On paper, the concept and production process of the $40 million Agrifuels ethanol plant seemed practical and, in many ways, technologically innovative: build an ethanol refinery adjacent to a sugar mill and use the mill's water supply (from the nearby Bayou Teche) for steam; permanently employ eighty to ninety refinery workers; purchase sugar cane and sweet sorghum from local farmers; use bagasse (sugar cane by-product) to fuel the boilers; refine the raw materials to make ethanol; transport the finished ethanol by barge, truck, or rail to be blended into gasohol at a blending facility; and sell the product to consumers. Four hundred thousand tons of sugar cane molasses, syrup, and sweet sorghum furnished by area farmers and a dozen sugar co-ops would produce 35 million gallons of ethanol a year (100,000 gallons a day).

Agrifuels would be the first major ethanol company in the U.S. to use sugarcane by-products and sweet sorghum as feedstock. Although the technology for using sweet sorghum to produce ethanol on a commercial level had not yet been realized, molasses, on the other hand, held well-known advantages over other biomass products, such as grain, because there was no feedstock preparation involved. Corn, for example, needed to be converted to sugar before beginning the ethanol fermentation process. Sugar cane also offered a high yield of sugar per acre in addition to a high yield of bagasse.[19] What made the plant truly revolutionary was that it would use bagasse, a renewable energy resource not fossil fuels, to power the plant and generate ethanol fuel. The National Alcohol Fuels Commission noted in its formal 1981 report that the bagasse-for-boiler-fuel concept "creates a positive energy balance, which attributes to conservation."[20] Using bagasse instead of natural gas or fuel oil for a plant's industrial requirements produced a net energy balance ratio (output to input) of 1.8:1, according to one report. Given this technological summary, it is not surprising that by 1980, according to Berard, the DOE labeled the Agrifuels project "the most viable in the U.S."[21]

Agrifuels continued on its development timeline. In late September 1980, representatives from Oasis Petroleum Corporation, including the president, and the four local entrepreneurs officially launched the $40 million Agrifuels plant at its 28-acre site in New Iberia, Louisiana, with a groundbreaking ceremony attended by three hundred invited guests, including Louisiana governor David Treen. Agrifuels president Carlos Toca stated, "We can all be proud of the contribution to the overall energy supply and the infusion of the capital investment of this project into the local economy." Donald P. Segura, president of the Cajun Sugar

Cooperative, hailed the project as the "salvation of the sugar industry." An editorial that appeared in the local newspaper the day after the ceremony pronounced that the experiment was "bound to be a resounding success."[22]

More encouraging news came the following year. Researchers from Louisiana State University in Baton Rouge and the University of Southwestern Louisiana in Lafayette and area farmers who participated in test projects reported for the first time that sweet sorghum could be grown commercially and in large quantity throughout the sugar parishes. Moreover, after submitting a loan application and required feasibility study to the DOE, the corporation received a "conditional commitment" for a federal loan agreement, which guaranteed repayment of 90 percent of the original loan to build the plant.[23]

By 1981 it appeared that the Agrifuels plant had real potential to become the nation's first sugar cane ethanol production facility. The project neared its final planning phase with construction completion and startup date slated for early 1983. Looking back, decades later, Berard shared his misgivings about the plant's unforeseen outcome: "This was truly the real beginning of an undertaking that had started as a sound idea that would ultimately develop into a mind-boggling nightmare."

Multiple economic and political factors on both the federal and state levels largely influenced the development of the Agrifuels ethanol plant. Skyrocketing oil prices in the 1970s led to government support for alternative fuel technologies across the country. However, the shifts in administrations and political ideals (from Carter to Reagan and Edwards to Roemer, for example) in the 1980s, along with a sharp decline in oil prices, ultimately led to the plant's failure. Several key elements of this story must be examined to explain the demise of the Agrifuels Refining Corporation. The combination of increased construction costs, federal energy policies, rapid decline in oil prices, and state politics created an unintended financial sinkhole from which the fledgling company could not survive.

With the federal government's involvement and the DOE's conditional commitment for a loan guarantee, additional costs and setbacks, as to be expected, began to mount. For starters, the loan program required a lengthy application and negotiation period, usually up to two years.[24] Administrative problems at the bureaucratic level contributed to the project's complexity and delay in construction. The National Alcohol Fuels Commission reported that federal efforts to develop and administer the program "have been not only slow but also fragmented."[25] Compliance with numerous federal regulations, along with rising interest rates, more than doubled the cost of the plant. Nevertheless, with guaranteed government financing and subsidies, the $105 million Agrifuels project moved forward.

Ronald Reagan's entry into the White House created additional problems for the upstart ethanol industry. His administration's threats to defund and ultimately terminate the alcohol fuel program compounded ethanol's questionable future. Ultimately, numerous plants across the country had enormous difficulty sifting through the maze of mixed signals from the government, deciphering changes to existing policies, and convincing investors and financial institutions to stick with and gamble on this government-backed program. With millions of dollars already on the line and contracts and construction of plants moving forward, developers, pro-ethanol lobbyists, and their political allies continued to champion Carter's alternative energy policies from the late 1970s. The subsidies held on, and even increased (by one cent, along with overall gas taxes), although other economic factors stymied ethanol's growth. Agrifuels, one of the nation's top alternative energy projects, continued to negotiate its uncertain future with creditors and contractors.[26]

Beginning in 1982 an unexpected, yet steady decline in oil prices had multiple consequences for the fledgling ethanol industry. As the oil glut progressed throughout the decade, the price differential between ethanol (gasohol) and cheap, regular gasoline continued to expand, thereby bringing into question the economics of the alcohol fuel. Oil companies and consumers slowly began to pull away from the gasohol blend. At that point, Berard recognized, "the [ethanol] market began to crater in Louisiana." Nevertheless, the Agrifuels plant still held promise.[27]

By 1985, Edgington Oil, a refining interest in California and subsidiary of Triad America Corporation, had purchased the plant from Oasis Petroleum, also a Triad subsidiary.[28] It is uncertain why or how this transaction took place. Records indicate that Edgington, a small refinery, faced a financial dilemma when in 1984 the Environmental Protection Agency began enforcing new regulations on refiners to reduce drastically the lead additive they used in gasoline.[29] By the mid-1980s ethanol became a likely substitute for lead as an octane booster in gasoline. In addition to purchasing the Agrifuels plant, according to Berard, Edgington also built a $21 million gasohol blending facility in California. In September of that year, Agrifuels finally entered the all-important loan agreement with the DOE, and construction began soon thereafter.

Throughout this period of oil price decline and uncertainty, Louisiana actually increased its support for ethanol. With a political eye on waning national support for ethanol and with four ethanol plants already under construction in his state and at least four more pending, in 1984 Governor Edwin Edwards attempted to buoy the struggling industry with new and improved tax incentives for ethanol production. Backed by Edwards and the pro-ethanol camp, the Louisiana legislature increased gasohol's 8-cents-a-gallon tax exemption to a whopping 16-cent exemption. Two years later when the oil market completely crashed and pressure for a policy change began to mount, state lawmakers ended the tax exemption and created a new direct subsidy program for ethanol producers. The new plan called for subsidies of 14 cents a gallon to be paid to ethanol producers.[30] But even with this policy reshuffling, state politicians seemed to be placating the ethanol and farm lobby at the expense of the state highway funds, taxpayers, and gasoline customers.[31] In 1979, Louisiana ranked fifth among states with tax exemptions for gasohol (8 cents a gallon). By 1986 the state had soared to the top of that shrinking list, with New Mexico (11 cents a gallon), North Dakota (8 cents a gallon), Virginia (8 cents a gallon), and a dozen other states (averaging 3–5 cents a gallon) trailing far behind.[32] Berard argued that the new increase in state subsidies was not necessary; even with uncertainties lingering in Washington over energy policy, the loan guarantee and existing federal and state exemptions provided more than enough incentives to make Agrifuels and the state ethanol industry viable. He feared, correctly, that the new wave of state subsidies might torpedo the ethanol industry when the new governor, Buddy Roemer, took office in early 1988, reduced the state budget, and canceled the program.

Construction of the $105 million plant continued throughout 1986, although as oil prices hit rock bottom its future remained on shaky ground. In January 1987 the company's financial situation worsened and the project began to unravel. Edgington Oil, the owner of Agrifuels, began experiencing financial troubles. Its parent company, Triad America, declared bankruptcy in February to protect itself from creditors. Apparently Adnan Khashoggi, owner of the Triad financial empire, had invested heavily in oil and real estate before the market for both industries began to collapse simultaneously.[33]

By late March, Edgington Oil could not come up with the required $6.5 million in working capital to begin operations and abandoned the Agrifuels plant. The president of Edgington explained to the press that Agrifuels proved no longer profitable: "Based on economic projections prepared by Agrifuels," Mark Newgard explained, "we formally advised the banks and contractors that it would be a mistake to put any additional funds into Agrifuels."[34] Newgard noted that state subsidies for ethanol were financed by a state motor fuels tax on gasohol sales, but when major distributors, including Texaco, pulled out of the gasohol market and several new plants came on line those subsidies would be depleted. By mid-April, workers at the nearly completed Agrifuels plant were sent home while the DOE and the local governing board—which included Pat Hamilton, the president of Agrifuels, and Berard, the vice president—contemplated the plant's fate. The DOE, which had guaranteed 90 percent of the loan agreement, cut all remaining financial disbursements to the plant and formally took over the remaining assets once the company defaulted on its loan payment.[35] Throughout that summer and into early fall, Hamilton scrambled to find the required operating capital to stall a DOE foreclosure. Berard read the writing on the wall and resigned from the board in August.

In the end, the DOE paid off the loan, a total of $78.9 million, and pumped in an additional $100,000 a month just to keep the equipment from corroding.[36] In 1988 the DOE spent additional funds mothballing the plant. And just as Berard feared, incoming governor Buddy Roemer did not renew the state alcohol fuel program, effectively terminating the ethanol industry in Louisiana. Prior to the drastic shifts in energy prices and policies of the 1980s, Louisiana's ethanol industry certainly had held promise; production across the state reached a peak of 32 million gallons in 1986. However, when Governor Roemer ended the state ethanol subsidy program, nearly all production stopped, as did the completion of several plants, including Agrifuels. Governors Edwards and Roemer had used the ethanol subsidies as a political tool, Berard believed, and the private investors and farming community were ultimately shortchanged. State politics, were not, however, solely to blame. The oil price collapse of the mid-1980s certainly contributed to the downfall of ethanol across the nation.

Beginning in 1988, Berard and local investors joined forces to purchase the mothballed Agrifuels plant from the government. They offered the DOE $25 million for the plant, but the agency turned them down, holding out for an offer of at least $50 million. Later the partners made a second offer: $15 million. But again the government held out. In the final chapter, the DOE dismantled the plant a few years later and sold the equipment to a firm from the Midwest. Agrifuels, a technological innovation arguably ahead of its time, never produced a gallon of ethanol.[37]

U.S. Ethanol Program: National Debates and Final Assessment

Beginning in 1978, the federal government through legislative action provided a variety of tax incentives for developing a domestic ethanol industry. By 1980 the program enjoyed public and private support.[38] But as the program transitioned from the planning to the application phase its weaknesses became apparent, especially to the growing anti-gasohol critics and the larger investment community. A program had been created and funded, yet the management of that program and the institutional constraints of the federal bureaucracy created problems for business models and decision making. President Reagan's ideas about limited government and reducing federal subsidies for various programs created uncertainties within the business community about whether or not support for the newly created ethanol industry would continue and for how long. Policymakers and program administrators apparently did not take into account the emerging economic recession driven in part by declining oil prices. Nevertheless, the federally backed crash program continued, regardless of its deficiencies, shifts in policy, and changing market conditions that threatened the future of this alternative fuel.

For much of the 1980s, factions from both inside and outside of government debated the benefits and drawbacks of the ethanol industry. With the Environmental Protect Agency's mandatory lead "phasedown" program, pro-ethanol interests argued for substituting clean-burning alcohol fuel for the noxious tetraethyl lead as an octane booster in gasoline.[39] Many in the agricultural sector also touted the benefits of ethanol production related to the sagging farm economy in the United States.[40] Opponents of the fuel alcohol plan argued that expanding the industry might severely impact food prices and food supplies. Energy experts and researchers questioned the "energy balance" and efficiency of converting fuel from crops to be used to power cars.[41] Environmentalists, moreover, contemplated excessive soil erosion and the environmental impact of using more herbicides and pesticides through large-scale land use.[42] Proponents noted the reduction in carbon dioxide emission from car engines that used gasohol.[43] Consumers often complained about gasohol's product performance in their vehicles. Above all, critics of ethanol pointed to the high costs to the economy and the loss of revenue to the states' highway trust funds, which depended on the federal gasoline sales tax (gasohol received exemptions) to repair and maintain the nation's interstate highway system.[44]

The internal problems that hampered the federal alcohol fuel program led some critics to argue that the alternative energy policies were, as one congressman stated, "alarmingly unbalanced." The pundits cited a disproportionate level of funds available to the program; less than $2 billion of the $20 billion authorized by the Energy Security Act for alternative energy had been "earmarked for biomass energy."[45] In addition, the DOE and USDA decided not to implement the price guarantees and purchase agreements for ethanol that were initially authorized by Congress and further recommended by the National Alcohol Fuels Commission. "This restrained approach," the Commission stated, "dampens the fullest private sector involvement in the program."[46] Looking back at the program and into ethanol's future, Congressman Ed Madison of Illinois stated in 1988, "It seems strange that we do not implement policies that will maximize the benefits we can obtain from this domestically produced, renewable fuel."[47]

Private firms across the country experienced the frustrating reality of dealing with an uncertain energy policy and unstable ethanol program in the wake of a downturn in oil markets. For example, American Energy, Inc., of North Dakota spent five years building an ethanol plant capable of producing 50 million gallons a year. By the summer 1984 the project had been "shelved," because one of the joint-venture partners, Ashland Oil, pulled out citing the fact that the energy investment tax credit had not been extended past 1985. Edward Wilkinson, chairman of National Alternative Energy Associates and president of Agri-Energy, Inc., of Minnesota complained, "We never know from one session of congress to the next what program or incentive is in danger of being eliminated." Wilkinson noted a loss of $4.3 million in equity to his firm because the "current administration changed their policy midstream." Another Minnesota firm's partner walked out in 1982 because of threats to the existing energy policy. In addition, that firm had to dish out half a million dollars to have attorneys decipher the complex energy tax credit law. "The tragedy is that we're four years in and here we are seeking clarification on acts of Congress that occurred back in 1978 and '79," the firm's representative stated.[48]

These frustrations speak to the faults that lay within this crash program initiated during the energy crisis. Throughout the 1970s and '80s, policymakers chose a short-term, reactive approach to expanding new technologies quickly and with little emphasis on research and development and long-term strategy to solve one of the nation's major problems. In the haste to develop these crash programs, bureaucratic decisions and institutional arrangements became even more ambiguous and disorganized, which created barriers to communication and cooperation with the private sector. Moreover, federal policymakers overextended their power by setting overly ambitious goals for these crash programs without providing the necessary tools and vision to sustain the program through abrupt changes in market conditions.

By contrast, the Brazilian alcohol fuel program, Pró-Álcool, proved successful at obtaining, and in some cases surpassing, near-term policy goals during the oil crises of 1970s and '80s. What started out in Brazil in 1975 as a "sugar industry bailout scheme" became a major national effort to stabilize the country's economy by producing and marketing ethanol from sugar cane on a grand scale. Brazil achieved an unprecedented level of production, 3 billion liters (about 800 million gallons) annually by 1979, by implementing long-term policy measures and building a coalition between government agencies, financial institutions, automakers, consumers, small- and large-scale farmers and ranchers, and Petrobras, the national oil company.[49] Over time, as the economics of sugar cane ethanol improved and Brazilian society accepted the shift to an alternative fuel, the government gradually reduced the subsidies, allowing the industry to continue to develop a fully integrated renewable energy economy.

In the United States, the Reagan administration, along with an economic recession, constantly worked against expanding the ethanol industry in America. By mid-decade Congress, which had originally initiated the incentive-based policy, had rescinded a majority of the operating funds to two of the three alternative fuel programs authorized in 1980. The Office of Alcohol Fuels, also created in 1980, barely existed by 1986. Consecutive funding cuts to the program began to limit the range of available financing for loan guarantees, the only biomass program still viable, incidentally, by 1986. Indeed, by 1985 the DOE had a budget of only $234 million for the alcohol program, a far cry from the $1.2 billion it was originally authorized. Ultimately the DOE offered grants to only three firms: New Energy of Indiana, Tennol of Tennessee, and Agrifuels Refining of Louisiana.[50] The New Energy plant defaulted on its federally guaranteed $126 million loan in 1987. Tennol followed suit by defaulting on its $64 loan guarantee and closed down in 1989.[51]

The individual states' subsidy programs proved problematic for the industry at times. State lawmakers had the ability and opportunity to increase or decrease their respective ethanol subsidy programs without consulting the federal government. Short of nationalizing the industry, or at least regulating prices, the federal government had absolutely no control over the states' subsidy programs and policy decisions. This institutional constraint actually proved beneficial to businesses during the industry's early years, since additional state subsidies meant additional profits for ethanol producers. But when market conditions changed and state policy-makers exercised their political power to alter the subsidies of *their* in-state industry, businesses and investors had to contend with additional problems and uncertainties.

The initial supporters of the economic and free-market potential of ethanol argued early on that with high oil prices alcohol fuels would eventually stabilize and become competitive. Analysts predicted a steady increase in oil prices throughout the 1980s; some estimates reached as high a $100 a barrel. Thus, much of the support for and impending success of ethanol rested with the market forces of global oil economics. "The market, not Congress," should dictate ethanol's role in the economy, the *Oil & Gas Journal* complained.[52] When the oil market collapsed, oil companies and their lobbyists began to pull away from the gasohol market and pulled back their support for the alcohol fuel program. After crude oil prices dropped below $10 a barrel in mid-1986, Tenneco, Texaco, and other national distributors discontinued sales of ethanol blended gasoline across the country. "Consumers have told us they don't want gasohol for their automobiles," a Tenneco spokesman stated, "and since we were losing market share we made this decision." Roughly half of the 165 distilleries that produced ethanol went out of business, and twenty closed their doors in 1986 alone.[53] The return of cheap gasoline at the pump and the economic recession left Americans and their political institutions less interested in energy conservation and less supportive of alternative fuels. The need for energy independence had faded, for the time being, along with high gas prices and the biomass alcohol fuels industry.

The development and demise of the alternative energy sources are definitive examples of the consequences of a complex political system that is, as one scholar explained, "uniquely ill-suited to handle energy policy."[54] Deficiency in government planning and implementation become compounded during crisis situations and often result in crash programs, as the case for Agrifuels clearly illustrates. With government serving the role as regulator and subsidizer, and with business producing energy products for the nation, it is vital that these two entities coordinate their efforts and communicate effectively during planning. When government programs involve long-term business planning, long construction periods, and financing from the private sector, it is incumbent on policymakers to ensure that the policies remain anchored and are not subject to change at midstream. The alcohol fuels crash program of the late 1970s and '80s illustrates the risks involved in cultivating a new industry that relies on government-backed loan guarantees and subsidies to compete with traditional domestic sources. Some of the policy issues that arose from that era are still being addressed today in government circles and company boardrooms. For historians, this period represents a watershed in energy history and government policy, where many of the questions have still gone unanswered.

Notes

1. August W. Giebelhaus, "Resistance to Long-Term Energy Transition: The Case of Power Alcohol in the 1930s," in *Energy Transitions: Long-Term Perspectives*, ed. Lewis J. Perelman, August W. Giebelhaus, and Michael D. Yokell (Boulder: Westview Press, 1981), 36.

2. In addition to sufficient feedstock supplies, other "economic feasibility factors" included labor, process water supplies, transportation, and an adequate market; see V. Daniel Hunt, *The Gasohol Handbook* (New York: Industrial Press, 1981), 183.

3. For excellent treatments of energy policy in postwar America, see Richard H. K. Vietor, *Energy Policy in America since 1945: A Study of Business-Government Relations* (Cambridge: Cambridge University Press, 1984); and Martin V. Melosi, *Coping with Abundance: Energy and Environment in Industrial America* (New York: Knopf, 1985).

4. Daniel Horowitz, *Jimmy Carter and the Energy Crisis of the 1970s: The "Crisis of Confidence" Speech of July 15, 1979; A Brief History with Documents* (Boston: Bedford/St. Martin's, 2005), 36, 41–42.

5. See Daniel Yergin's chapter, "Conservation: The Key Energy Sources," in *Energy Future: Report of the Energy Project at the Harvard Business School*, ed. Robert Stobaugh and Daniel Yergin (New York: Random House, 1979), 136–82.

6. Horowitz, *Jimmy Carter*, 99, 116.

7. U.S. National Alcohol Fuels Commission, *Fuel Alcohol: An Energy Alternative for the 1980s* (Washington D.C.: Government Printing Office, 1981), 26.

8. The goal of the program was 920 million gallons of ethanol, or 60,000 barrels of ethanol a day, produced by the end of 1982. A more ambitious, if not unrealistic, goal of the Carter administration's energy policy was to produce enough alternative fuels to meet 10 percent (roughly 10 billion gallons) of the nation's gasoline consumption by 1990, thereby replacing 2.5 million barrels of oil a day; Hunt, *Gasohol Handbook*, 18; U.S. Congress, House, Subcommittee on Fossil and Synthetic Fuels of the Committee on Energy and Commerce, *Alcohol Fuels and Lead Phasedown* (Washington, D.C.: Government Printing Office, 1986), 73; "Ethanol Buildup Continues but Shortfall Is Predicted," *Chemical Week*, September 3, 1980, 17. The Energy Security Act also authorized the DOE and USDA to enter into purchase agreements and price guarantees with firms to encourage further interest in ethanol developments; see U.S. National Alcohol Fuels Commission, *Fuel Alcohol*, 15.

9. U.S. National Alcohol Fuels Commission, *Alcohol Fuels Tax Incentive, A Summary: Alcohol Fuels Provisions of the Crude Oil Windfall Profit Tax Act* (Washington, D.C.: Government Printing Office, 1980); see also Hunt, *Gasohol Handbook*, 42. The Energy Security Act also created the U.S. Synthfuels Corporation for production of methane from coal. Funds from the Windfall Profit Tax Act would be used to finance and operate the Synthfuels Corporation.

10. U.S. National Alcohol Fuels Commission, *Fuel Alcohol*, 50; see also U.S. National Alcohol Fuels Commission, *State Initiatives on Alcohol Fuels: A State-by-State Compendium of Laws, Regulations, and Other Activities Involving Alcohol Fuels* (Washington, D.C.: Government Printing Office, 1980).

11. *Environment* 20, no. 6 (1979), 21; U.S. Congress, House, Subcommittee on Fossil and Synthetic Fuels, *Alcohol Fuels and Lead Phasedown*, 72; Hunt, *Gasohol Handbook*, 18; U.S. National Alcohol Fuels Commission, *Fuel Alcohol*, 50.

12. U.S. National Alcohol Fuels Commission, *Fuel Alcohol*, 6–11. The Commission further recommended the implementation of purchase agreements and price guarantees, in addition to design and production of pure-alcohol vehicles by U.S. automakers backed by government incentives. By the late 1980s, however, the government decided not to follow through with these policies and recommendations.

13. "Sugarcane: Old Problems, New Opportunities," *Acadian Profile*, 6, no. 3 (1977), 62. The report also stated that sugar prices fell to 10 cents a pound in 1977, where 17 cents a pound was needed to make a profit.

14. "Gasohol: Blending Agriculture and Oil," *Acadian Profile* 7, no. 3 (1979), 50. Sugar cane acreage in south Louisiana had dropped 36 percent from 1978/79. The Agribusiness

review in this section is based on the author's personal communication with Dailey J. Berard, July 24 and 26, 2007, Cankton, Louisiana, along with other cited sources.

15. "Gasohol: Blending Agriculture and Oil," 50.

16. *State of Louisiana: Acts of the Legislature, Vol. 1, Regular Session 1979* (Baton Rouge: State of Louisiana, 1979), 2284–86.

17. "Gasohol," *Acadian Profile* 7, no. 5 (1979), 73.

18. Oasis Petroleum, formed in 1977, had in 1980 an estimated 120 service stations in twenty-one states where it planned to market gasohol. Berard first met with representatives from Oasis Petroleum at an ethanol conference in New Orleans, and there they discussed plans for the joint business venture into Agrifuels.

19. Hunt, *Gasohol Handbook*, 113, 143–44.

20. U.S. National Alcohol Fuels Commission, *Fuel Alcohol*, 6. A second congressional committee hearing, one critical of Carter's alternative energy plan, agreed with the positive energy balance argument from biomass fuel "as long as crop residues are the major fuel for conversion of biomass to ethanol." U.S. Congress, Subcommittee on Energy of the Joint Economic Committee, *Farm and Forest Produced Alcohol: The Key to Liquid Fuel Independence* (Washington, D.C.: Government Printing Office, 1980), 13.

21. C. S. Hopkinson and J. W. Day, "Net Energy Analysis of Alcohol Production from Sugarcane," *Science* n.s. 207, no. 4428 (1980), 302–304; "Alternative Energy Resources Being Developed," *Acadian Profile* 8, no. 6 (1980), 39. As a by-product of ethanol manufacturing, the plant would also produce an estimated 148,000 tons of solubles for animal feed.

22. "$40 Million Ethanol Refinery Begun Officially at Morbihan," *Daily Iberian*, Oct. 1, 1980, 33; "Sugar as Fuel," *Daily Iberian*, Oct. 1 1980, 14.

23. "Sorghum Experiment Could Change Face of Acadiana Agriculture," *Acadian Profile* 9, no. 4 (1981), 58. Bagasse from sweet sorghum could be used as an additional source for boiler fuel. The 90 percent loan guarantee was enacted by the Biomass Energy and Alcohol Fuels Act 1980 (Title 11).

24. The DOE used a "batch review" process to evaluate each loan proposal. This process included meeting very specific criteria during a limited time period. However, priority was given to projects that used biomass as the primary boiler fuel and "new technologies that use different biomass feedstocks" such as cane molasses and sweet sorghum; see Hunt, *Gasohol Handbook*, 27–28.

25. The report also noted that, prior to the passage of the Energy Security Act of 1980, six different departments and agencies administered nine different financial support programs for alcohol fuels. Most of these programs remained intact with little change in their bureaucratic functions after the passage of the new act; see U.S. National Alcohol Fuels Commission, *Fuel Alcohol*, 140–41.

26. One 1984 report noted that as much as $1 billion in capital investment had been spent on the new industry since 1979. However, the program apparently reduced the trade deficit by $210 million and created nearly twenty thousand jobs in that five-year period; see U.S. Congress, House, Subcommittee on the Oversight of the Committee on Ways and Means, *Tax Incentives for the Production and Use of Ethanol Fuels* (Washington D.C.: Government Printing Office, 1984), 1–5, 141.

27. "Federal Subsidies Foster Ethanol Industry Growth," *Washington Post*, Dec. 8, 1985, H6. The report noted that even with the depressed corn prices a gallon of ethanol cost 40–55 cents more than gasoline. As in the 1930s, oil companies launched an anti-ethanol campaign using brochures and billboards to discourage the public's support for gasohol.

28. Triad America was a holding corporation of Adnan Khashoggi, a wealthy Saudi arms dealer.

29. "If Leaded Gas Goes, So Will Some Small Refiners," *Business Week*, July 23, 1984, 77.

30. Louisiana Department of Natural Resources, *Summary: Fuel Alcohol in Louisiana* (Baton Rouge: Department of Natural Resources, 1987). The plan called to reduce the subsidy even further to 12 cents from 1988/89.

31. In 1985 the state exemption for ethanol in Louisiana amounted to $28 million. In accordance with the Agricultural Ethanol Production Law (1985), payments to state ethanol producers were capped at $52 million. However, if the five plants in operation and the five plants under construction during that time produced to their capacity, the total payments for subsidies would exceed $217 million. It is not clear how the Louisiana government planned to subsidize the plants once they became fully operational; see Louisiana Department of Natural Resources, *Summary: Fuel Alcohol in Louisiana*.

32. "Ethanol Tax Exemptions under Fire," *Oil and Gas Journal*, Jan. 20, 1986, 35.

33. In early 1987 national news outlets began reporting that Khashoggi owed millions to creditors; "Creditors Putting Pressure on Saudi," *New York Times*, Jan. 17, 1987, 6; "Payouts on Loans to Khashoggi Firm Stopped by U.S.," *Wall Street Journal*, Apr. 13, 1987, 1; "Khashoggi's U.S. Company Files for Bankruptcy," *Washington Post*, Jan. 29, 1987, A30.

34. Newgard in "Edgington Abandoned Agrifuels in Late March," *Daily Iberian*, May 2, 1987, 1. "U.S. Pays Project's Creditors; Fuels Firm in Default Owned by Khashoggi," *Washington Post*, Sept. 4, 1987, F1.

35. "Workers Off at Agrifuels; Talks Begin," *Daily Iberian*, Apr. 15, 1987, 1–2.

36. "New Iberia Journal; Tanks Are Tombstones of Ruined Fuel Policy," *New York Times*, June 16, 1988, 16.

37. "Ethanol . . . Ahead of Times," *Daily Iberian*, Apr. 16, 2006, 2.

38. The National Alcohol Fuels Commission noted the following benefits of an ethanol industry: creates jobs, disposal of waste products (e.g., bagasse), and environmentally friendly (meaning less carbon emissions). The report confirmed that gasohol's performance matched that of gasoline in cars; see U.S. National Alcohol Fuels Commission, *Fuel Alcohol*, 2.

39. See U.S. Congress, *Alcohol Fuels and Lead Phasedown*.

40. "Federal Subsidies Foster Ethanol Industry Growth," *Washington Post*, Dec. 8, 1985, H6. The article reports that ethanol production increased the price of corn by 10 cents a bushel, which "pumped an extra $849 million into the depressed farm economy." A congressional subcommittee further noted that farm-based ethanol offered opportunities to the agricultural sector, in which "3,700 family farms a month are disappearing from the map of rural America"; see U.S. Congress, *Farm and Forest Produced Alcohol*, 5.

41. Nicholas Wade, "Gasohol: A Choice That May Buy Grief," *Science* n.s. 207, no. 4438 (1980), 1450–51; C. S. Hopkinson, "Net Energy Analysis of Alcohol Production from Sugarcane," *Science* n.s. 207, no. 4428 (1980), 302–304. Automobile efficiency tests from

Illinois, Nebraska, and Iowa in the late 1970s indicated that gasohol performed as well as if not better than gasoline; see Hunt, *Gasohol Handbook*, 31.

42. "America's Home-Grown Fuel," *Nation*, Sept. 27, 1980, 275–76.

43. U.S. Congress, House, Subcommittee on Forests, Family Farms, and Energy and the Subcommittee on Wheat, Soybeans, and Feed Grains of the Committee on Agriculture and the Subcommittee on Energy and Power of the Committee on Energy and Commerce, *Review of the Role of Ethanol in the 1990's* (Washington D.C.: Government Printing Office, 1988), 19. The report noted that gasohol reduced carbon dioxide emission by 25 percent.

44. "Ethanol Industry Booms amid New Controversy," *Washington Post*, Dec. 8, 1985, H1. The federal and state tax incentives for gasohol cost the federal government an estimated $420 million a year and a "drainoff" of $280 million annually from the federal highway fund. The revenue loss, "with no end in sight, is simply counterproductive in meeting the highway needs of this nation," a representative from the auto industry lamented; see U.S. Congress, *Tax Incentives for the Production and Use of Ethanol Fuels*, 77.

45. U.S. Congress, House, Subcommittee on Energy of the Joint Economic Committee, *Farm and Forest Produced Alcohol*, 1.

46. Hunt, *Gasohol Handbook*, 24–25; U.S. National Alcohol Fuels Commission, *Fuel Alcohol*, 15.

47. U.S. Congress, House, *Review of the Role of Ethanol in the 1990's*, 1.

48. U.S. Congress, *Tax Incentives for the Production and Use of Ethanol Fuels*, 42–45, 59, 101. This incentive allowed businesses to claim a tax credit on their income taxes for purchasing equipment that was used in the production of ethanol from crops.

49. Michael Barzelay, *The Politicized Market Economy: Alcohol in Brazil's Energy Strategy* (Berkeley: University of California Press, 1986), 22, 172.

50. U.S. Congress, *Alcohol Fuels and Lead Phasedown*, 73, 76–77.

51. "Payouts on Loans to Khashoggi Firm Stopped by US," *Wall Street Journal*, Apr. 13, 1987, 1.

52. U.S. National Alcohol Fuels Commission, *Fuel Alcohol*, 29, 43; "The Market, not Congress, Should Determine Role for Ethanol in Fuel," *Oil and Gas Journal*, May 23, 1988, 13.

53. "Demand Dries Up for Gasohol," *Chemical Week*, Mar. 4, 1987, 16; "Alcohol Fuels Move off the Back Burner," *Business Week*, June 29, 1987, 100.

54. Personal communication with Joseph A. Pratt, Aug. 1, 2007.

Chapter 8

Pipe Dreams for Powering Paradise
Solar Power Satellites and the Energy Crisis

Jeff Womack

In October 2007, just as crude oil prices soared to record highs, the U.S. National Security Space Office (NSSO) released a study, titled "Space-Based Solar Power as an Opportunity for Strategic Security," that pitched solar power satellite (SPS) technology as a "game changer" that would allow the United States to send power anywhere in the world for purposes as diverse as powering armies and restoring electricity in disaster zones. In the following spring, the National Space Society's official magazine, *adAstra*, published a special issue touting "Space-Based Solar Power: Inexhaustible Energy from Orbit." CNN.com ran a story in May 2008 under the headline "How to harvest solar power? Beam it down from space!" that focused on entrepreneurs seeking to bring electricity to rural parts of India using "massive sun-gathering satellites in geosynchronous orbits"; the article also described Japan as aggressively pursuing space power systems. A year later, the Pacific Gas and Electric Company promised to do the same thing by 2016 in the United States, signing a contract with Solaren Corp. to purchase 200 megawatts of electricity from that company's proposed solar satellite.[1]

The various reports and stories described space-based solar power as an exciting new technology. In reality, however, the concept has its roots in the energy crisis. The National Aeronautics and Space Administration (NASA) began its flirtation with SPSs in the early 1970s, shortly after Neil Armstrong took his first steps on the moon. In the intervening years, however, these satellites have neither appeared nor disappeared. No country has launched one, but governments and entrepreneurs continue to talk about them, and the talk has changed very little in half a century. To understand this strange state of affairs, we must look back to the beginnings of the SPS discussion. In spite of its persistence, however, space-based solar power has received little attention from historians interested in the energy crisis, alternative energy, or solar power. Once unraveled, the story offers an interesting set of insights into the factors that affect the success—or failure—of alternative energy technologies.

Space-based solar power began as a science fiction concept. In 1941, Isaac Asimov published a short story in *Astounding Science Fiction*, "Reason," that took place on a "solar space station" that collected solar radiation and fed it to Earth using "energy beams." "Reason" eventually appeared in the author's influential *I, Robot* anthology, and other science fiction authors also picked up on the concept. Arthur C. Clarke, for instance, wrote a speculative article in the October 1945 issue of *Wireless World* describing solar-powered space stations that would eventually make possible the creation of a worldwide communications network.[2]

By 1968, *I, Robot* had reached bestseller status and gone through no fewer than seven printings. In that year, however, Peter E. Glaser—a scientist working for the Arthur D. Little consulting company—brought space-based solar power out of the world of science fiction and into a rapidly expanding debate about the future of energy in the United States when he delivered a paper on the topic at the Engineering Energy Conversion Conference. After the conference, Glaser's research appeared as an article, "Power from the Sun: Its Future," in the journal *Science.*[3]

According to Glaser, it would "be increasingly more difficult and more costly" to control "the environmental deterioration" created as a side effect of energy production. As examples, he cited air and water pollution associated with burning fossil fuels, but he also emphasized the problem of thermal pollution from nuclear power plants. Glaser advocated major research funding for alternative energy sources, suggesting that such research would prepare the world for a rapidly approaching point where "the adverse effects on the human environment of present energy sources become intolerable."[4]

Among alternative energy technologies, Glaser favored solar energy as an environmentally friendly methods of producing electricity. He did not, however, simply want to see government support for sun-powered hot water heaters or rooftop photovoltaic panels. Instead, he had a much more ambitious proposal in mind: a "satellite conversion system" designed to meet the electricity needs of the United States. Glaser described a system based on satellites positioned in geosynchronous orbit (where they would receive near-continuous sunlight and remain stable with respect to a receiving station on the ground), using highly efficient photovoltaic cells to convert solar energy into electricity. The satellites would beam their energy back to the Earth in the form of microwave radiation—the same energy used in a common household appliance to heat food and drinks— where ground-based receiving stations would convert it into electricity and send it out to customers. Glaser described his satellites as massive constructions; the illustration accompanying the article showed a circular solar array 6 kilometers in diameter.[5]

Peter Glaser's article arrived at an interesting turning point for solar power advocacy. Solar enthusiasts had a long, if unsuccessful, history in the United States. As early as 1833, John Adolphus Etzler had attempted to attract investors to finance the construction of a utopian community that would feature steam engines fired by mirror-focused sunlight and plowing machines driven by windmills. John Ericcson, the Swedish American inventor famous for designing the ironclad USS *Monitor*, fearing the day when "Europe must stop her mills for want of coal," spent the final two decades of his life working on solar-powered steam engines. The popular Philadelphia entrepreneur Frank Shuman began working on a solar engine in 1907 as a replacement for coal-fired water pumps, and he eventually succeeded in building a steam engine in Egypt powered by water heated to the boiling point in a series of glass-covered "hot boxes."[6]

Although many of those previous solar advocates had warned of a coming age of energy scarcity, their claims tended to fall flat in the face of abundant fossil fuel supplies. Published in 1968, however, Glaser's "Power from the Sun" appeared just as electricity companies, buffeted by a perfect storm of technological and economic factors, began to face serious challenges to their ability to meet the need for power in the United States.[7] Speaking to a *New York Times* interviewer in July 1969, the outgoing chairman of the Federal Power Commission, Lee C. White, explicitly warned of that the nation's largest cities had "inadequate capacity to meet peak demands, which are more and more coming in the summertime with increased use of air-conditioning." White foresaw "repeated electric power shortages" in the near future—a prediction that came true in New York City in 1970 when two generators failed and left the city without enough capacity to meet daytime electricity demands.[8] As members of the government cast around for ways to address these shortages, it should come as little surprise that "Power from the Sun" found an audience.

On June 4, 1971, President Richard Nixon gave an address to Congress billed in a White House press release as "the first time that a message on energy of a comprehensive nature has been sent to the Congress by a President of the United States." Responding to "the brownouts . . . the possible shortages of fuel . . . the sharp increases in certain fuel prices, and our growing awareness of the environmental consequences of energy production," President Nixon focused most of his remarks on proposals for new coal and oil policies and accelerated research into "fast breeder" nuclear reactor technology, but he also talked about solar energy, saying that "the sun offers an almost unlimited supply of energy" for those who could "learn to use it economically." To jump-start that learning process, President Nixon declared that both NASA and the National Science Foundation (NSF) had begun "re-examining their efforts in this area." The president instructed his scientific advisors to "make a detailed assessment of all of the technological opportunities in this area and to recommend additional projects which should receive priority attention."[9]

In the service of these directives the administration commissioned a special investigation, jointly run by NASA and the NSF, into potential applications for solar energy technologies in the United States. The two agencies assembled a "Solar Energy Panel" and charged the panelists with "assessing the potential of solar energy as a national energy resource and the state of the technology in the various solar energy application areas" and "recommending necessary research and development programs to develop the potential in those areas considered important." The panel began its work in January 1972 and delivered its report, *An Assessment of Solar Energy as a National Energy Resource*, in December.[10]

The presence of officials from NASA on the panel made sense; NASA, more than any other government agency, had experience with solar technology, which it used to power spacecraft. NASA also had a reputation as the government's premier engineering and science agency—the legacy of its successes during the Apollo program. NASA officials, however, came to the project at a difficult time for the agency.

Many think of 1969, the year of the first moon landing, as NASA's heyday. The agency's budget, however, told a different story. According to the *New York Times*, between 1965 (the beginning of the fiscal 1966 year) and 1969, the NASA budget was "cut more sharply than that of any other Federal agency," having shrunk by 35 percent, from $5.9 billion to $3.8 billion, in just four years. In a 1969 interview the agency's new administrator, Thomas Paine, expressed his fear that "the space program, caught between the pressure on the right for ever-larger defense appropriations and on the left for massive social projects," would find itself "ground down to insignificance."[11]

Caught in the budgetary vice, NASA officials found themselves casting about for a project that would help to justify the agency's existence to its critics. Glaser's SPS article, which proposed to solve a pressing national energy problem with a massive investment in the space agency's area of expertise, must have looked like a plan of salvation for officials from the agency. It therefore comes as little surprise that Glaser received an invitation to join the Solar Energy Panel. Two other members of the nine-person Photovoltaic subcommittee of the Solar Energy Panel also had aerospace ties: Paul Goldsmith, of TRW Systems, and Nicholas F. Yannoni, of the Air Force's Cambridge Research Lab.[12]

The appearance of "Power from the Sun" at the hour of NASA's need was probably more than just a coincidence. Glaser's employer, Arthur D. Little, actually had a long-term relationship with NASA; according to Christopher Kraft, the engineering firm belonged to a stable of "think-tank organizations" that existed "to furnish the brainpower" when NASA wanted to investigate a new idea. Given this relationship, the contractor was well positioned to understand both the space agency's troubles and the benefits, for NASA contractors, of a large new aerospace project. This is not to say that Glaser wrote his "Power from the Sun" article exclusively with an eye toward generating new government contracts. Kraft remembered Glaser as "always a big advocate" of SPSs, and Glaser spent most of his life personally promoting the idea.[13] But Glaser's work definitely had a tangible benefit for Arthur D. Little. In 1972 that company headed up a coalition of government contractors, including the Grumman Aerospace Corporation, Textron, and the Raytheon Corporation, that received a six-month, $197,000 contract from NASA "to explore technical problems of a satellite solar power station" as part of the larger Solar Energy Panel study.[14]

Readers can easily spot the influence of the aerospace coalition on the *Assessment of Solar Energy.* The report described a variety of technologies and proposed several avenues of investment for the government. One proposal, however, stands out: the section "Electric Power from Space," tucked into the "Photovoltaic Solar Energy Conversion Systems" section of the report. "Electric Power from Space" essentially contains a more detailed version of Glaser's *Science* article.[15]

At this point, readers familiar with the history of public policy and federal politics may recognize the beginnings of a phenomenon often referred to by historians as the "iron triangle." In *Chain Reaction,* his book chronicling the public and private debate over nuclear power, historian Brian Balogh offers this excellent encapsulation of the "iron triangle" concept: "Since the Progressive Era, federal programs have depended on the tripartite support of an interest group, a congressional oversight committee (or subcommittee), and an administrative bureau." As Balogh points out, these combinations of reinforcing support come in a variety of iterations, but they all demonstrate a single truth: government programs live and die based more on their ability to muster political support than on their intrinsic value to the nation.[16]

At first glance, the SPS program seems to have built two legs of the iron triangle: an administrative bureau and the support of powerful interests in industry. These developments did not go unnoticed at the time; contemporary critics of the government's solar energy policies went so far as to accuse industry interests of hijacking government policy. Published in 1979, Ray Reece's *The Sun Betrayed: A Report on the Corporate Seizure of U.S. Solar Energy Development* epitomizes this type of critique. Begun in 1977 as an investigative piece for the *Texas Observer, The Sun Betrayed* described the SPS project as a perfect example of "the politics of corporate hegemony," with "deceit, vested interest, and collusion in the highest echelons of U.S. industry and government." Reece claimed in particular that members of NASA and the aerospace industry, facing the end of the Apollo program, used the Solar Energy Panel's report to serve their "direct financial and professional interests in finding (or creating) a successor to the shrinking [government] aerospace program."[17]

It originally appeared that congressional support for space-based solar power—the third leg of the iron triangle—might materialize. The House Committee on Science and Astronautics released a report stating its support for solar power as the best strategy for meeting the nation's "long-term energy needs" and recommending that the federal government begin spending $150 million per year on research and development. The committee's enthusiasm, however, looks rather tepid when set against the "Electric Power from Space" section of the NSF/NASA proposal, which had called for a national commitment "comparable to the 'Peaceful Uses of Atomic Energy' and 'Man on the Moon' programs."[18] Such a commitment would have amounted to billions of dollars a year in funding. A huge gap existed, moreover, between the House committee's recommendations and what it delivered. Actual spending on solar power in the 1974 fiscal year amounted to approximately $12 million, less than a tenth of the committee's request.[19] Despite the supposed "collusion in the highest echelons," space-based solar power had failed to make much of an impact in either Congress or the White House.[20] The iron triangle remained incomplete.

In a sign of things to come, however, the lack of congressional interest in space-based solar power did not signal the end for the scheme. Instead, administrators at NASA used money from the space agency's own budget to commission further studies of a possible SPS system, and in 1974 Glaser headed a group that produced a "Feasibility Study of a Satellite Solar Power Station." This and other studies contemplated the specific details of an SPS system, including spacecraft needs, costs, and a project schedule. According to Kraft, however, the ongoing research did not amount to much; the space agency engaged think tanks to produce such long-range reports each year, but they represented only a tiny fraction of NASA's budget and rarely make it past the dream phase. The SPS program thus fell into a kind of program limbo, awaiting someone to rescue it from obscurity.

If Glaser brought SPSs out of science fiction, Kraft, the charismatic director of NASA's Manned Spaceflight Center (later renamed the Johnson Space Center), rescued them from the bureaucratic netherworld. In

an interview on the subject, Kraft remembered that the SPS studies came
to his attention because "Bob Pylan and the people in the Engineering
Directorate at JSC were thinking about what I wanted: an expanded
use of space," and specifically a follow-up program for the space shuttle
that NASA was in the process of developing. The director described his
motives as "ulterior"; although Kraft liked the idea of helping to solve
"the big energy problem," he essentially "saw the solar power satellite as a
test program that would allow us to produce other types of vehicles than
just the Space Shuttle," including "vehicles like space tugs and smaller
space stations with the capability to do things in orbit and go to and from
geosynchronous orbit." A new spacecraft program would also have paid
obvious dividends for NASA's major contractors, and Kraft found a ready
ally in Gordon Woodcock from Boeing.

Boeing Aerospace Company's "Solar Power Satellite System Definition
Study," released in 1979 and overseen by Woodcock, demonstrates Kraft's
point; the study's description of the system reads like a NASA and Boeing
wishlist. Since the SPS project envisioned using extremely large satellites,
with many square kilometers of solar panels affixed to a truss or frame,
NASA would have needed to get a lot of materials into orbit. The System
Definition Study concluded, therefore, that the space agency would need a
new vehicle (the study refers to it as a "heavy lift launch vehicle") capable
of lifting 230 tons of payload into near-Earth orbit.[21]

According to the report, NASA would also have had to construct a new
infrastructure in space. As the report pointed out, the components of an
SPS would have to travel into orbit in some disassembled or compacted
form. An operational program would therefore have included "not only
the satellite, but also space construction and support systems: a base in
low Earth orbit (LEO base) for construction . . . and for service as a space
transportation and logistics base" and "a base in geosynchronous orbit
(GEO base) that constructs [the satellites] and serves as a maintenance
support base," as well as "mobile maintenance systems that visit operating
SPSs to provide periodic maintenance" (the "space tugs" mentioned by
Kraft).[22] In short, the program requirements described in Boeing's System
Definition Study would require no less than two new vehicles and two new
space stations.

To Kraft, it felt like a good time to pitch a big, new project. The incoming presidential administration of Jimmy Carter expressed far more interest

in solar energy than its immediate predecessors. President Carter spoke repeatedly about an energy crisis facing the nation in speeches delivered both during the presidential campaign and after he took office. In April 1977 he stated in a televised address that, with the exception of war, the energy crisis represented "the greatest challenge our country will face during our lifetimes" and declared a new U.S. commitment to develop "the new, unconventional sources of energy we will rely on in the next century." President Carter included solar power in this assessment, suggesting that the country should make it a goal to "use solar energy" in more than 2.5 million houses by 1985.[23]

SPSs seemed poised to take off in 1977. Kraft put an enormous amount of energy into promoting the project. He managed to get a meeting with two officials from the incoming Carter administration: Bert Lance, director of the Office of Management and the Budget, and James R. Schlesinger, Carter's "special advisor on energy" and eventually the first secretary of the Department of Energy. The meeting bore promising fruit; Kraft got the go-ahead to put together a working group of people from NASA and the Energy Research and Development Administration (folded into the Department of Energy in the fall of 1977) "to canvas the energy world and see what technology needed to be developed or expanded upon" to make large-scale solar power feasible. Kraft also promoted the idea in other ways, such as giving lectures in the United States on the subject of space-based solar power and helping to organize an international symposium on alternative energy, headlined by Prince Charles, that featured SPSs.

Unfortunately for Kraft, not everyone shared his excitement. Although he met with incoming White House officials to pitch the concept, he ultimately had to win over not the White House but a new boss in his own agency—Robert A. Frosch, appointed by President Carter as the new administrator of NASA. In many ways, the new administrator looked like a potential ally. He had experience with big government research programs and an interest in environmental issues, having previously served as assistant secretary of the navy for research and development and as the assistant executive director of the United Nations Environment Programme. Frosch's experience could have made him an influential advocate. When Kraft briefed the new administrator on the program, however, he found an unreceptive audience. In Kraft's words, "[Frosch] thought I was 'nuts,'" and the administrator told his subordinate that, "if I had a thought of going to the Congress or the President [to ask for support for the SPS plan], I was crazy."

Given its potential benefits for NASA, why did Frosch react so badly to the proposal? Frosch came from outside the space agency, and the president who appointed him, Jimmy Carter, had serious reservations about the value of the manned spaceflight program.[24] Kraft may have looked at the proposal and seen possibilities for new funding, but Frosch saw something else: the technical, economic, and environmental barriers to success. Given the enormity of these challenges, Frosch concluded that he could not sell the program to a skeptical, and potentially hostile, audience in the White House or Congress.

From its inception, major technical barriers bedeviled the SPS concept. In his original article, Glaser had recognized that the photovoltaic technology of his time had a maximum theoretical efficiency of only 24 percent, meaning that solar cells could convert only 24 percent of the energy they received into electricity. For the system he envisioned, however, Glaser assumed that the development of "organic compounds" would allow for cells with at least 80 percent efficiency and significantly lower production costs.[25] Glaser's prediction failed to materialize; cheap, high-efficiency cell technology remained elusive throughout the twentieth century.

Another technical barrier involves the price of solar arrays. Writing in 1972, the Solar Energy Panel ascribed "special importance" to reducing the cost of solar arrays (the large, electricity-generating sheets of solar cells) to an approximate unit cost of $32 per square foot. As of 1972, according to the report, "prices for space flight qualified solar arrays range from $1000 to $8000" per square foot, meaning that the report predicated the success of an SPS program on a 97 percent reduction in the cost of its core component. Moreover, even if the program could achieve this price reduction in the cost of solar panels, the report's authors still speculated that the system would generate electricity at a unit cost of 38 mills per kilowatt-hour (mills/kWh; expressed in monetary terms as $.038/kWh).[26] Electricity for residential use cost, on average, $.024/kWh in 1972, so the price of electricity would have had to increase by half before the system envisioned by the report could compete economically with other forms of electricity generation. Writing at a time of relative scarcity, the authors may have assumed that the price of electricity would go up, but that assumption was incorrect. Inflation-adjusted electricity prices between

1972 and 2008 never rose to so high a level as to make the projected cost of satellite-delivered power competitive.[27]

Of course, a substantial government investment might have overcome the potential technical and economic challenges of the project. The United States had, after all, successfully undertaken other major technological challenges, such as the Manhattan Project and the race to put a man on the moon, which cost phenomenal sums of money, faced major technical barriers, and yielded relatively little immediate economic benefit. Moreover, both of these programs were sold to the public and to Congress at least partially as matters of national security and prestige—the same concerns raised by President Carter at the beginning of his term in office. But the various studies revealed that the SPS concept had problems beyond just technical barriers and cost. These power-generating satellites, it turned out, had potentially serious consequences for the environment.

The "SPS Concept Development and Evaluation Program Plan," a working paper released in February 1977 (just four months before Frosch took over at NASA), exemplified the types of environmental questions that arose with regard to the program. The paper was part of a larger effort to provide "a 'best' reference conceptual system design" for an SPS system and to assess barriers facing the program, including "environmental impacts."[28] In the section "Atmospheric, Ionospheric, and Magnetospheric Degradations" the authors tackled one of the least-understood technologies envisioned by the system: the process of transferring energy from orbiting satellites to ground stations using beams of microwave radiation.

To be clear, the authors of the study repeatedly and forcefully asserted that the proposed system would be safe. Unfortunately, the details of the report belied those assertions; the authors could not, for example, rule out the possibility that microwave beams would cause atmospheric changes. In particular, they found that microwave radiation could "affect the stratosphere's ozone layer density with the risks of increased ultraviolet transmission," and they concluded that the beams could "cause weather modification by heating of the troposphere." Moreover, they buried one of the biggest questions about the safety of the program at the end of the paragraph on atmospheric heating. Their study looked at the possible consequences of a single satellite, but by 1977 the solar power plan called for an eventual total of two hundred or more satellites beaming down radiation—a fact the study's authors admitted might "exacerbate" any problems.[29]

The working paper's "Biological/Ecological Effects" section also raised questions. As the report noted, considerable controversy existed in the 1970s around the level of microwave radiation exposure acceptable for human safety. The United States defined the acceptable level as 65 mW/in^2, basing its standard on the point at which microwave radiation has "tissue heating potential," but the study's authors noted that microwave ovens have to meet a much lower threshold, 1 mW/cm^2, "to account more conservatively for long-term exposures." Obviously, this standard might also apply to radiation beamed to Earth by satellites. The Soviet Union, moreover, had publicly proposed a much more stringent standard of 0.065 mW/in^2, which, according to the Soviets, reflected the radiation's "additional presumed effects on the nervous system." Other countries had adopted a variety of standards between the extremes. Aside from the issue of human exposure, the report's authors also had to grapple with the question of the system's impact on local animal populations. They anticipated that birds, for instance, would "avoid the higher power density region as a consequence of the heating effects they would experience." They worried, however, that some animals might find themselves attracted, "with possibly long-term deleterious effects, to the pleasant warmth created at some distance from the beam center."[30]

In their report, the writers offered several responses to the radiation concern. They concluded that the beam of microwave radiation passing between SPSs and their ground stations might "require large exclusion areas . . . to protect other satellites and, perhaps, high altitude aircraft." According to the study, however, "a problem would exist in only a 10 mile (15 km) radius about the beam," and the United States possessed enough land to accommodate power stations set in remote locations with large exclusion zones around their borders. In fact, the enthusiastic authors went so far as to suggest that "mesh shielding could be employed within at least a portion of this radius to reduce ecological and biological effects and enable use of the land beneath for agriculture or industry."[31]

The authors again had to concede, however, that the proposed scope of the program—to reiterate, "200+ beams" of microwave radiation—raised some concerns, since "the low level radiation from the beams of many satellites could cumulate." Incidental scattering from the beams, it turned out, would perpetually bathe most of the hemisphere beneath the satellites in low-level microwave radiation, at a level of roughly 0.03 mW/cm^2. The degree to which Canadians, Central and South Americans, and Europeans would appreciate this side-effect of the system remains unconsidered in the report, but its authors did concede that 0.03 mW/cm^2 exceeded the Soviet standard.[32]

When added to the technical and economic barriers, the environmental concerns about radiation exposure and atmospheric alteration produced a trifecta of problems that would have challenged the survival of any program. It had a particularly lethal impact on the SPS scheme, however, because the particular combination of technical and economic concerns on the one hand and environmental concerns on the other essentially cut the program off from every constituency that might have supported it. To understand why, it helps to consider the larger debates taking place at the time over the issue of solar power.

Prior to World War II, solar research mostly took place in the laboratories of individual inventors like John Ericcson and Frank Shuman, but by the middle of the twentieth century solar power technology had moved beyond the purview of a few famous inventors and developed a larger constituency. That community incorporated a variety of people—academics and industrial scientists, environmental lobbyists and private enthusiasts, think tanks and technical journals—who came together because of their common interest in solar energy. Together, these people formed what John Kingdon has described as a "policy community."[33]

At the time of the energy crisis, according to historian Frank Laird, the solar energy policy community divided roughly into two camps, with the cleavage based on differing beliefs about the viability of solar technology and the type of society that solar energy would support. The "conventional advocates" regarded solar energy skeptically, believing that it would make only a small contribution to society in the near future. These advocates generally took a long view of solar power, casting it as a replacement technology for a future point when existing energy sources ran out, and they worried about "[compromising] their credibility as serious scientists and engineers" with support for "technologically immature systems." What Laird describes as "nonconventional advocates," by contrast, had a much more ambitious and hopeful belief that solar energy technologies, given massively increased financial support, could have a substantial impact on the energy economy within a few decades. They believed, moreover, that the adoption of solar technologies would create a world very different "from the status quo, in political and social as well as technological terms." In the 1970s, in particular, newcomers to the solar power policy community with ties to the rapidly growing environmental movement dramatically increased the ranks, and the clout, of the nonconventionalists.[34]

The particular combination of challenges facing the SPS system proposed in the NASA studies managed to accomplish the impressive feat of making the idea simultaneously unpalatable to both camps in the solar power community. For the skeptical conventionalists, the project's unanswered technical questions made it an unacceptably expensive and risky proposition. The nonconventionalists, by contrast, found the program's potential environmental costs unbearable. This unanimity of disapproval helps to explain the depth of negativity in Frosch's reaction to the scheme. He looked at the NASA studies and concluded two things: he could not sell the program to a skeptical White House and Congress, and Kraft's vocal support for a scheme with so many detractors might ultimately attract criticism for the agency. According to Kraft, that internal rejection at NASA eventually spelled the end of the SPS program as a going concern.

Looking back at the program, it might strike some readers as strange that it took so long for the leadership at NASA to pull the plug on such a questionable scheme and even stranger that the idea has experienced its recent resurgence in popularity. As Kraft himself pointed out, some people supported the project because they saw it as a means to an end, but that analysis does not do enough to explain the persistence of an idea that had made little or no progress by the end of the century. To explain that longevity, one must turn away from the motivations of individual supporters of SPSs and think instead about the general arguments made in favor of space-based solar power. Three things stand out in particular.

First, advocates sold SPSs as magic bullets—technology to replace a complicated scarcity problem with a new age of abundance. At a time when the federal power commissioner warned of electricity shortages, Peter Glaser promised "an era in which an abundance of power could free man from his dependence on fire," and most of the periodic resurgences of interest in space-based solar power have coincided with other moments of scarcity.[35] The technological promise of abundance speaks directly to what

historians of technology have described as a "widespread American confidence in the 'technological fix'—faith in technology to solve all problems
with less effort and cost than required to change social behavior or political
realities."[36]

Space-based solar power also stayed around because the idea sounded
less revolutionary than it was. The key components of the proposed system, such as photovoltaic cells, spacecraft, and microwave transmitters, all
existed in 1968, even though none of them could live up to the needs of the
proposed system. Glaser emphasized this point when he argued that space-
based solar power would not "require the discovery or development of new
physical principles" and that solutions for the systems' biggest problems
lay "within the projected capabilities of systems engineering."[37] Pitched
as a simple process of upgrading existing technology, technical challenges
often sound less daunting. In the area of nuclear power, for example, "fast-
breeder" reactors operated in test facilities as early as the 1950s and generated a good deal of interest during the energy crisis, but they never quite
made the transition to commercial viability despite continued government
support.

A third important advantage for this program lay in its association with
the popularity of solar energy technology. In 1977, just as Kraft began
his big push for the SPS program, the magazine *Science* captured popular
sentiment about solar energy when it proclaimed photovoltaic cells "at
once the most sophisticated solar technology and the simplest, most environmentally benign source of electricity yet conceived." As demonstrated
by a 2008 CNN story describing satellites that could "electromagnetically
beam gigawatts" of "clean, renewable" energy back to Earth, even in the
face of contrary evidence, space-based solar power continued to benefit
from the perception of environmental cleanliness.[38]

SPSs as real-world technology failed to make much progress in the
face of major challenges to implementation, but space-based solar power
as an idea has enjoyed a measure of success based in part on technological optimism, its connection to existing technology, and positive popular
sentiment about solar power. This success might provide a useful example
for advocates of other alternative energy technologies to use in selling their
ideas. It also suggests, however, that Americans should regard promises of
technological deliverance with a healthy dose of skepticism. The apparent
elegant simplicity of the SPS idea—just take existing photovoltaic panels

and put lots of them into orbit, where the sun always shines!—concealed a variety of problems which, taken together, made space-based solar power an expensive, questionable strategy for solving U.S. electricity needs.

This critique does not apply only to alternative energy technologies. Advocates of oil often tout the low cost of petroleum without attempting to account for the environmental cost of carbon emissions or the social cost of doing business with unsavory regimes, and the continued viability of the nuclear power industry is based in part on utilities' expectation that they will eventually pass along the economic and environmental burdens of radioactive waste to taxpayers. The most important lesson that we can learn from the history of space-based solar power is one that applies to most forms of energy production: the highest costs in a system are often the least obvious ones.

Notes

I thank the Johnson Space Center History Office for its generous support of my research, some of which is reported here.

1. National Security Space Office (NSSO), "Space-Based Solar Power as an Opportunity for Strategic Security: Phase 0 Architecture Feasibility Study," Release 0.1, National Security Space Office (Oct. 10, 2007), www.nss.org/settlement/ssp/library/final-sbsp-interim-assessment-release-01.pdf; NSSO, "Special Report—Space-Based Solar Power: Inexhaustible Energy from Orbit," *adAstra*, Spring 2008; Lara Farrar, "How to Harvest Solar Power? Beam It Down from Space!" *CNN.com*, May 30, 2008, www.cnn.com/2008/TECH/science/05/30/space.solar/index.html; David R. Baker, "California's New Power Source a Solar Farm," *San Francisco Chronicle*, Apr. 14, 2009, *SFGate.com*, www.sfgate.com/cgi-bin/article.cgi?f=/c/a/2009/04/14/MN7S171PSL.DTL.

2. Isaac Asimov, *I, Robot* (New York: Bantam Books, 1991 [Gnome Press, 1950]), 56–59; Arthur C. Clarke, "Extra-Terrestrial Relays: Can Rocket Stations Give World-Wide Radio Coverage?" *Wireless World* 51 (Oct. 1945): 305–308. Clarke's prediction essentially came true with the launch of the first communications satellites.

3. William Ledbetter, "An Energy Pioneer Looks Back," *in* NSSO, Special Report, 26–27; Peter Glaser, "Power from the Sun," *Science* 162, no. 3856 (1968).

4. Glaser, "Power from the Sun," 857–58.

5. Ibid., 859–61.

6. On Etzler, see Lamont C. Hempel, "The Original Blueprint for a Solar America," *Environment* 24, no. 2 (1982): 25–32. John Ericsson, *Contributions to the Centennial Exposition* (New York: Printed by the author, 1876), 577, quoted by John Perlin, *From Space to Earth: The Story of Solar Electricity* (Ann Arbor: aatec, 1999), 5. For the complete story of Frank Shuman, see Frank T. Kryza's book *The Power of Light: The Epic Story of Man's Quest to Harness the Sun* (New York: McGraw-Hill, 2003).

7. For further information on the technical challenges faced by electricity providers, see Robert H. Williams, "The New Problems of Energy Supply," in *The Hazards of the International Energy Crisis: Studies of the Coming Struggle for Energy and Strategic Raw Materials,* ed. David Carlton and Carlo Schaerf (London: Macmillan, 1982), 19–25.

8. Eileen Shanahan, "F.P.C. Chief Sees More Power Cuts," *New York Times,* July 28, 1969; see also David Bird, "Fuel Shortage May Force Rationing, Swindler Warns," *New York Times,* Sept. 19, 1970.

9. "Special Message to the Congress on Energy Resources," and "Remarks About a Special Message to the Congress on Energy Resources," both accessible via the American Presidency Project website, University of California–Santa Barbara, www.presidency.ucsb.edu/ws/?pid=3038, and www.presidency.ucsb.edu/ws/?pid=3037.

10. National Science Foundation and National Aeronautics and Space Administration, *An Assessment of Solar Energy As a National Energy Resource,* NSF/RA/N-73–001 (Washington, D.C.: Government Printing Office, 1972), 1.

11. Tom Buckley, "NASA's Tom Paine—Is This a Job for a Prudent Man?" *New York Times,* June 8, 1969.

12. NSF/NASA, *Assessment of Solar Energy,* 84.

13. This and following reference to Kraft's comments are drawn from the author's interview with Kraft, Jan. 8, 2009.

14. "Solar Energy Study Is Planned by NASA," *New York Times,* July 9, 1972.

15. NSF/NASA, *Assessment of Solar Energy,* 60–62.

16. Brian Balogh, *Chain Reaction: Expert Debate and Public Participation in American Commercial Nuclear Power* (New York: Cambridge University Press, 1991), 62–63.

17. Ray Reece, *The Sun Betrayed: A Report on the Corporate Seizure of U.S. Solar Energy Development* (Montreal: Black Rose Books, 1979), 4, 31.

18. U.S. House of Representatives, Committee on Science and Astronautics, "Solar Energy Research," Staff report, 92d Congress, 2d Session, Serial Z, Dec. 1972, 3–4, 8–9; NSF/NASA, *Assessment of Solar Energy;* 63.

19. U.S. Congress, House, Committee on Science and Astronautics, Subcommittee on energy, "Solar Energy for the Terrestrial Generation of Electricity," Hearings, 93d Congress, 1st Session (June 5, 1973), 1.

20. For a more complete discussion of the numbers and the testimony, see Frank N. Laird, *Solar Energy, Technological Policy, and Institutional Values* (Cambridge: Cambridge University Press, 2001), 159–60.

21. National Aeronautics and Space Administration, "Solar Power Satellite System Definition Study," Vol. 1, Phase 2, Final Report, Seattle: Boeing Aerospace Company (Nov. 1979), 5, [IA 1285, Box 177, Center Series, New Initiatives Office] JSC History Collection, University of Houston–Clear Lake. By comparison, the space shuttle orbiter only carried payloads of 65,000 lbs., or approximately 32.5 tons, into orbit.

22. NASA, "SPS System Definition Study," 7.

23. James Carter, televised speech on the president's proposed energy policy, Apr. 18, 1977. Full text and video can be accessed online from the Miller Center of Public Affairs, University of Virginia, at http://millercenter.org/scripps/archive/speeches/detail/3398.

24. Andrew J. Dunar and Stephen P. Waring, *Power to Explore: A History of Marshall Space Flight Center, 1960–1990* (Washington, D.C.: NASA History Office, 1999), 310. Carter's vice president, Walter Mondale, had an even less charitable view of the space administration and inserted several like-minded individuals into the Office of Management and Budget, which oversaw the agency's funding.

25. Glaser, "Power from the Sun," 860–61.

26. NSF/NASA, *Assessment of Solar Energy*, 46–63.

27. All electricity price figures taken from the Energy Information Administration, www.eia.doe.gov/emeu/aer/elect.html.

28. National Aeronautics and Space Administration, "SPS Concept Development and Evaluation Program Plan," Vol. 1, Working Paper (Feb. 1977), 46 [IA 1246, Box 176, Center Series, New Initiatives Office], JSC History Collection, University of Houston–Clear Lake.

29. Ibid., 64.

30. Ibid., 65, 67. It did not escape the paper's authors that the Soviet standard might have more to do with intentional obstructionism than with actual concerns; they noted in a footnote that the Soviet number "may in fact be less formal than a legal 'standard.'"

31. Ibid., 65.

32. Ibid., 67.

33. John W. Kingdon, *Agendas, Alternatives, and Public Policies*, 2nd ed. (New York: HarperCollins, 1995).

34. Frank Laird, "Advocating Energy Technologies in the Cold War," *Technology and Culture* 44, no. 1 (2003), 32. The full article offers a much deeper analysis of the two camps and the differences that separated them.

35. Eileen Shanahan, "F.P.C. Chief Sees More Power Cuts," *New York Times*, July 28, 1969; Glaser, "Power from the Sun," 861.

36. Gary Cross and Rick Szostak, *Technology and American Society: A History* (Englewood Cliffs, N.J.: Prentice-Hall, 1995), 311. For an insightful discussion of the issues of abundance, scarcity, and energy use in American society, see Martin Melosi, *Coping with Abundance: Energy and Environment in Industrial America* (New York: Knopf, 1985).

37. Glaser, "Power from the Sun," 861.

38. Allen Hammond, "Photovoltaics: The Semiconductor Revolution Comes to Solar," *Science* 197, no. 4302 (1977), 445; Farrar, "How to Harvest Solar Power?

Chapter 9

The Nuclear Power Debate of the 1970s

J. Samuel Walker

K imberly Wells, a reporter for a Los Angeles television station, was an unlikely source for a major story on the hazards of nuclear power. She specialized in lightweight human interest pieces, such as a feature on singing telegrams, rather than issues as controversial and complex as nuclear power during the 1970s. But after inadvertently witnessing a near accident at the Ventana nuclear power station in California, she learned about the dangers of a core meltdown. Greg Minor, a nuclear engineer, told her not only that she was "lucky to be alive" but that "we might say the same for the rest of southern California." Elliot Lowell, a physics professor who opposed nuclear power, affirmed this frightening assessment. He claimed that, if the fuel rods in the core of a nuclear reactor overheated, they would melt through the floor of the plant in a matter of minutes and release enough radioactivity to "render an area the size of Pennsylvania permanently uninhabitable."

Spurred by her new knowledge of the dangers of nuclear power, Wells, at considerable risk to her own career, joined forces with a television cameraman, Richard Adams, to find out what happened at Ventana the day she had visited. Eventually she gained the confidence of Jack Goodell, a nuclear engineer and control room supervisor, who had discovered shoddy construction practices that imperiled the plant. After officials of California Gas and Electric, the utility that owned the plant, refused to heed Goodell's warnings and publicly dismissed the near accident as a routine malfunction, Goodell decided that drastic measures were necessary. He took over the control room by grabbing a guard's pistol and then threatened to "flood containment with radiation." Utility executives were informed that this action would destroy the plant. While Goodell enlisted Wells to broadcast the dangerous conditions at the plant and a throng of reporters descended on the site, the utility sent an armed police squad to force its way into the locked control room. As the tension escalated, the police team rushed into the control room and killed Goodell in a burst of gunfire. Wells inherited the responsibility of revealing the causes of the showdown at the Ventana reactor.

This was the plot of the motion picture *The China Syndrome*, which opened in more than six hundred theaters across the United States on March 16, 1979.[1] The title referred to a tongue-in-cheek term that nuclear experts had coined during the 1960s to describe an accident in which an overheated core would melt through the bottom of a plant and presumably through the earth's core toward China. The film starred Jane Fonda as Kimberly Wells, Jack Lemmon as Jack Goodell, and Michael Douglas as Richard Adams. It received many favorable reviews as a suspenseful and entertaining thriller and quickly became a box office smash. Within two weeks, Columbia Pictures reported that *The China Syndrome* had produced the highest income of any film it had ever released in a nonholiday period.[2]

Although *The China Syndrome* commanded a great deal of attention because of its entertainment value, it also received wide notice because of its unflattering depiction of nuclear power. It was the latest salvo in a bitter national debate that had intensified throughout the 1970s. The film makers denied that they set out to present a strongly anti-nuclear message. Fonda claimed that the movie was "intended as an attack on greed, not on nuclear energy." Nevertheless, director James Bridges conceded that it was "not impartial," and Mike Gray, who conceived the project and coauthored the screenplay, wrote privately that it explained "the fundamental horrors of nuclear technology." A reviewer in *The Progressive* predicted that the film's treatment of nuclear power would have a major influence on public opinion: "Simply put, *The China Syndrome* is an incendiary piece of work that promises to cripple if not destroy whatever effect two decades of

Nice Mr. Nuclear ads have had in making the public receptive to nuclear power."[3]

While nuclear critics hailed the movie as a way to alert the public to the hazards of the technology, nuclear proponents complained that it greatly exaggerated the risks. Supporters of nuclear power agreed that a major reactor accident could, in the worst case, release large amounts of radiation into the environment. But they took issue with the film's suggestion that the China syndrome would inevitably result if the core of a plant overheated. Although such an accident was possible, they insisted that it was highly unlikely. Pro-nuclear groups sent information packets to news organizations to counter the manner in which the technology was presented in *The China Syndrome*. An executive for a nuclear utility, perhaps with Jack Goodell's nonsensical threat to "flood containment with radiation" in mind, charged that the movie had no scientific credibility. The divisions over *The China Syndrome* strikingly demonstrated the passions of the controversy over nuclear power that played a prominent role in discussions of energy policies in the United States during the 1970s.[4]

The Growth of Commercial Nuclear Power

The spirited contest over nuclear power during the decade before the release of *The China Syndrome* contrasted sharply with the widespread support that prevailed when the technology was introduced commercially during the 1950s. In 1954, Congress passed the Atomic Energy Act, a law that made the development of nuclear power possible by allowing for the first time the dissemination of basic information about atomic energy for civilian applications. It hoped to encourage the rapid growth of a nuclear industry that would provide a new source of electrical power. But privately owned utilities, though interested in exploring the use of nuclear technology to meet future energy needs, did not, for the most part, rush to build nuclear plants. Their response was lukewarm for several reasons. First, supplies of conventional fuel for the production of electricity, especially coal and oil, were plentiful and cheap. There was no pressing need for nuclear power to meet energy requirements for the near future. Further, many technical and economic questions about the technology remained to be answered. Although experiments with government reactors had established the technical feasibility of using nuclear energy to generate electricity, those tests did not prove that the technology could meet the demands of commercial power production. Finally, utilities were cautious because of nuclear power's potential hazards. Experts regarded the chances of a disastrous nuclear accident to be remote, but they did not dismiss them entirely.[5]

Burdened by those drawbacks, the nuclear power industry did not grow as quickly as its advocates had hoped. Nevertheless, the electrical power industry viewed nuclear technology as an important long-term source of energy, and many utilities took steps to investigate its potential. By 1962, six relatively small privately owned nuclear plants were generating power on the commercial grid, and several more reactors were under construction or on order. Although those plants were far from being technologically proven or commercially competitive, they signaled substantial progress since passage of the Atomic Energy Act. Despite some lingering uncertainty about the prospects for the industry, the omens were good. Public attitudes toward nuclear power were strongly favorable. A public opinion poll taken in February 1956 showed that 69 percent of those questioned had "no fear" of having a nuclear plant located in their community; only 20 percent expressed concern. In the spring of 1960, among responses to a national telephone survey on the proposition that "atomic power should be used to produce electricity," 64 percent were positive and only 6 percent were negative. In October 1963, *Nucleonics*, a monthly trade journal, suggested that "public acceptance of nuclear power would not be a significant problem."[6]

The first few years of the nuclear industry's existence were generally tranquil; the growth of nuclear power occurred at a measured pace in a stable and supportive political environment. In the mid-1960s the trends suddenly and unexpectedly changed. The industry experienced a surge in reactor orders, which Philip Sporn, past president of the American Electric Power Service Corporation, described in 1967 as the "great bandwagon market." The boom in the nuclear industry was accompanied by increasingly visible and vocal opposition to nuclear power that soon created a major national controversy.[7]

The bandwagon market was an outgrowth of several developments that enhanced nuclear power's appeal to utilities in the mid- and late 1960s. One was the intense competition between the two leading builders of nuclear plants, the General Electric Company and the Westinghouse Electric

Corporation. In 1963, General Electric made a daring move to increase its reactor sales and to convince utilities that nuclear power had arrived as a safe, reliable, and cost-competitive alternative to fossil fuel. It offered a "turnkey" contract to Jersey Central Power and Light Company, a subsidiary of the General Public Utilities Corporation, to build a 515-megawatt nuclear plant near Toms River, New Jersey. For a fixed cost of $66 million, General Electric agreed to supply the entire Oyster Creek plant to the utility (the term "turnkey" suggested that the utility would merely have to turn a key to start operating the facility). The company's bid was successful, winning out over not only Westinghouse but also manufacturers of coal-fired units. General Electric expected to lose money on the Oyster Creek contract but hoped that the plant would help stimulate the market for nuclear power.

The Oyster Creek contract opened the "turnkey era" of commercial nuclear power and came to symbolize the competitive debut of the technology. Westinghouse followed General Electric's lead by offering turnkey contracts for nuclear plants, setting off a fierce corporate battle. The turnkey plants were a financial drain for both companies; their losses ran into the hundreds of millions of dollars before they stopped making turnkey arrangements. One General Electric official commented: "It's going to take a long time to restore to the treasury the demands we put on it to establish ourselves in the nuclear business." But the turnkey contracts fulfilled General Electric's hopes of stirring interest among utilities and played a major role in triggering the bandwagon market.[8]

Other important considerations helped convince a growing number of utilities to buy nuclear plants during the latter part of the 1960s. One was the spread of power-pooling arrangements among utilities, which encouraged the construction of larger generating stations by easing fears of excess capacity and overexpansion. A utility with extra or reserve power could sell it to other companies through a widening network of interconnections. The desirability and feasibility of using larger individual plants benefited nuclear vendors, who emphasized that bigger plants would produce "economies of scale" by cutting capital costs per unit of power and improving efficiency. This helped to overcome a major disadvantage of nuclear power reactors relative to fossil fuel: the significantly heavier capital requirements for building them. During the late 1960s, designs for nuclear facilities jumped from the 500 to 800 to 1,100 electrical megawatt range even though operating experience was still limited to units of 200 megawatts or less. The practice of design by extrapolation, or scaling up the size of plants based on experience with smaller facilities, had been employed for fossil-fuel units since the 1950s. It had appeared to work well and, therefore, it was natural for vendors to apply the same procedures to nuclear power plants.

In addition to turnkey contracts, system interconnections, and increasing unit size, growing national concern about air pollution made nuclear power more attractive to utilities. During the 1960s, the deteriorating quality of the environment, including visible evidence of foul air, took on increasing urgency as a public policy issue. Plants that burned fossil fuels provided more than 85 percent of the nation's electricity and contributed heavily to air pollution. Coal, by far the most commonly used fuel for producing power, placed a much greater burden on the environment than other fossil fuels, releasing millions of tons of noxious chemicals into the atmosphere annually. But the demand for electricity was steadily rising, and experts predicted that it would continue to grow at a high rate of 7 percent or more per year. An article in *Fortune* magazine vividly explained the predicament: "Americans do not seem willing to let the utilities continue devouring . . . ever increasing quantities of water, air, and land. And yet clearly they also are not willing to contemplate doing without all the electricity they want. These two wishes are incompatible. That is the dilemma faced by utilities." After the mid-1960s, utilities increasingly viewed nuclear power as the answer to that dilemma. Because nuclear plants did not burn fossil fuels, they did not contribute to air pollution. The trade publication *Nucleonics Week* commented in 1965 that, in comparison with coal, "the one issue on which nuclear power can make an invincible case is the air pollution issue."[9]

A combination of technological, economic, and environmental developments launched the bandwagon market for nuclear power plants. Between 1966 and 1968, utilities committed to purchasing sixty-eight nuclear units, in contrast to the twenty-two they had ordered between 1955 and 1965. After a modest slowdown, the boom in orders resumed in the early 1970s. Utilities bought twenty-three nuclear stations in 1971, thirty-nine in 1972, and a record forty-four in 1973. By the end of 1973, thirty-seven nuclear plants were producing commercial power or had recently received operating licenses, and this number was dwarfed by projections of future requirements. In early 1973 the U.S. Atomic Energy Commission (AEC), the government agency that the 1954 Atomic Energy Act had made responsible for encouraging the growth and regulating the safety of the nuclear industry, estimated that by the year 2000 nuclear units in the United States would increase their capacity to 1.2 million megawatts. This would mean that the total number of plants would grow to more than one thousand. The future looked so bright for nuclear proponents that the trade magazine *Nuclear Industry* suggested the industry had witnessed the "virtual collapse of competition from fossil fuels." *Nucleonics Week* reported in December 1972 that the prevalent attitude among utility officials was "one of enthusiastic optimism." It added: "Utility executive after executive repeats the theme that nuclear power is the only way to go for future generation needs."[10]

The Nuclear Power Slump

The soaring optimism of the nuclear industry soon proved to be short-lived. The immediate cause of the downturn in the industry's fortunes, ironically, was an energy crisis that emerged as a prominent national issue in the fall of 1973. As early as 1971, President Richard M. Nixon publicly expressed concern about the long-range energy needs of the United States, and in June 1973 he issued a statement on energy in which he declared, "America faces a serious energy problem." A short time later a war between Israel and its Arab neighbors led to an oil embargo by the Organization of Arab Petroleum Exporting Countries, which gave much greater urgency to Nixon's efforts to increase the nation's energy supplies. In a televised speech on November 7, 1973, he called for a program he named Project Independence, so that "by the end of this decade we will have developed the potential to meet our own energy needs without depending on any foreign energy sources." A key component of Project Independence was a major expansion in nuclear power. After succeeding Nixon as president, Gerald R. Ford placed the same emphasis on boosting nuclear power production to keep up with national energy requirements. In his State of the Union address on January 15, 1975, he set a goal of having "200 major nuclear power plants" in operation within ten years, an ambitious objective even if less lofty than the AEC's projections two years earlier.[11]

The energy crisis initially seemed likely to spur further growth in nuclear power. Industry executives expressed confidence in early 1974 that "the energy crisis will result in a net increase in utility commitment to nuclear power." But this prediction turned out to be woefully inaccurate. The energy crisis severely damaged the electric power industry and, by extension, the nuclear industry in two ways. It quickly and sharply drove up the price of oil and other fuels that utilities purchased to run their plants, which drained their financial resources. It also exacerbated the already serious problem of inflation, which greatly increased the cost of borrowing money for plant construction. At the same time, an economic slump and increasing unemployment curtailed demand for electricity, which grew at a substantially slower rate than experts had anticipated. As expenses skyrocketed and markets diminished, utilities postponed or canceled plans to build many new plants. Although utilities cut back on both coal and nuclear projects, the blow fell disproportionately on builders of nuclear units because of higher capital costs. By September 1974, 57 of 191 nuclear plants under construction, under licensing review, on order, or announced by utilities had been delayed, generally by a year or two but sometimes by several years, and a few had been canceled altogether. Fourteen months later, 122 nuclear projects had been deferred and nine canceled. Sales of new plants declined sharply from the peak levels of the early 1970s; between 1975 and 1978, U.S. utilities ordered only eleven nuclear units.[12]

The nuclear industry's suddenly worsening prospects shook its confidence and, within a short time, threatened its financial well-being. *Nucleonics Week* informed its readers in January 1975 that the industry was in a state of "utter chaos," and that utilities had "no idea how to finance nuclear plants." A few months later *Nuclear Industry* found a "scene of almost unrelieved gloom and anger" at a meeting sponsored by the Atomic Industrial Forum, an organization that promoted industrial applications of nuclear energy. The mood had not improved by December 1977, when the same magazine reported that the "collective frame of mind" of delegates at the Atomic Industrial Forum's annual conference was "as appropriate for wringing hands as for shaking hands." Nuclear industry officials were deeply troubled by evidence that a growing number of utilities, especially those in precarious financial condition, were turning away from the nuclear option in favor of building coal-fired plants. The forecasts for the expansion of nuclear power were much less promising than those of the early

1970s. Whereas in 1973 the AEC had predicted that more than a thousand nuclear plants would be operating by the year 2000, just five years later the recently created U.S. Department of Energy reduced the estimate to five hundred at most and perhaps only about two hundred. In March 1978, *Business Week* suggested that the outlook for the nuclear industry was so bleak that within ten years it was "apt to contract dramatically and it may collapse altogether."[13]

Expanding Opposition to Nuclear Power

The rising costs and slowing demand that plagued the nuclear industry on the financial front were compounded by increasing controversy on the political front. To industry officials, the emerging anti-nuclear movement—described by one executive as populated with "latter-day Luddites"—was the "root of the political problem." Eroding support for and growing protests against nuclear power were closely tied to increasing public fear of exposure to radiation. This, in turn, was a direct result of a major scientific debate during the late 1950s and early 1960s over the effects of radioactive fallout from aboveground nuclear bomb tests by the United States, the Soviet Union, and Great Britain. The tests spread fallout to populated areas far from the test sites and ignited a highly publicized controversy over the hazards of low levels of radiation. Although it was clear that exposure to heavy doses of radiation was harmful, the risk of exposure to low doses was a source of uncertainty and sometimes sharp disagreement among scientists. The fallout debate moved the issue of low-level radiation hazards from the realms of scientific and medical discourse to the popular realms of newspaper reports, magazine stories, and political campaigns. For the first time, it became a matter of sustained public concern.[14]

The Limited Test Ban Treaty of 1963, which prohibited nuclear atmospheric testing by its signatories, effectively ended the fallout debate as a prominent public issue. But it did not dispel public concern about low-level radiation that the fallout controversy had fostered. This was evident in public protests against the construction of several nuclear power plants in the early 1960s. Critics of the proposed Ravenswood plant in New York City and the proposed Bodega Bay and Malibu plants in California cited the dangers of radiological contamination that might result if the reactors were built. Citizen objections to those projects played an important role in their eventual termination. Organized opposition to nuclear power remained sporadic and localized; it focused on conditions that applied to a particular area, such as the density of the population in New York and the threat of earthquakes at the California sites. Nevertheless, it was disquieting for nuclear power proponents. In June 1963, *Nucleonics Week* advised the nuclear industry and the government to deal "fully and forthrightly" with public concern about the technology. "If the public does not accept nuclear power," it warned, "there will be no nuclear power."[15]

As the nuclear power industry expanded during the late 1960s, so too did challenges to the construction of new plants. The growing objections focused not only on plans for specific sites but also on concerns over the implications of building a large number of reactors. The increase in anti-nuclear activism went hand in hand with the expansion of the industry. Most of the plants built during the bandwagon market years met with little or no opposition, but several triggered strong dissent. Although there was no organized, broad-based movement against nuclear power, the cumulative effect of anti-nuclear activities called attention to reservations about nuclear technology in general.

The growth of the nuclear industry occurred simultaneously, if coincidentally, with the rise of the environmental movement in the United States. Environmentalists recognized the advantages of nuclear power in reducing air pollution, but they became increasingly critical of the technology on other grounds. The view of nuclear power as better for the environment than conventional fuels was undermined in the late 1960s by a major controversy over the effects of waste heat from nuclear plants on water quality, widely known as thermal pollution. The nuclear industry gradually and reluctantly took action to combat thermal pollution by building cooling towers or cooling ponds for plants that lay on inland waterways, but not before it sustained a barrage of attacks that aroused public doubts about the environmental effects of nuclear power. As the thermal pollution question generated criticism, an even more bitter debate over radiation emissions from nuclear plants gained prominence across the nation. Several scientists challenged the prevailing view that the small amounts of radiation released by nuclear plants during normal operation were not a serious problem. They charged that the routine releases were a severe threat to public health that could cause tens of thousands of deaths from cancer every year. The exchange of views over radiation risks stirred further uneasiness about nuclear power, especially among those unable to evaluate the conflicting claims.

Within a short time, the concern about radiation exposure from routine emissions was intensified by even more potent apprehensions about reactor safety. The larger size of individual plants ordered during the late 1960s raised new safety questions and provoked fears of a severe reactor accident that would spew large quantities of radiation into the environment. In the early 1970s, a highly contentious controversy over the performance of emergency core cooling systems in nuclear plants, designed to prevent a core meltdown that could lead to the China syndrome, received coverage in the popular media as well as in technical journals. By highlighting the uncertainties about the technology that its supporters acknowledged, sometimes involuntarily, the controversy enhanced the stature of critics who questioned the safety of nuclear power.[16]

Those problems, along with a series of other environmental, technical, and public health questions, emerged within a short period of a few years and made nuclear power the source of acute controversy. Public support, which had seemed so strong in the early 1960s, was shaken. As the combination of key issues that arose in rapid sequence fed public misgivings, the arguments of nuclear critics substantially affected the fortunes of the industry. "The antinuclear coalition has been remarkably successful," commented *Forbes* in September 1975. "It has certainly slowed the expansion of nuclear power."[17]

The individuals and groups who opposed nuclear power, or at least objected to specific nuclear projects, did not constitute a monolithic front in their tactics or motivation. By the mid-1970s, anti-nuclear activism had moved beyond localized protests and politics to gain wider appeal and influence. Although it lacked a single coordinating organization and uniform goals, it emerged as a movement sharply focused on fighting nuclear power. Its efforts attracted a great deal of national attention. According to one estimate, coverage of nuclear power issues in the print media grew by 400 percent between 1972 and 1976. Anti-nuclear activism was spearheaded by leaders who, as even their opponents conceded, were well informed, articulate, and increasingly media-savvy. They were frequently brash and sometimes belligerent in taking on nuclear proponents, whom they regarded with attitudes ranging from skepticism to contempt. Few nuclear critics were nuclear physicists or engineers, but many were scientifically literate. Even those not trained in science often were skillful at learning the fundamentals of nuclear technology and challenging the positions of its supporters. Myron M. Cherry, an attorney who argued against the licensing of several nuclear plants, commented in 1977: "There are some things I'm not so good at, but I'm absolutely *fantastic* at asking questions." None of his adversaries in the hearings in which he participated would have disagreed. The nuclear industry discovered that the anti-nuclear movement was too tenacious to be dismissed and too influential to be ignored.[18]

Some prominent critics of nuclear power first became involved in opposing local projects and then branched out to gain recognition as national anti-nuclear leaders. As a young attorney in a large Chicago law firm, Cherry had received an assignment from a senior partner in 1969 to give the utility constructing the Palisades nuclear station in Michigan a "hard time." Although at first he knew little about nuclear power, he participated in licensing proceedings on the plant and raised questions about the environmental and health effects of nuclear power. After playing the same role in a variety of other hearings, he commented privately in 1976 that he had become so "seriously afraid of nuclear accidents" that he wanted to "put people in jail," presumably people from the nuclear industry.[19]

Cherry's opposition to specific nuclear projects often received support from David Dinsmore Comey, whom *Nuclear Industry* described in early 1973 as "probably the most formidable . . . foe of nuclear power." As a faculty member in Soviet studies at Cornell University, Comey organized a successful campaign in 1968 to halt plans to build a nuclear plant on nearby Lake Cayuga, largely because of concerns about thermal pollution. He was then hired by a Chicago-based nonprofit group called Business and Professional People for the Public Interest to carry out extensive anti-nuclear activities. Although Comey was regarded by industry officials as less confrontational and doctrinaire than many other environmentalists, he was relentless in citing objections to the design and performance of nuclear plants. Another well-known critic, Anthony Z. Roisman, was a lawyer who had made a name for himself by representing environmental groups in petitioning the AEC to greatly expand its consideration of the environmental impact of two plants under construction on the Chesapeake Bay in Maryland. When the issue went to court, Roisman won a milestone victory in 1971 that applied not only to the twin Calvert Cliffs facilities but to all nuclear plants under

construction or under licensing review. The Calvert Cliffs decision established Roisman as a leading voice among anti-nuclear forces.[20]

The efforts of Cherry, Comey, Roisman, and a dedicated but disparate
cadre of other nuclear critics were often combined with and enhanced by
a growing number of anti-nuclear organizations. Although many of those
groups pursued issues unrelated to nuclear power, they formed a crucial
component of the anti-nuclear coalition. The Sierra Club, for example,
was an organization of 140,000 members that addressed a wide range of
environmental questions. Despite the appeals of some of its leaders and a
serious internal rift, it had refused to take a blanket stand against nuclear
power during the 1960s and early 1970s. In 1974, however, its board of
directors voted in favor of a nuclear moratorium. Other organizations were
less ambivalent in their opposition to nuclear power. The Natural Resources
Defense Council (NRDC) was founded in 1969 as a "public interest law
firm" to take legal action on environmental issues. Its goal was to advance
environmental protection through "responsible militancy." By 1976 it
had enlisted about 15,000 members and litigated on matters ranging from
administration of anti-pollution legislation to cleanup of industrial sites. It
also "invested a steadily increasing portion of its resources in the nuclear
energy issue" because, the group concluded, "for too long, we let the glim-
mering promise of nuclear energy blind us to the fact that we may have
opted for an unforgiving and potentially unmanageable technology."[21]

The smallest of the national organizations that contested nuclear power
was perhaps the most influential. The Union of Concerned Scientists
(UCS) was formed at the Massachusetts Institute of Technology in late
1968. It began largely as a faculty organization that published an appeal
in March 1969 to use science and technology for addressing social and
environmental problems rather than for building nuclear, chemical, and
biological weapons. It soon turned its attention to environmental issues
and increasingly became involved in the debate over nuclear power. Key
leaders of the UCS were Henry W. Kendall, a high-energy physicist at
the Massachusetts Institute of Technology who later won a Nobel Prize in
physics, and Daniel F. Ford, a graduate student in economics at Harvard
University before devoting his efforts to the nuclear controversy. The
organization had only two hundred to three hundred members, but Kendall
and Ford together provided it with energy, commitment, and credibility.
Science magazine reported in 1975 that Kendall and Ford, "in large part, *are*

the UCS." The UCS first joined the nuclear debate when it intervened in a licensing hearing for the Pilgrim plant near Plymouth, Massachusetts, and it won increased recognition after it challenged the AEC and the industry on the performance of emergency core cooling systems in 1971.[22]

The UCS was disappointed, however, that its warnings about the risks of a core meltdown during the emergency-cooling debate did not receive more media coverage. Therefore, in November 1972, Kendall and Ford approached the consumer advocate and corporate watchdog Ralph Nader to ask that he join their campaign to resolve the "number one public safety problem in the country today." Nader had earned enormous public respect for his well-publicized efforts to improve automobile safety and to lobby for other regulatory, environmental, and tax reforms. Kendall and Ford concluded that, with Nader's assistance, "the reactor safety issue that the UCS has been pursuing can finally achieve widespread public attention and that important remedial changes can result." Nader accepted the invitation of the UCS and within a short time emerged as the leading, or at least the best-known, critic of nuclear power. In a speech on November 21, 1972, he called nuclear power "a terrible hazard" that could cause "the greatest destruction that this country has ever known." A short time later, he told approximately 3 million viewers of the nationally televised *Dick Cavett Show* that "the risks of something going wrong with these nuclear power plants are so catastrophic that they are not worth the benefit."[23]

On January 3, 1973, Nader held a press conference with Kendall and Ford that was attended by about forty reporters from the print media and television. He told them he wanted to raise the debate over the role of nuclear power "to the moral level that is so important." He further suggested that the issue was more urgent than the "so-called energy crisis" because, "if we don't deal with it now, we'll have perhaps a radioactive crisis 10 years hence." The alliance with Nader went a long way toward accomplishing the goals that Kendall and Ford had sought when they recruited him. A consultant for the nuclear industry thought "Nader's entry into the anti-nuclear business" was "ample cause for concern," and *Nuclear Industry* commented that "the support of a public figure such as Nader gives [UCS] demands public exposure to a degree that Kendall and Ford could not achieve on their own." Nader's activities clearly boosted the visibility, credibility, and morale of the anti-nuclear movement. In November 1974 he organized a three-day conference called "Critical

Mass '74" that attracted 750 enthusiastic participants and about eighty press representatives. "The tone of the meeting, in sharp contrast to the nuclear industry meetings a month earlier, was one of buoyant optimism," observed *Nucleonics Week.* "The lasting impression of the affair was that the antinuclear movement is growing in size and dedication."[24]

Nuclear power critics did not fully agree on their objectives, which ranged along a spectrum from appealing for an immediate shutdown of plants to demanding major improvements. Although Nader had not actively opposed the technology until Kendall and Ford approached him, he gradually took a more hard-nosed position than that of the UCS. In their joint press conference of January 3, 1973, he supported the UCS's call for a moratorium on the completion of nuclear plants already in the early stages of construction until safety issues were resolved and for "derating," or reducing the power output, of facilities in operation or nearing completion by as much as 50 percent. By the time the "Critical Mass '74" conference took place, Nader had toughened his stance. "There has to be a moratorium on all construction of nuclear power plants and the most expeditious shutdown of existing plants," he declared. "I don't think that a position as to the hazards of nuclear power is consistent with any other position than a moratorium and a shutdown." Nader's hard-line position was echoed by other nuclear opponents. Cherry made clear that he was "firmly committed to stopping nuclear power in the U.S.," and a staff member of the national environmental organization Friends of the Earth announced, "We oppose nuclear power completely. Something that produces radioactivity cannot be made safe any more than war can be made safe."[25]

Other nuclear critics took a more moderate position while continuing to emphasize their deep reservations about nuclear power. The UCS did not support an immediate moratorium on nuclear plants. It suggested that nuclear power could be made acceptable, at least on a temporary basis, if its "grave weaknesses" were corrected. When Kendall told the "Critical Mass '74" meeting that he opposed an immediate moratorium and favored "sharp restrictions" on nuclear plants and a gradual phase-out instead, the audience groaned audibly. Comey remarked in 1977 that "nuclear energy can probably be made safe," though he added that it was "nowhere near that point now" and that the costs of improving its safety would make it economically uncompetitive. Anthony Roisman, an effective adversary of the industry in several licensing and rule-making hearings, told a group of staff members from utilities and the AEC that publicists for both sides in the nuclear debate had misrepresented their opponents. He urged his listeners to take the views of nuclear critics seriously but also advised them: "Don't treat us as your enemies. We both have the same goals." Frank von Hippel, a physicist at Princeton University who had expressed serious misgivings about the performances of the nuclear industry and the AEC, regretted that the controversy focused on the question of whether nuclear power should be abandoned completely or pushed "full steam ahead." He hoped that the debate would be couched in less categorical terms and would deal responsibly with the issue of improving existing safety programs.[26]

The Anti-nuclear Position

Despite differences in goals and priorities, leading critics marshaled similar complaints against nuclear power, which were, in turn, echoed by local citizens and organizations that took an anti-nuclear position or objected to plans for specific sites. Grassroots opposition reflected concerns ranging from aesthetic sensibilities to scientific uncertainties, but by the mid-1970s the anti-nuclear movement highlighted several major arguments in its indictments of the technology. Convinced that the risks of nuclear power far outweighed the benefits, it placed its greatest emphasis on the issue of nuclear safety. Nuclear opponents asserted that, despite the efforts of the industry and the AEC to ensure the safe operation of nuclear plants, a core meltdown was possible, if not probable. They pointed out that nuclear plants contained as much radioactivity as thousands of Hiroshima-type atomic bombs and estimated that the large amount of radiation, if released to the environment in an accident, would cause tens of thousands of deaths. Nuclear critics argued that a system as large and complex as a nuclear power plant was vulnerable to human errors. They claimed that even if plants were well designed and well built, which they regarded as unlikely, nuclear safety depended on flawless operating performance.

Moreover, from the perspective of nuclear opponents, the routine operation of plants was a dire threat to public health even if severe accidents were avoided. Drawing on the views of several scientists who dissented from the prevailing consensus, one writer alleged in 1977, for example, that the death toll from normal releases of low levels of radiation from nuclear units "may now be in the thousands" and would "in time climb into the hundreds of thousands." In addition, nuclear critics pointed out that the nuclear industry and the government had not found a satisfactory method of disposing of the radioactive wastes produced by nuclear power, which they claimed would pose a grave danger to the public for generations. Concerns about the safety of nuclear plant operation and the risks of population exposure to radiation were intensified by the hazards that the projected use of plutonium for reactors seemed to present. Although the slightly enriched uranium employed in existing nuclear power plants was not suitable for nuclear weapons, most experts believed that plutonium would be widely used in the future to fuel reactors. Many nuclear opponents cited the potential catastrophe that could occur if terrorists acquired enough plutonium to build an atomic bomb. They further suggested that protecting plants from terrorist activities would require measures so extensive and intrusive that they would undermine American civil liberties. One anti-nuclear group, the Citizens Energy Project, contended in 1978 that "nuclear power and civil liberties cannot co-exist; to the extent that nuclear power is expanded, civil liberties must be restricted."[27]

In the calculus of anti-nuclear activists, the risks of nuclear power generation were unacceptably large whereas the benefits were slight. They argued that nuclear power was unnecessary to meet the energy requirements of the United States. Nader, for example, told a congressional committee in January 1974 that by 1985 geothermal energy could produce "the equivalent of one half of the electric energy which is now produced in our economy." The idea that nuclear power could be replaced by more benign sources of energy was a staple of anti-nuclear literature. The foremost champion of this view, Amory B. Lovins, attracted a great deal of notice for an article he published in the journal *Foreign Affairs* in October 1976. Lovins was a twenty-eight-year-old "consultant physicist" who worked for the Friends of the Earth in London. He urged the adoption of a "soft path" to energy sufficiency rather than continuing along the "hard path" of dependence on fossil fuels and nuclear power. He called for the immediate abandonment of nuclear power and a more gradual retreat from fossil fuels in favor of conservation and development of alternative sources of power from solar, wind, and geothermal energy. "Enterprises like nuclear power are not only unnecessary but a positive encumbrance," he wrote, "for they prevent us

. . . from pursuing the tasks of a soft path at a high enough priority to make them work."[28]

In addition to submitting that nuclear power was unsafe and unnecessary, nuclear critics maintained that it was unreliable. They pointed to AEC statistics showing that nuclear plants fell short of targets for generating power. Their average "capacity factor," which was the ratio of the actual power a plant produced compared to its capacity if it operated all the time, usually ran in the 50–60 percent range during the mid-1970s because of a variety of equipment and operating problems. Most of the problems were minor, but many required shutdowns in order to be corrected or repaired.[29]

The Pro-nuclear Position

Supporters of nuclear power took strong exception to the arguments of its opponents. They acknowledged that developing the technology imposed risks on the population, but they insisted that the benefits far exceeded those risks. Although they admitted that a serious accident was conceivable and that the loss of life and property could be severe if one occurred, they contended that the chances of an accident releasing large amounts of radiation to the environment were remote. Alvin M. Weinberg, director of Oak Ridge National Laboratory in Tennessee, who was widely regarded as one of the most thoughtful nuclear power advocates, wrote in 1972, "Nuclear people have made a Faustian bargain with society." He suggested that advanced reactor designs could provide a cheap, clean, and virtually inexhaustible source of energy but required "a vigilance and a longevity of our social institutions that we are quite unaccustomed to." Weinberg concluded that the advantages of nuclear power made the bargain "well worth the price." Another prominent defender of nuclear power, Hans A. Bethe, a professor of theoretical physics at Cornell University, made the same point about nuclear safety. He had been a pioneer in the field of nuclear physics during the 1930s, a leading scientist on the Manhattan Project during World War II, and a recipient of the Nobel Prize in physics in 1967. Physicist Frank von Hippel thought that "a large fraction of the scientific community" saw Bethe as a "model of independence, incisive analysis, and public responsibility." In a series of articles and interviews,

Bethe asserted that even in the "extremely unlikely" event of a core melt-down in a nuclear plant the probability of a major release of radiation was "extremely small."[30]

Nuclear proponents countered the critics' argument that a nuclear plant contained more radiation than an atomic bomb by emphasizing that a reactor could not explode like a bomb. Nearly all the nuclear power reactors in operation or on order used uranium fuel enriched to a level of about 3 percent of the fissionable isotope uranium-235. A nuclear bomb, by contrast, needed fuel enriched to 80 percent or more of uranium-235 (as in the atomic bomb dropped on Hiroshima) or fissionable plutonium (as in the atomic bomb dropped on Nagasaki and in U.S. nuclear weapons tested after World War II). "Comparing a reactor to an A-bomb is a popular scare tactic," declared A. David Rossin, an official with the Commonwealth Edison Company of Chicago. He pointed out that, whereas radiation from a bomb was released when it exploded, nuclear plants were designed to keep the radioactivity "isolated and sealed." Supporters of nuclear power also denied the charge that plants were safe only if their design, construction, and operation were infallible. They insisted that the redundant safety systems and multiple barriers to a large release of radiation provided ample protection from the consequences of equipment failures and human errors.[31]

Nuclear advocates sharply disputed the claim that even routine operation of plants would cause large numbers of cancer deaths annually. This issue, which won headline treatment in the late 1960s and early 1970s, centered on the effects of exposure to low levels of radiation, and the scientific evidence did not provide definitive guidance. But most radiation protection professionals, while cautioning against unnecessary exposure to any amount of radiation, believed the available data strongly indicated that the risks of low-level exposure were slight. They maintained that the charge that normal emissions of radiation from nuclear plants would greatly increase the incidence of cancer were exaggerated beyond evidence or reason. Nuclear supporters dismissed other health and safety concerns raised by critics as overstated or alarmist. They admitted that the disposition of nuclear waste was a problem, but they expressed confidence that satisfactory solutions would be found. They acknowledged that the future use of plutonium, which experts believed would largely replace uranium as reactor fuel by the end of the century, would require adequate safeguards against terrorist threats. But nuclear proponents argued that power reactors were a poor target for terrorists, and that newly strengthened regulations offered sufficient protection against them. They denied that measures taken to guard nuclear plants against terrorist activities would undermine American civil liberties.[32]

In the minds of its supporters, nuclear power was essential to meet the energy requirements of the United States. They dismissed the conclusion of Lovins, Nader, and other critics that alternative sources of energy along with conservation could replace nuclear power. Ralph E. Lapp, a nuclear physicist and freelance writer, played a prominent role in attacking this point of view. Lapp was a veteran of the Manhattan Project who had challenged the AEC's assurances during the 1950s that radioactive fallout from nuclear weapons testing did not present a significant threat to public health. He had later criticized the AEC and the nuclear industry on reactor safety issues, but by the mid-1970s he concluded that nuclear opponents, especially Nader, had overestimated the risks of nuclear power and undervalued its benefits. He commented in 1975 on Nader's views on energy supplies: "It is extremely difficult to critique Mr. Nader's proposals for alternative energy sources because he has never put them together in anything approaching a coherent framework. His emotional attachment to the sun seems profound, but on questioning about its value, he usually concedes it is a future source."[33]

Nuclear supporters agreed on the need for conservation and new energy sources, but they discounted the "soft" path to energy sufficiency that Lovins recommended. Bethe responded skeptically to Lovins's article in *Foreign Affairs:* "The energy problem of the United States and other industrial countries is extremely serious. We need to combine many different techniques to solve it. But it cannot be solved by combining 'soft' arithmetic with wishful thinking." Although Bethe and other nuclear proponents favored the development of solar, wind, and geothermal energy, they denied that those alternatives to nuclear power could satisfy either the short-term or long-term energy requirements of the United States. Advocates contended that nuclear power was reasonably reliable and predicted that it would improve with more operating experience. They presented data showing that the industry-wide capacity factor for nuclear units was comparable to that of large coal facilities and that for some nuclear plants it was better.[34]

The Nuclear Power Controversy

The issues surrounding the safety, necessity, and reliability of nuclear power had erupted into a full-fledged national controversy by the mid-1970s. One indication of the growing uncertainty about the technology was the ambivalence of President Jimmy Carter, who took office in January 1977. During his election campaign in 1976, he had emphasized the need for energy conservation and the development of solar power and clean-burning coal; he described the use of nuclear power as "a last resort only." Nevertheless, Carter regarded nuclear power as an essential component of his energy policies. "When I say 'last resort,'" he explained to news editors in January 1978, "it doesn't mean that it's a necessary evil. . . . There's a legitimate role for atomic power to play."[35]

The nuclear debate that had emerged by that time was intense, dogmatic, and highly polarized; representatives of both sides described it as a "religious war." With the fervor of holy warriors, partisans used emotional appeals to win public support. "The result has been a flood of advertising and pamphlets," observed reporter Joanne Omang in the *Washington Post*, "either scaring us about the horrors of [a] nuclear holocaust or scaring us about the horrors of inadequate electricity." Howard K. Smith, a commentator for *ABC Evening News*, repeated a favorite theme of nuclear proponents when he told viewers in 1975 that, without an expansion of nuclear power, "the day will come, probably in the early 1980s, when the home will grow cold, auto traffic turned to a trickle, and industries go on two days a week, with lots of unemployment, for lack of fuel." Although supporters of the technology seldom drew the issue in terms as stark and alarming as Smith's commentary, they insisted that nuclear power development was necessary to avoid a serious energy shortage.[36]

Nuclear advocates used emotional appeals in advancing their arguments, but their critics were even more inclined to make their case by evoking strong sentiments. Perhaps the most arresting example was a poster that anti-nuclear protesters often carried at rallies. It asked the question "What Do You Do in Case of a Nuclear Accident?" and provided a hauntingly apocalyptic answer: "Kiss Your Children Goodbye." Students from Archbishop Carroll High School in Washington, D.C., made a similar argument in a letter to President Carter: "People often say, 'I don't want my kids to sit around in the cold twenty years from now,'" they wrote.

"But the proliferation of nuclear power raises the more important question of whether there will be anyone alive twenty years from now." Hans Bethe complained that trying to explain the advantages of nuclear power to opponents was like "carving a cubic foot out of a lake." He related an incident that occurred when he spoke to a largely anti-nuclear audience at a meeting in Berkeley, California. After he had presented his position on the need for nuclear power, a woman in the audience stood up, turned her back on him, and shouted "Save the Earth!" The crowd reacted, he said, with "thunderous applause."[37]

In keeping with its heavy emotional content, the contest over nuclear power featured a strong element of gamesmanship. This was evident in competing petitions that each side publicized to show authoritative scientific support for its position. In January 1975 a group of thirty-four eminent American scientists, including eleven Nobel laureates, released a statement on energy policy that had been drafted primarily by Bethe and Lapp. Contending that the energy crisis confronted the United States with "the most serious situation since World War II," it maintained that there was "no reasonable alternative to an increased use of nuclear power to satisfy our energy needs." The petition faulted nuclear critics for a lack of "perspective as to the feasibility of non-nuclear power sources and the gravity of the fuel crisis." When Bethe and Lapp issued the statement at a press conference attended by about a hundred reporters, Nader countered by attending the event and handing out an appeal of his own. It was a letter to President Ford signed by eight prominent scientists, including five Nobel laureates, that opposed a "massive speedup of nuclear power plant construction." *Science* magazine scored this exchange as "Nuclear Advocates 34, Opponents 8."[38]

A short time later, the Union of Concerned Scientists circulated yet another petition that urged a "drastic reduction" in new construction of reactors. Of the approximately sixteen thousand people who received the statement from the UCS, about twenty-three hundred signed. The UCS then delivered the petition to the White House and Congress on August 6, 1975, the thirtieth anniversary of the atomic bombing of Hiroshima. This initiative, in turn, prompted the American Nuclear Society, an organization of nuclear professionals in industry, government, and academic institutions, to launch its own drive. Eventually it secured more than thirty-two thousand signatures on a statement that underlined the need for both coal and nuclear power and asserted that there were "no technical problems incapable of being solved" in the use of either technology.[39]

As the adversaries in the nuclear debate attempted to win public favor by citing the numbers and professional qualifications of their supporters, each also acclaimed defectors from the competing side. Nuclear proponents pointed to the views expressed by Ian A. Forbes, a former member of the UCS who had coauthored a stinging reproach of the AEC's treatment of the emergency core cooling question in 1971. By 1974 he had concluded that the issue had been satisfactorily resolved, and he became a vocal backer of nuclear power. He rebuked Nader and the UCS for polarizing discussion in a way that "made reasoned debate almost impossible." The effect of Forbes's changeover from a critic to a defender of nuclear power was modest, however, compared to the highly publicized resignation of three midlevel engineers from their positions with General Electric's nuclear power division in 1976. The three men, Gregory C. Minor, Richard B. Hubbard, and Dale C. Bridenbaugh, had a total of fifty-four years of experience with General Electric, families to support, and no immediate job prospects. They resigned with a flourish by announcing that "nuclear power is a technological monster that threatens all future generations." General Electric sought to deflate the effect of their action by pointing out that they were a small portion of the hundreds of nuclear engineers it employed. Nevertheless, as *Time* reported, the "trio's defection seemed like a major victory for the antinuke forces in the great nuclear debate."[40]

Although the battle over nuclear power was usually fought in press conferences, hearings, meetings, petitions, articles, and television appearances, it occasionally was joined in more direct confrontations. In May 1977 a demonstration against two proposed nuclear plants in Seabrook, New Hampshire, attracted about two thousand poster-carrying, slogan-chanting, nonviolent protesters. "We feel Seabrook in particular and nuclear power in general are life and death issues," explained one of their leaders. After the demonstrators occupied the construction site, police arrested more than fourteen hundred of them for trespassing. The conflict over Seabrook commanded a great deal of attention and suggested that citizen protests against nuclear plants would grow. Nader predicted that direct action "will spread all over the country as needed . . . if there is no more formal way to protest." Opponents of nuclear plants in Indiana, Oklahoma, Missouri, Alabama, California, and elsewhere adopted similar nonviolent tactics in efforts to halt or slow construction, promote their views, and win sympathy for their cause. The result was to amplify the emotional and uncompromising quality of the nuclear power debate. "Increasingly, the debate is constituted less of reason and logic and more of emotion," lamented Jon Payne, editor of *Nuclear News*, the monthly publication of the American Nuclear Society. "And its outcome is based less on the accuracy of the arguments than on the number of voices behind them."[41]

The influence of the nuclear controversy on public attitudes toward the technology was difficult to assess. Public opinion surveys showed strong support for nuclear power. A poll conducted by Louis Harris and Associates in August 1975 indicated that 63 percent of the public favored the expansion of nuclear power in the United States, whereas 19 percent opposed and 18 percent were not sure. Later polls yielded similar results. Nevertheless, the polls were not unequivocally favorable to nuclear power. The 1975 Harris survey, for example, found that support for nuclear construction was substantially weaker if Nader opposed it. A Gallup poll conducted in June 1976 showed that 71 percent of those interviewed thought it was "extremely important" or "somewhat important" to build more nuclear power plants, but it also showed that 40 percent of the respondents believed that operations of existing plants should be cut back until stricter safety regulations were imposed.[42]

The outcome of anti-nuclear initiatives that appeared on ballots in seven states in 1976 was similarly ambiguous. The campaign with the broadest potential consequences occurred in California, where nuclear opponents collected 400,000 signatures to place a proposition on the ballot for elections held in June. The initiative, known as Proposition 15, was intended to stop construction of new plants, reduce operation of existing ones, and eventually close down nuclear power in the state. The battle over the initiative underscored the bitter and emotional nature of the nuclear debate. Foes of Proposition 15 claimed it would cause the economy to collapse. Supporters of the measure asserted that nuclear power was a severe public health threat; one flyer proclaimed, "We are irreversibly committed to one million deaths from nuclear radiation." The nuclear industry spent heavily to defeat Proposition 15, and California voters rejected it by a margin of two to one. Even so, the message of the outcome was mixed. As David Pesonen, the leader of the initiative drive, commented, "A million and a half people were willing to vote to shut down nuclear power. Those people are firm and will not go away."[43]

Five months later, voters in six other states turned down proposals to place restrictions on nuclear power by decisive margins. Nuclear supporters welcomed the election results but recognized that they did not represent final victories. Anti-nuclear opposition and reservations about nuclear technology among a substantial segment of the public had become permanent parts of the political landscape. "It's a funny situation, where we're losing all the battles but winning the war," said one nuclear opponent. "Even when these proposals go down to defeat, we've educated more millions of people about the problems we see." In 1978, 63 percent of the voters in Montana provided support for that view when they approved a referendum imposing sharp restrictions on nuclear power, even though the state had no nuclear plants. Other samplings of public opinion also produced some ominous indications for nuclear advocates. A poll of college students and members of the League of Women Voters in Oregon in 1978, for example, offered startling information about the public's fear of nuclear power. Asked to rank thirty sources of risk "according to the present risk of death from each," both groups rated nuclear power as number one, ahead of smoking, motor vehicles, motorcycles, handguns, and alcoholic beverages.[44]

There were several reasons for the intensity and polarization of the nuclear power controversy; it was not simply a debate over energy sufficiency. Like a religious controversy, the nuclear power issue was so emotional in part because it could not be resolved with available information. All the key questions surrounding the technology—the probability of a severe accident, the consequences of a severe accident, the effects of low-level radiation, the dangers of radioactive waste disposal, the level of threat from terrorist attacks on nuclear facilities, and the costs and reliability of nuclear power—were subjects of dispute among experts. Operating experience and scientific evidence were still too limited to provide conclusive answers. When Bethe acknowledged at a public hearing that he could not say for certain that safety systems in reactors would work as designed, a woman in the audience audibly murmured, "My God!—they really don't have the answers, do they?" Further, many political and social questions regarding the risks and benefits of nuclear power compared to other energy sources remained to be addressed. John P. Holdren, a physicist at the University of California, Berkeley, suggested in 1976 that "the disagreement among experts on major aspects of nuclear power is not a temporary condition."[45]

In addition to the lack of definitive evidence in crucial matters, long-standing public attitudes toward nuclear energy in general and cultural trends in the United States during the 1970s contributed in critical ways to the temper of the nuclear power debate. Perhaps the key issue was the connection between nuclear power and nuclear bombs. Despite the efforts of nuclear proponents to dispel popular misconceptions, a significant percentage of the public feared that a nuclear plant could explode like an atomic bomb. A Harris poll conducted in 1975, for example, showed that 39 percent of those surveyed believed that a failure in a nuclear power plant could produce a "massive nuclear explosion" (24 percent thought this could not occur, and 37 percent were not sure). Nuclear critics claimed that nuclear power could not be separated from nuclear weapons because they used the same materials for fuel and posed the same threat of radioactive contamination. "If you're against nuclear warfare, you're also against nuclear power," declared one environmentalist. This argument was vital in shaping public attitudes toward nuclear power. Three scholars who studied growing public concern about the technology during the 1970s concluded that "distrust of nuclear power is . . . rooted in the fear of nuclear weapons."[46]

Public misgivings that arose from the stigma of nuclear weapons were reinforced by deep-seated fear of radiation. Although public apprehension about radiation predated World War II, it was greatly increased by the development and use of atomic weapons. Within a short time after the atomic bombings of Hiroshima and Nagasaki, accounts of the effects of radiation, embellished in science fiction books and articles, comics, and films, combined to heighten existing anxieties. During the 1950s and 1960s, debates over radioactive fallout and the effects of radioactive emissions from nuclear plants made radiation safety a bitterly contested issue. The allegation that routine radiation released from nuclear plants would cause thousands of cases of cancer every year among the population was a staple of nuclear opponents. In 1973, E. F. Schumacher, an economist and technology critic, wrote in his influential *Small Is Beautiful* that radiation from nuclear power was perhaps a greater menace to humanity than the atomic bomb. It was, he argued, "the most serious agent of pollution of the environment and the greatest threat to man's survival on earth." The hazards of low-level radiation were a source of sustained publicity during the postwar period and, as a result, of uniquely intense public fears that were important in setting the tone of the debate over nuclear power.[47]

The controversy was further inflamed by suggestions that whether to use the technology was not only a technical question but also a serious moral issue. Some national church organizations were prominent in opposing nuclear power on moral grounds. The General Conference of the United Methodist Church passed a resolution in May 1976, for example, that stated, "In our opinion, no generation has a right to assume risks in its decision making which bear heavily upon the potential destruction of the earth as a habitable place for future generations." At about the same time, the National Council of Churches, an ecumenical organization of thirty Protestant and Orthodox denominations, considered an even sharper condemnation of nuclear power. Drafted by a committee chaired by Margaret Mead, the eminent anthropologist, and Rene Dubos, a professor emeritus of pathology at Rockefeller University, it focused on the dangers of using plutonium for nuclear fuel, a prospect it called "morally indefensible and technically objectionable." To the Council leaders' surprise, the report stirred many protests from, as one staff member put it, "good church people working in the industry who said, 'How dare you say that what I'm doing is immoral.'" After nuclear industry representatives and some theologians vigorously complained that nuclear power should not be "prejudged as intrinsically evil," the Council softened the draft statement. Its deliberations called attention to differing moral assessments that greatly reduced the likelihood of compromises on technical, economic, and political issues.[48]

Like the continuing disagreements among experts, fears of atomic explosions and radiation, and conflicting moral positions, the impassioned tone of the nuclear power debate was a result of cultural and philosophical trends in America during the 1970s. A growing chorus of social critics claimed that technological development and economic growth threatened to undermine democratic freedoms, moral values, environmental resources, public health, and eventually economic well-being. Schumacher, a leading advocate of this view, warned, "In the excitement over the unfolding of his scientific and technical powers, modern man has built a system of production that ravishes nature and a type of society that mutilates man." Condemning bigness and centralization in industry and government, Schumacher and other critics urged alternative systems that were modest in scale and decentralized in authority. Adherents to a "small is beautiful" outlook identified nuclear power as a powerful threat to their vision, and although they did not represent a large percentage of the population they provided leadership, commitment, energy, and often money to anti-nuclear campaigns. A group called the Creative Initiative Foundation, whose numbers included the three engineers who created a stir by resigning from General Electric, was instrumental in placing the anti-nuclear initiative on the ballot in California in 1976. Opposition to nuclear power on the basis of its violation of the principles of smallness and decentralization added another dimension to the nuclear power debate that made accommodations between the competing sides unlikely.[49]

Many observers complained about the prevalence of emotional, moral, and philosophical appeals that polarized the nuclear controversy. Representatives of both sides of the debate called for a calm, reasoned discussion of the topic, and each expressed confidence that its arguments would prevail if the American people were well informed about the issues. Nader claimed in 1973, "If the country knew what the facts were and if they had to choose between nuclear reactors and candles, they would choose candles." Bethe affirmed that he welcomed "a factual public discussion on a broad basis," but he wanted it "conducted with rational arguments, not vast exaggerations." The nature of the nuclear power debate, however, made a dispassionate exchange of views difficult, if not impossible. The arguments used by both sides contained an ample measure of theoretical projections and unprovable assertions, and nuclear critics added a liberal portion of frightening associations. The adversaries in the debate contended that the stakes were very high, ranging from the economic welfare of the nation to the survival of the human race. *Fortune* commented on March 12, 1979, that the nuclear controversy, "the bitterest environmental confrontation of the Seventies," was "complex, confusing, and muddied by overstatements from both sides."[50]

The contention over nuclear power showed no signs of abating as the end of the 1970s neared. The issues were divisive, emotions were high, and opportunities for compromise were meager. In that atmosphere, *The China*

Syndrome was more than simply an exciting motion picture. A screenplay in which utility executives were willing to gun down an honorable man for trying to correct unsafe conditions seemed disturbingly plausible to the growing number of Americans who believed the benefits that nuclear power provided were a poor tradeoff for the risks it imposed. Less than two weeks after the film opened, the debate over the technology took on even greater urgency when the worst accident in the history of commercial nuclear power in the United States occurred at the Three Mile Island plant in Pennsylvania.

Notes

A previous version of this essay was published as chapter 1 of J. Samuel Walker, *Three Mile Island: A Nuclear Crisis in Historical Perspective* (Berkeley: University of California Press, 2004).

1. *The China Syndrome*, written by Mike Gray, T. S. Cook, and James Bridges, Columbia Pictures, 1979.

2. *New York Times*, Mar. 16, 1979, sec. 3, 16, Apr. 4, 1979, sec. 3, 18; *Baltimore Sun*, Mar. 26, 1979, B5; *Washington Post*, Mar. 29, 1979, 7; David Ansen, "Nuclear Politics," *Newsweek*, Mar. 19, 1979, 103; Stanley Kauffmann, "What Price Thrills?" *New Republic*, Apr. 7, 1979, 24–25; Mike Gray and Ira Rosen, *The Warning: Accident at Three Mile Island* (New York: W. W. Norton, 1982), 142.

3. *New York Times*, Mar. 11, 1979, sec. 2, 1; Kenneth Turan, "Anchored to Reality," *Progressive* 43 (May 1979): 52–53; Mike Gray to Henry Kendall, Feb. 16, 1979, Box 60 [no folder], Papers of the Union of Concerned Scientists (hereafter UCS Papers), Massachusetts Institute of Technology Archives, Cambridge, Mass.

4. *New York Times*, Mar. 18, 1979, sec. 2, 1; *Boston Globe*, Mar. 29, 1979, 53; *Washington Post*, Mar. 30, 1979, B1.

5. George T. Mazuzan and J. Samuel Walker, *Controlling the Atom: The Beginnings of Nuclear Regulation, 1946–1962* (Berkeley: University of California Press, 1984), 22–23, 77–82, 92–96, 203–208.

6. Hazel Gaudet Erskine, "The Polls: Atomic Weapons and Nuclear Energy," *Public Opinion Quarterly* 27 (Summer 1963): 164; "Big Hurdle for A-power: Gaining Public Acceptance," *Nucleonics* 21 (Oct. 1963): 17–24; Stanley M. Nealey, Barbara D. Melber, and William L. Rankin, *Public Opinion and Nuclear Energy* (Lexington, Mass.: D.C. Heath, 1983), 4; Mazuzan and Walker, *Controlling the Atom*, 418.

7. U.S. Congress, Joint Committee on Atomic Energy, "Nuclear Power Economics—1962 through 1967," 90th Cong., 2nd Sess., 1968, 15; J. Samuel Walker, *Containing the Atom: Nuclear Regulation in a Changing Environment, 1963–1971* (Berkeley: University of California Press, 1992), 392–98.

8. "When the Music Stops," *Forbes*, Feb. 1, 1969, 3; Walker, *Containing the Atom*, 30–31.

9. *Nucleonics Week*, Feb. 25, 1965; Jeremy Main, "A Peak Load of Trouble for the Utilities," *Fortune* 80 (Nov. 1969): 116–19, 205; Walker, *Containing the Atom*, 32–36, 267–70.

10. "Fossil Competition Fades in U.S.," *Nuclear Industry* 19 (Jan. 1972): 6–9; "'72 Sets Sales Record," *Nuclear Industry* 19 (Nov.–Dec. 1972): 7–10; "AEC Capacity Forecast to 2000," *Nuclear Industry* 20 (Mar. 1973): 20–21; *Nucleonics Week*, Dec. 14, 1972, Jan. 10, 1974.

11. White House Press Release, "Statement by the President," June 29, 1973, Staff Member Office Files, Energy Policy Office, Richard M. Nixon Papers, Nixon Presidential Materials Project, National Archives, College Park, Md.; Glenn Schleede to Jim Cannon, Nov. 18, 1975, Domestic Council—Glenn R. Schleede Files, Box 25 (Nuclear Energy 1975, General), Gerald R. Ford Papers (hereafter Ford Papers), Gerald R. Ford Library, Ann Arbor, Mich.; Richard Nixon, "Address to the Nation about Policies to Deal with the Energy Shortages," Nov. 7, 1973, available via the University of California–Santa Barbara's American Presidency Project, www.presidency.ucsb.edu/ws/index.php?pid=4034&st=&st1=; "President's Report on the State of the Union," House Document 94–1, 94th Cong., 1st Sess., 1975, 5–6 ; David E. Nye, *Consuming Power: A Social History of American Energies* (Cambridge, Mass.: MIT Press, 1998), 217–19.

12. Dick Livingston to Jim Cannon, Nov. 18, 1975, Domestic Council—Glenn R. Schleede Files, Box 25 (Nuclear Energy 1975, Overview Memo), Ford Papers; "Why Atomic Power Dims Today," *Business Week*, Nov. 17, 1975, 98–106; "Nuclear Dilemma," *Business Week*, Mar. 25, 1978, 54–68; "Another Peak Year," *Nuclear Industry* 21 (Jan. 1974): 3–6; "Deferrals of Nuclear Power Plant Construction Continue," *Nuclear Industry* 21 (Sept. 1974): 17; "Industry Groping for Answers to Utility Financing Crisis," *Nuclear Industry* 21 (Oct. 1974): 17–19; *Nucleonics Week*, Sept. 19, 1974; Richard F. Hirsh and Adam H. Serchuk, "Momentum Shifts in the American Electric Utility System: Catastrophic Change—or No Change at All?" *Technology and Culture* 37 (Apr. 1996): 280–311; Richard F. Hirsh, *Technology and Transformation in the American Electric Utility Industry* (Cambridge: Cambridge University Press, 1989), 110–13; John Robert Greene, *The Presidency of Gerald R. Ford* (Lawrence: University Press of Kansas, 1995), 67–81.

13. Ken Pedersen to Commissioner Ahearne, Jan. 31, 1979, Subject File, Box 630 (Energy—Nuclear, 1977–78), Victor Gilinsky Papers (hereafter Gilinksky Papers), Hoover Institution on War, Revolution and Peace Archives, Stanford University, Palo Alto, Calif.; "A Call to Arms," *Nuclear Industry* 22 (Aug. 1975): 7–10; "Record Turnout, Marked Concern," *Nuclear Industry* 25 (Jan. 1978): 3–4; *Nucleonics Week*, Jan. 16, 1975, July 10, 1975, Nov. 24, 1977; Peter Stoler, "The Irrational Fight against Nuclear Power," *Time*, Sept. 25, 1978, 71–72; "Nuclear Dilemma," *Business Week*, Mar. 25, 1978, 54.

14. "Is Nuclear's Defense Worth the Effort?" *Nuclear Industry* 25 (Jan. 1978): 26–27; J. Samuel Walker, *Permissible Dose: A History of Radiation Protection in the Twentieth Century* (Berkeley: University of California Press, 2000), 18–22.

15. *Nucleonics Week*, June 20, 1963; Walker, *Containing the Atom*, 387–93.

16. The issues of thermal pollution, radiation emissions, and reactor safety in the late 1960s and early 1970s are discussed in detail in Walker, *Containing the Atom*, 139–202, 267–362, 388–407.

17. "Don't Confuse Us with Facts," *Forbes*, Sept. 1, 1975, 20.

18. Roger Smith, "The Nuclear Critic: What Lies Ahead?" attachment to *Nucleonics Week*, Apr. 25, 1974; Frank Graham, Jr., "The Outrageous Mr. Cherry and the Underachieving Nukes," *Audubon* 79 (Sept. 1977): 50–66; President's Commission on the Accident at Three Mile Island, *Report of the Public's Right to Information Task Force* (Washington, D.C.: Government Printing Office, 1979), 1. For excellent studies of the anti-nuclear movement, see Brian Balogh, *Chain Reaction: Expert Debate and Public Participation in American Commercial Nuclear Power, 1945–1975* (Cambridge: Cambridge University Press, 1991), and Thomas Raymond Wellock, *Critical Masses: Opposition to Nuclear Power in California, 1958–1978* (Madison: University of Wisconsin Press, 1998).

19. Victor Gilinsky, Memorandum to the Files, Dec. 16, 1976, Personal NRC File, Box 13 (F.13.9), Gilinsky Papers; Graham, "Outrageous Mr. Cherry," 56–57.

20. "David Comey Combines Dedication with a Thirst for Knowledge," *Nuclear Industry* 20 (Apr. 1973): 27–28; McKinley C. Olson, "Nuclear Energy: It Costs Too Much," *Nation* 219 (Oct. 12, 1974): 331–33; Dorothy Nelkin, *Nuclear Power and Its Critics: The Cayuga Lake Controversy* (Ithaca, N.Y.: Cornell University Press, 1971), 58–64; Walker, *Containing the Atom*, 378–86.

21. *San Francisco Sunday Examiner and Chronicle*, Mar. 24, 1974, California Living Magazine, 14–19; "Squeaky Wheel," *New Yorker* 48 (Dec. 16, 1972): 27–28; John H. Adams, "Responsible Militancy: The Anatomy of a Public Interest Law Firm," *Record of the Association of the Bar of the City of New York* 29 (1974): 631–46; *Five-Year Report, 1970–1975* (New York: Natural Resources Defense Council, 1976), 5, 15; Wellock, *Critical Masses*, 101–13.

22. Robert Gillette, "Nuclear Power: Hard Times and a Questioning Congress," *Science* 187 (Mar. 21, 1975): 1058–62; Dorothy Nelkin, *The University and Military Research: Moral Politics at M.I.T.* (Ithaca, N.Y.: Cornell University Press, 1972), 56–65; Smith, "Nuclear Critic," 2.

23. Daniel F. Ford and Henry W. Kendall to Richard Sandler, Nov. 4, 1972, Box 37, and Transcript of Ralph Nader's Remarks on Nuclear Reactor Safety on the ABC-TV Dick Cavett Show, Dec. 13, 1972, Box 37 (Nader-Lippmann Correspondence), Nader Transcript, Cavett Show, UCS Papers; "Nader's Conglomerate," *Time*, June 11, 1973, 82; *Nucleonics Week*, Dec. 21, 1972.

24. Pat Fogarty to Hal Stroube, Jan. 15, 1973, Box 32 (Atomic Industrial Forum), UCS Papers; John A. Harris to the Commissioners, Jan. 3, 1973, Records of the Atomic Energy Commission, History Division, Department of Energy, Germantown, Md. (hereafter AEC/DOE); "Effort to Put ACRS on the Spot Highlights New AEC Opposition," *Nuclear Industry* 20 (Jan. 1973): 15–17; *Nucleonics Week*, Nov. 21, 1974.

25. Fogarty to Stroub, Jan. 15, 1973, UCS Papers; *Nucleonics Week*, Jan. 21, Oct. 3, 1974; *Christian Science Monitor*, May 24, 1977, 6.

26. "Some Scientists Question Nuclear Power Safety," *National Wildlife* 10 (Aug.–Sept. 1972): 19; "Fuel Cycle Rulemaking," *Nuclear Industry* 20 (Feb. 1973): 15–20; "Effort to Put ACRS on the Spot," *Nuclear Industry* 20 (Feb. 1973): 16; "Nuclear Licensing Detente?" *Nuclear News* 17 (Sept. 1974): 42–44; *Nucleonics Week*, Oct. 3, Nov. 21, 1974; *Christian Science Monitor*, May 24, 1977, 6; *New York Times*, Oct. 5, 1977, 27; Frank von Hippel to Morris K. Udall, Feb. 6, 1975, Box 338 (Interior, E&E Subcommittee—Nuclear Correspondence), Papers of Morris K. Udall (hereafter Udall Papers), University of Arizona, Tucson; Frank

von Hippel, "A Perspective on the Debate," *Bulletin of the Atomic Scientists* 31 (Sept. 1975): 37–39.

27. *Wall Street Journal*, Jan. 26, 1972, 1; "Just How Safe Is a Nuclear Power Plant?" *Reader's Digest* 100 (June 1972): 95–100; *Nucleonics Week*, Apr. 24, 1975; "Nuclear Energy Alert," *Nation* 222 (Mar. 13, 1976): 302; David Dinsmore Comey, "The Perfect Trojan Horse," *Bulletin of the Atomic Scientists* 32 (June 1976): 33–34; Susan Schiefelbein, "Is Nuclear Power a License to Kill?" *Saturday Review*, June 24, 1977, 10; Donna Warnock, *Nuclear Power and Civil Liberties: Can We Have Both?* (Washington, D.C.: Citizens' Energy Project, 1978), 2; John G. Fuller, *We Almost Lost Detroit* (New York: Reader's Digest Press, 1975), 18, 64; J. Samuel Walker, "Regulating against Nuclear Terrorism: The Domestic Safeguards Issue, 1970–1979," *Technology and Culture* 42 (Jan. 2001): 107–32.

28. John Abbotts to Gene L. Woodruff, July 26, 1974, Henry W. Kendall to Woodruff, July 25, 1974, Box 8 (Nuclear Energy Controversy, 1974–1980), Papers of Fred H. Schmidt, University of Washington, Seattle; U.S. Congress, Joint Committee on Atomic Energy, "Hearings on Nuclear Reactor Safety," 93rd Cong., 2d Sess., 1974, 514; Amory B. Lovins, "Energy Strategy: The Road Not Taken?" *Foreign Affairs* 55 (Oct. 1976): 65–96; Allen L. Hammond, "'Soft Technology' Energy Debate: *Limits to Growth* Revisited?" *Science* 196 (May 27, 1977): 959–61; *New York Times*, July 24, 1977, sec. 4, 17; Oct. 16, 1977, sec. 3, 1.

29. Transcript of CBS News Special, "Nuclear Power—What Price Salvation?" March 31, 1974, Box 32 (CBS), UCS Papers; *Washington Star-News*, Apr. 2, 1974, 1; *New York Times*, Mar. 9, 1975, 42; "Kouts Says Data Inadequate for Thorough Reliability Analysis, Raps Nader on Leaks," *Nuclear Industry* 21 (May 1974): 35; "Proposed NRC Steps to Stimulate Higher Plant Reliability," *Nuclear Industry* 22 (Mar. 1975): 32–33; Deborah Shapley, "Nuclear Power Plants: Why Do Some Work Better Than Others?" *Science* 195 (Mar. 25, 1977): 1311–13; Gladwin Hill, "Power Play," *National Wildlife* 14 (Oct.–Nov. 1976): 39; SECY-75-33 (Feb. 7, 1975), SECY-75-377 (July 21, 1975), SECY-75-178 (Apr. 21, 1975), Nuclear Regulatory Commission (NRC) Records, NRC Public Document Room, Rockville, Md. "SECY" papers were prepared by AEC, and later NRC, staff to provide information or outline policy options for the commissioners, who used them as a basis for making decisions.

30. Alvin M. Weinberg, "Social Institutions and Nuclear Energy," *Science* 177 (July 7, 1972): 27–34; "As I See It," *Forbes*, July 15, 1972, 30; Ralph E. Lapp, "Nuclear Power Safety—1973," *New Republic* 168 (Apr. 28, 1973): 17–19; "Forbes: A Call to Reason," *Nuclear News* 17 (Aug. 1974): 42–43; H. A. Bethe, "The Necessity of Fission Power," *Scientific American* 234 (Jan. 1976): 31; Don G. Meighan (pseudonym), "How Safe Is Safe Enough?" *New York Times Magazine*, June 20, 1976, 8–9ff; Frank von Hippel to Morris K. Udall, Feb. 6, 1975, Box 338 (Interior, E&E Subcommittee—Nuclear Correspondence), Udall Papers.

31. A. David Rossin, "Let's Set the Record Straight," typescript, June 17, 1976, Box 346 (Interior Committee Files—Nuclear Issues, David Rossin Correspondence), Udall Papers.

32. Weinberg, "Social Institutions and Nuclear Energy," 32–33; Bethe, "Necessity of Fission Power," 27–28; Walker, *Permissible Dose*, 36–66; Walker, "Regulating against Nuclear Terrorism," 108–24.

33. Ralph E. Lapp, *Nader's Nuclear Issues* (Greenwich, Conn.: Fact Systems, 1975), 47; Petr Beckmann, *The Health Hazards of Not Going Nuclear* (Boulder, Colo.: Golem Press, 1976), 152–54.

34. Hans A. Bethe, Letter to the Editor, *Foreign Affairs* 55 (Apr. 1977): 636–37; A. D. Rossin and T. A. Rieck, "Economics of Nuclear Power," *Science* 201 (Aug. 18, 1978): 582–89; Dennis J. Chase, "Clouding the Nuclear Reactor Debate," *Bulletin of the Atomic Scientists* 31 (Feb. 1975): 39–40.

35. Frank Press and John Deutch to the Vice President and others, Nov. 21, 1979, Staff Offices—Counsel (Cutler), Box 100 (Nuclear Regulatory Commission), Papers of Jimmy Carter, Jimmy Carter Library, Atlanta, Ga.; *Wall Street Journal*, Oct. 19, 1977, 22.

36. "Pulling the Plug on A-power," *Newsweek*, Feb. 24, 1975, 23–24; "How Safe Is Nuclear Power?" *Newsweek*, Apr. 12, 1976, 70; "Meeting the Media: Industry Tells Its Side," *Nuclear News* 20 (Aug. 1977): 34–35; "It's Time to End the Holy War over Nuclear Power," *Fortune*, Mar. 12, 1979, 81; *Wall Street Journal*, July 21, 1977, 32; *Washington Post*, Mar. 12, 1978, C5; Commentary by Howard K. Smith, *ABC Evening News*, Sept. 9, 1975.

37. Statement by Students of Archbishop Carroll High School, n.d., Public Document Room, NRC Records; "Hans Bethe Recounts Some Frustrations as a Nuclear Spokesman," *Nuclear Industry* 22 (Nov. 1975): 20–21.

38. Robert Gillette, "Nuclear Advocates 34, Opponents 8," *Science* 187 (Jan. 31, 1975): 331; Robert Gillette, "Nuclear Critics Escalate the War of Numbers," *Science* 189 (Aug. 22, 1975): 621; Philip M. Boffey, "Nuclear Power Debate," *Science* 192 (Apr. 9, 1976): 120–22.

39. "Scientists Speak Out: 'No Alternative to Nuclear Power,'" *Bulletin of the Atomic Scientists* 31 (Mar. 1975): 4–5; "Nuclear Hazards," *Bulletin of the Atomic Scientists* 31 (Apr. 1975): 3.

40. "Five MIT Nuclear PhD's Wearing Environmental Tags, Blast Nader," *Nuclear Industry* 21 (June 1974): 11–12; "Nader Responds to Forbes Group's 'Call to Reason,'" *Nuclear News* 17 (Nov. 1974): 60; *New York Times*, Feb. 3, 1976, 12; "The San Jose Three," *Time*, Feb. 16, 1976, p. 78; "The Apostates," *Newsweek*, Feb. 16, 1976, p. 64; "Forbes: A Call to Reason," *Nuclear News* 17 (Aug. 1974): 42–43; "Don't Confuse Us with Facts," *Forbes*, Sept. 1, 1975, 28; Wellock, *Critical Masses*, 165–66.

41. "The Siege of Seabrook," *Time*, May 16, 1977, 59; "The No-Nuke Movement," *Newsweek*, May 23, 1977, 25–26; "Antinuclear Protests Are Busting Out All Over," *Science* 200 (May 19, 1978): 746; Jon Payne, "Seabrook: A Cause Celebre," *Nuclear News* 21 (Sept. 1978): 45; *Washington Post*, Oct. 8, 1978, 49; "Three Demonstrations Test Non-violent Tactics," *Nuclear Industry* 25 (Nov. 1978): 26; Henry F. Bedford, *Seabrook Station: Citizen Politics and Nuclear Power* (Amherst: University of Massachusetts Press, 1990), 76–93.

42. "Majority Favors Nuclear—Harris Survey," *Nuclear News* 18 (Sept. 1975): 31–34; "Second Harris Poll Finds Public Still Favors Nuclear," *Nuclear News* 20 (Jan. 1977): 31–35; Gallup Poll Press Release, July 22, 1976, Box 34 (Gallup Poll), UCS Papers; Nealy, Melber, and Rankin, *Public Opinion*, 15–60.

43. Wellock, *Critical Masses*, 147–72.

44. "Voters Reject Six Initiatives," *Nuclear Industry* 23 (Nov. 1976): 5; "Nuclear Initiatives: Two Sides Disagree on Meaning of Defeat," *Science* 194 (Nov. 19, 1976): 811–12; "Montana Passes a Nuclear Initiative," *Science* 202 (Nov. 24, 1978): 850; *Christian Science Monitor*, Nov. 5, 1976, 6; *New York Times*, Nov. 9, 1976, 18; *Washington Post*, Nov. 9, 1978, 7; Paul Slovic, Baruch Fischhoff, and Sarah Lichtenstein, "Rating the Risks," *Environment* 21 (Apr. 1979): 14–21, 36–39.

45. John P. Holdren, "The Nuclear Controversy and the Limitations of Decision-Making by Experts," *Bulletin of the Atomic Scientists* 32 (Mar. 1976): 20–22; Wellock, *Critical Masses*, 163.

46. Robert Jay Lifton, "Nuclear Energy and the Wisdom of the Body," *Bulletin of the Atomic Scientists* 32 (Sept. 1976): 16–20; Roger M. Williams, "Massing at the Grass Roots," *Saturday Review* 4 (Jan. 22, 1977): 14–18; Christoph Hohenemser, Roger Kasperson, and Robert Kates, "The Distrust of Nuclear Power," *Science* 196 (Apr. 1, 1977): 25–34; "Anti-atom Alliance," *Newsweek*, June 5, 1978; Nealy, Melber, and Rankin, *Public Opinion*, 83.

47. E. F. Schumacher, *Small Is Beautiful: Economics As If People Mattered* (New York: HarperCollins [1973], 1989), 143; Spencer R. Weart, *Nuclear Fear: A History of Images* (Cambridge, Mass.: Harvard University Press, 1988), 54; Walker, *Permissible Dose*, 145–53.

48. Alvin M. Weinberg, "The Moral Imperatives of Nuclear Energy," *Nuclear News* 14 (Dec. 1971): 33–37; "Clerics Soften Nuclear Posture," *Nuclear Industry* 23 (Mar. 1976): 28; Philip M. Boffey, "Plutonium: Its Morality Questioned by National Council of Churches," *Science* 192 (Apr. 23, 1976): 356–59; Margaret N. Maxey, "Nuclear Energy Debates: Liberation or Development?" *Christian Century* 93 (July 21–28, 1976): 656–61; Extracts from General Conference Resolution on Energy, May 6, 1976, Box 25 (Subject File—Nuclear, General), Papers of Carlton Neville, Carter Library.

49. Richard Rovere, "Letter from Washington," *New Yorker* 53 (Jan. 2, 1978): 54–58; Carroll Pursell, "The Rise and Fall of the Appropriate Technology Movement in the United States," *Technology and Culture* 34 (July 1993): 629–37; Schumacher, *Small Is Beautiful*, 313; Wellock, *Critical Masses*, 157–67; Weart, *Nuclear Fear*, 341–43.

50. Hans A. Bethe, "The Controversy about Nuclear Power," typescript, c. Jan. 1977, Box 4 (Bethe, Hans A. and Related Correspondence), Papers of Fred H. Schmidt, University of Washington, Seattle; C. Hosmer, "The Anatomy of Nuclear Dissent," draft, Mar. 31, 1976, Domestic Council—Glenn R. Schleede Files, Box 26 (Nuclear Energy 1976, General), Ford Papers; Lapp, *Nader's Nuclear Issues*, 3; "It's Time to End the Holy War over Nuclear Power," *Fortune*, Mar. 12, 1979, 81.

Part IV

Demand

Chapter 10

The Consumer's Hand Made Visible
Consumer Culture in American Petroleum Consumption of the 1970s

Brian Black

When Chrysler unveiled a reorganized automobile engine in 1937, it took the predominant design of a previous era and literally turned it. With a hemispherical orientation of the engine's cylinder head, the Hemi as it became known to consumers, offered new levels of power. The reconfigured engine design provided impressive air flow, torque, power, and, most important, a throaty rumble. With no concern about efficiency or conserving fuel, designers, first at Chrysler and later at General Motors and Ford Motor Company, kept upping the ante in the "horsepower wars" that lasted from the early 1950s until the early 1970s. In hindsight, the era of "muscle cars," as they became known, might mark the golden era of petroleum-powered automobiling in the United States.

Such distinctions—drawn from a single engine design but made legend in films such as *American Graffiti*—reveal themselves if our history does not resist limiting its scope and instead embraces the need to investigate certain specific topics over a relatively long period of time. This chapter approaches its organization from a few ideas based primarily in environmental history. First, when naturalist Aldo Leopold encouraged us to adopt an "ecological interpretation of history" and see "that man is, in fact, only a member of a biotic team," he guided historians to view human history through the same biological communities that drive other species' communities—to tell human stories wrapped in the natural systems that support and determine their survival and living patterns (their culture).[1] Second, particularly for the modern human, energy, drawn from a variety of sources and used for a host of activities, would be one of the critical natural systems on which our species relies. And, finally, in particular for contemporary Americans, our living patterns are predicated on the use of petroleum.

The use and consumption of petroleum defined much of American life in the twentieth century. Within the petroleum story, the 1970s operate as a hinge era to introduce an energy transition away from petroleum reliance. Although this shift is neither immediate nor particularly clear-cut, the global and domestic polices of the 1970s clearly constructed a new paradigm in Americans' relationship with petroleum. The most important lessons of this shift for historians, though, may come from the attempt by American manufacturers and consumers in ensuing decades to ignore or deny the lessons of the 1970s shift in petroleum culture.

Historian John McNeil refers to efforts to transform resources into work (broadly defined) as an "energy regime." He defines such a regime as "the collection of arrangements whereby energy is harvested from the sun (or uranium atoms), directed, stored, bought, sold, used for work or wasted, and ultimately dissipated." Historian of technology David Nye takes the large, systematic argument to a more individual level when he writes that "energy systems that a society adopt create the structures that underlie personal expectations and assumptions about what is normal and possible. . . . Each person lives within an envelope of such 'natural' assumptions about how fast and far one can go in a day, about how much work one can do, about what tools are available, about how that work fits into the community, and so forth."[2]

McNeill, using the macro-historical view, urges us to realize the centrality of energy in all of human life; Nye, using the micro-style, tells us that through our choices consumers help to support or alter existing energy regimes. When these changes in resource use and management reach such a significant level, they are referred to as an "energy transition."

Throughout the twentieth century, petroleum has operated as the dominant energy regime in the lives of most Americans. We came to live more closely with it than any energy resource in human history. The infrastructure of our business practices and laws functioned to inculcate it more deeply into consumers' lives—to make it cheaper, easier to get, and, indeed, essential to everyday life. This was not a diabolical charade; during the first two-thirds of the twentieth century, petroleum was one of the cheapest energy resources in history and its use in inexpensive innovations revolutionized personal travel and greatly expanded human capabilities. Although two-thirds of Americans' existence with petroleum was organized by an overarching premise of abundance, the 1970s introduced American consumers to the reality of petroleum's finite supply. As the hinge point from which a new paradigm of scarcity—based on awareness of its finite supply—remade Americans' petroleum reality, the 1970s merit closer inspection.[3]

Policy shifts during this decade mark a revolution in Americans' relationship with and conception of petroleum. In fact, by analyzing a few of these policies and the consumer patterns that they inspired, our view of the 1970s clears to form continuity with our present era in petroleum culture. That continuity is Americans' next great energy transition away from reliance on petroleum to a new energy regime. One indicator of this transition at the start of the twenty-first century, I believe, lay under the hood of vehicles such as Dodge's 1968 Charger.

When the Hemi design came into vogue during the 1960s, consumer interest grew from the fact that the reorganized layout of the engine enabled additional cylinders to be included, first a V-6 and later as large as a V-12. These experiments drew directly out of Chrysler's work making tank engines (which used a V-16) during World War II. No other car could match the 426 Hemi engine in terms of acceleration, which was similar to the power of a drag racing vehicle. From 1966 to 1971, Chrysler produced approximately 11,000 Hemi cars (only 475 of the 1968 Chargers). The engine—with neither effort to prioritize efficiency nor comfort and safety—provides us with a symbol of the culture of cheap gasoline. By the early 1970s, though, the engine had disappeared from new vehicles. Closer inspection of general trends in vehicle purchasing in the United States from the 1970s to 2005 allows us to use this

emblematic engine as a guide for understanding broader trends in the
culture of petroleum consumption.

The 1970s as Hinge in Petroleum History

A *New York Times* article from February 5, 1974, reads:

> Anxious motorists overwhelmed gasoline stations in the metropolitan
> [New York] area yesterday, with many stations running out of sup-
> plies early in the day, while dealers hoped incoming deliveries under
> February allocations would restore calm by mid-week.
>
> In Brooklyn, Murray Cohen, an owner of the AYS Service Station at
> Avenue Z and East 17th Street, said he had imposed a $3 maximum
> for each car's purchases, only to find that most people needed only 75
> cents' worth to fill up. One man, he said, waited in line for an hour
> and could use only 35 cents' worth.
>
> In Washington, William E. Simon, director of the Federal Energy
> Office, who had asked drivers not to buy more than 10 gallons at a
> time, yesterday issued an appeal to them to stay away from stations
> unless they bought at least $3 worth. . . . "Panic buying isn't helping
> the situation."[4]

Previously, many Americans were content to drive their cars until gas
gauges neared empty. By early 1974, though, prudent consumers topped-
off their tanks because they were uncertain about gasoline's future availability. Many states implemented odd-even gas purchasing based on the
car's license plate number.

Intermittently, motorists throughout 1973–74 had to wait in line for
an hour or more—ironically, of course, with their engines running the
entire time. In other regions, the worst harbinger became signs that read
"Sorry, No Gas Today." Expressway speeds were cut from 60–70 miles
per hour to 50. Many communities—as well as the White House—for-
went lighting public Christmas trees. Some tolls were suspended for
drivers who carpooled in urban areas. Rationing plans were leaked to
the public, even if they were not implemented. For instance, in the New
York City region the Federal Energy Office estimated that residents

eighteen years of age and older could expect to receive books of vouchers for 37 gallons per month.[5]

By the end of 1973, in fact, gas lines were plentiful throughout the nation. Supplies of petroleum were least disturbed on the West Coast, but by February even California had adopted odd/even-day rationing. Gas station operators were subjected to mistreatment, violence, and even death. Drivers also reacted with venom to other drivers attempting to cut into gas lines. At the root of such anger, of course, was the cruel reality that the events of our everyday lives—kids going to school, adults going to work or shopping, goods moving in every direction, and even cutting our grass—might be constrained, our choices limited. Nothing could seem more un-American.

The scene of winter 1974 was not the American relationship with petroleum with which Americans had gotten comfortable over the previous three decades. Based in priorities such as those at the root of the Hemi engine, petroleum was the lifeblood of America's renowned era of conspicuous consumption.[6] Figuratively and literally these moments in the gas line demonstrated a new phase in America's petroleum culture. The opulence of seeing the finite as infinite was suddenly overcome by the dark shadow brought by envisioning scarcity—the possibility of a world without cheap petroleum. The likely outcomes may seem obvious, but they were not just a product of fears about energy security. The 1970s role as a hinge point in our petroleum story also grows from new, emerging environmental sensibilities. Historical convergence made petroleum use—and specifically automobiles—a foil for this new ethic.

These petroleum shortages extended into 1974; the implications of them, though, extended through the rest of the decade. The shock was an abrupt lesson. Although few Americans understood why the price fluctuated so wildly, for the first time most Americans learned three valuable lessons: petroleum was a finite resource; the United States imported the bulk of its petroleum supply; and the nation was entirely reliant on this commodity. In fact, the reality of petroleum dependence had actually begun to emerge in many ways by the late 1960s. Whether rooted in new scientific understanding or a reconception of American patterns of consumption, though, there was no impetus for action while gas prices remained low. Such are the patterns of "successive intellectual paroxysms" that sociologists find defining the history of Americans' relationship with energy and particularly petroleum.[7] In this chronology, though, various trends converge on the 1970s to create the crucial catalyst for the energy transition from petroleum dependence.

Most often, historians boil these socioeconomic factors into the term "Arab oil embargo" to denote the hinge point of change in American energy consumption. This proves to be dangerous oversimplification when one attempts to trace each strand of these patterns. In fact, the actual event of the Organization of Arab Petroleum Exporting Countries (OAPEC) cutting its oil shipments to the West in the 1970s is merely one formative moment—albeit critical—in a decade-long remaking of the way American consumers viewed their petroleum supply. In this chapter I also briefly analyze the ways intellectual, economic, and political shifts of the 1970s influenced America's petroleum culture; the gas line, though, is entirely a product of supply. Although the shortage of 1973–74 was artificial, its value as a cultural spur grew from its interpretation as a symbolic precursor of a future certainty.

The moment of scarcity in the 1970s was, of course, a political construct (supplies of petroleum remained but were temporarily derailed). The historical processes that led to this crucial moment unfolded for more than five decades prior. From the dominant stranglehold of Western powers and the large petroleum corporations that dominate them, oil morphed into a tradable, ultravolatile commodity after World War II. Historian Daniel Yergin writes that this new era demonstrated that "oil was now clearly too important to be left to the oil men."[8] As political leaders in each oil nation assessed how best to leverage power for their nation from their supply of crude, it took little time for them to also realize the merit of joining forces with similarly endowed nations. Although OPEC was created at the Baghdad Conference in Iraq in September 1960, it took a decade to take full form.[9] OPEC's ability to manipulate prices did not become a reality until Egyptian leader Anwar Sadat urged his fellow members to "unsheath the oil weapon" in early 1973.[10]

Factoring in production increases elsewhere, the net loss of supplies when the embargo took hold in December 1973 was 4.4 million barrels per day (bpd), which accounted for approximately 9 percent of the total oil available previously. Although these numbers told of a genuine shortfall in the overall supply, the fickle petroleum market accentuated the embargo's importance by inserting a good bit of uncertainty and panic. It was the American consumers who felt the impact most. To provide oil to consumers, brokers began bidding for existing stores of petroleum. In November 1973, per-barrel prices had risen from around $5 to more than $16. Consuming nations bid against each other in order to ensure sufficient petroleum supplies. For American consumers, retail gasoline prices spiked by more than 40 percent. Although high costs were extremely disconcerting, scarcity also took the form of temporary outages of supply. The front on this new resource war could be found on the home front: the American gas station.

President Richard M. Nixon went before Americans on November 7, 1973, to declare an "energy emergency."[11] He spoke of temporary supply problems:

> We are heading toward the most acute shortages of energy since World War II. . . . In the short run, this course means that we must use less energy—that means less heat, less electricity, less gasoline. In the long run, it means that we must develop new sources of energy which will give us the capacity to meet our needs without relying on any foreign nation.

> The immediate shortage will affect the lives of each and every one of us. In our factories, our cars, our homes, our offices, we will have to use less fuel than we are accustomed to using. . . .

> This does not mean that we are going to run out of gasoline or that air travel will stop or that we will freeze in our homes or offices any place in America. The fuel crisis need not mean genuine suffering for any Americans. But it will require some sacrifice by all Americans.

Nixon went on to introduce Project Independence, which he viewed "in the spirit of Apollo, with the determination [that] the Manhattan Project, [would] . . . by the end of this decade" help the nation to develop "the potential to meet our own energy needs without depending on any foreign energy source."[12]

Of course, the argument for a conservation ethic to govern American consumers' use of energy was a radical departure from the postwar American urge to resist limits and flaunt the nation's decadent standard of living. The ethical concept of conserving natural resources is most often traced to roots in the Progressive era of the early 1900s and particularly the work of Gifford

Pinchot.[13] Although such an ethic of restraint could be applied to any natural resource, petroleum fell more into the category of an abundant resource that could fuel a standard of living that would enable us to conserve species and sites of particular merit.[14] Most important, gas was cheap. Therefore, historian Paul Sutter and others have demonstrated that petroleum actually became an important tool for enabling Americans to visit far-off areas where conservation and preservation were put into practice.[15]

But in the 1970s, higher prices inspired by temporary scarcity shed a new emphasis on energy conservation. Although this ethical shift did not take over the minds of all Americans in the 1970s, a large segment of the population began to consider a new paradigm of accounting for our energy use and needs. They became interested in energy-saving technologies, such as insulation materials and low-wattage light bulbs, as well as limits on driving speeds that might increase engine efficiency.[16] As a product of the 1970s crisis, some Americans were even ready and willing to consider less convenient ideas of power generation, such as alternative fuels.

Creating the other crucial plank leading to policy shifts of the 1970s, this new environmental ethic created new expectations by consumers on both the goods they purchased and the federal government to regulate their production.[17] The modern environmental movement came of age with wide-reaching policies beginning with the National Environmental Policy Act in 1969 and exploded on the public scene with Earth Day 1970, celebrated by approximately 20 million people worldwide. Ultimately, environmentalism was a cultural shift that became animate and even permanent through policy and regulation.[18] Consumption was a major emphasis of this new ethic toward resource use and management. Petroleum was the lifeblood and crowning symbol of American postwar consumption and excess. The stage was already set for dramatic changes in America's petroleum culture by the early 1970s when additional variables entered the scene.

American ideas of transportation received another jolt when events known as the "second oil shock" arrived in 1979. Just as some Americans might have begun to think problems with the petroleum supply were a thing of the past, relations with Iran in the late 1970s demonstrated the problems associated with relying on politically volatile nations to supply our lifeblood. Revolution in Iran halted that nation's oil exports temporarily. When Iranians took Americans hostage in 1979, Jimmy Carter from the Oval Office placed an embargo on the importation of Iranian oil into the United States and froze Iranian assets. And in 1980 the Iran-Iraq war abruptly removed almost 4 million daily barrels of oil from the world market—15 percent of total OPEC output and 8 percent of free-world demand.

Things had clearly changed in the consumptive nature of American life. In particular, the American idea of energy use—in its broadest sense—was brought under new scrutiny. This impact could be seen most clearly with Carter, a trained nuclear engineer. He reflected on revolutionary new ideas inspired by an environmental ethic such as that put forward by economist Amory Lovins in a 1976 *Foreign Affairs* article titled "Energy Strategy: The Road Not Taken?" In his subsequent book, Lovins contrasted the "hard energy path," as forecast at that time by most electrical utilities, and the "soft energy path," as advocated by Lovins and other utility critics:

> The energy problem, according to conventional wisdom, is how to increase energy supplies . . . to meet projected demands. . . . But how much energy we use to accomplish our social goals could instead be considered a measure less of our success than of our failure. . . . [A] soft [energy] path simultaneously offers jobs for the unemployed, capital for businesspeople, environmental protection for conservationists, enhanced national security for the military, opportunities for small business to innovate and for big business to recycle itself, exciting technologies for the secular, a rebirth of spiritual values for the religious, traditional virtues for the old, radical reforms for the young, world order and equity for globalists, energy independence for isolationists. . . . Thus, though present policy is consistent with the perceived short-term interests of a few powerful institutions, a soft path is consistent with far more strands of convergent social change at the grass roots.[19]

Carter took the ethic of energy conservation directly to the American people. His administration would be remembered for events such as the Iranian hostage crisis; however, when he controlled the agenda he steered American discourse to issues of energy. Often we assume that no previous president has had to consider the issue of energy—and, certainly, that no leader could resist the influence of "Big Oil." Carter, however, attempted to steer the nation toward a future of energy conservation and independence. In a 1977 speech he urged the nation:

Tonight I want to have an unpleasant talk with you about a problem unprecedented in our history. With the exception of preventing war, this is the greatest challenge our country will face during our lifetimes. The energy crisis has not yet overwhelmed us, but it will if we do not act quickly.

It is a problem we will not solve in the next few years, and it is likely to get progressively worse through the rest of this century.

We must not be selfish or timid if we hope to have a decent world for our children and grandchildren.

We simply must balance our demand for energy with our rapidly shrinking resources. By acting now, we can control our future instead of letting the future control us. . . .

Our decision about energy will test the character of the American people and the ability of the President and the Congress to govern. This difficult effort will be the "moral equivalent of war"—except that we will be uniting our efforts to build and not destroy.[20]

Carter would introduce wide-reaching policy initiatives mainly aimed at energy conservation.[21] During the late-1970s, though, some of his initial efforts pressed these new expectations onto American automobile manufacturers. It was a radical shift in petroleum culture because it involved new priorities placed on the resource's most obvious application, the automobile. Automobile historian Christopher Finch saw these new ethics play out in the cars driven by Americans:

Acceptance of the small car, whether American, European or Japanese, was a victory for reason, accompanied, in 1974, by the adoption of a nationwide 55-mile-per-hour speed limit dictated by the perceived need for fuel conservation. . . . For the first time in history, Americans were driving modest-sized, fuel-efficient cars, fitted with seat belts (the use of which became compulsory in some places), at speeds that verged on the sedate, on a well-engineered Interstate system that was now virtually complete. For most of the wrong reasons, the forces of responsibility had prevailed.[22]

Policy Changes Reflect New Expectations of Consumers

The application of new ideas about petroleum and, in particular, its use in automobiles primarily arrived at consumers through federal policies. These policies fall into three primary categories: safety, emissions, and supply conservation. Particularly the first two categories represent policy culminations of efforts that predate the 1970s. Here, I discuss them only briefly and, instead, draw out the implications of the third category for consumers.

The calls for vehicle safety represent the first front in the reconception of American ideas of the automobile. Spurred to action by Ralph Nader and others, Congress passed the 1966 National Traffic and Motor Vehicle Safety Act. This law established the U.S. Department of Transportation (DOT), with automobile safety as one of its purposes. Also in 1966, Congress held a series of highly publicized hearings regarding highway safety, which resulted in the passage of legislation that made installation of seat belts mandatory. In 1970 the National Highway Traffic Safety Administration (NHTSA) was officially established, and in 1972 the Motor Vehicle Information and Cost Savings Act expanded NHTSA's scope to include consumer information programs. Prioritizing vehicle safety also came to place new expectations on drivers. Seat belts were required in new cars after the 1968 model, but Americans often did not use them. Manufacturers, however, consistently argued that increased safety measures did not sell vehicles.[23]

The connection with pollution could also be included as a safety concern. The seminal event in the emergence of modern environmentalism, Earth Day 1970 contained many activities that related to air pollution. In one of the day's most dramatic and public displays, New York City's Fifth Avenue was transformed into an auto-free zone. Only pedestrian traffic was allowed to traverse the city's symbolic primary artery. Accomplishing its intention, this public display was meant to strip away the noise, congestion, and exhaust that the vehicles brought to the space. The founder of Earth Day, Sen. Gaylord Nelson, was even more specific, saying that "the automobile pollution problem must be met head on with the requirement that the internal combustion engine be replaced by January, 1, 1975."[24] One of the major proponents of clean air legislation was Sen. Edwin Muskie, a Democrat from Maine. He acted as a bridge between the new environmental nongovernmental organizations springing from middle-class America's Earth Day exuberance and the 1960s conception of using the federal government to regulate and ultimately solve the nation's various ills. Together, a conglomeration of concerns focused public opinion against the internal combustion engine as an inefficient, polluting threat to U.S. health and security. Although Nelson and others argued for banning the engine altogether, the most likely outcome appeared to be placing federal regulations (similar to those used in states such as California) on all American cars.

In terms of pollution, lead, which had been inserted into gasoline to improve engine performance in the 1920s, was the first target of environmentalists. In January 1971, the EPA's first administrator, William D. Ruckelshaus, declared that "an extensive body of information exists which indicates that the addition of alkyl lead to gasoline . . . results in lead particles that pose a threat to public health."[25] The resulting EPA study released on November 28, 1973, confirmed that lead from automobile exhaust posed a direct threat to public health. As a result, the EPA issued regulations calling for a gradual reduction in the lead content of the nation's total gasoline supply, which included all grades of gasoline.

For other pollutants, initial legislation passed in 1970 (national emission standards were contained in the Clean Air Act) required specific pollutants contained in vehicle exhaust, such as CO and HCl, to drop 90 percent from 1970 levels by 1975. The intention was to force manufacturers to create the technologies that could meet the new standards. In 1975 a California act required that vehicle exhaust systems be modified to include a device, the catalytic converter, prior to the muffler. Costing approximately $300, early converters ran the exhaust through a canister of pellets or honeycomb made of either stainless steel or ceramics. The converters offered a profound, cost-effective way of refashioning the existing fleet of vehicles to accommodate new expectations on auto emissions. With the fleet largely converted, 1989 brought Congress to finally ban the use of leaded gasoline.[26]

The automobile was no longer a mindless cultural good; the 1970s had led consumers to identify some of its downsides, and lawmakers had begun to address them. The truest measure of the new culture of scarcity, though, came with the effort to make cars use *less gas*. Creating a system of regulations for all vehicles needed to precede any such regulations. Therefore, the DOT first had to locate usable data. In a decision that would have significant implications for consumers and manufacturers, regulators chose the EPA's collection of measurements, which also required that any new regulations would be guided by the agency's definition of light trucks. When such regulations were first discussed in 1973, only AMC (maker of Jeep) argued for utility vehicles to be included with trucks; many other manufacturers joined this call by the mid-1970s. When the rules were included in the Federal Register in 1977, they used the EPA classification of utility vehicles as separate from cars and included them in a separate category with trucks.

After the initial emergence of new expectations on automobile performance, the issue came to a political head when 1977 elections brought in new Democrats into shake up Congress. Many of these politicians were not willing to allow the manufacturers to forestall further meeting the new requirements discussed earlier in the decade. In spite of calls from Detroit that new mileage requirements would "shut down" American plants, Congress passed a bill requiring American vehicles to meet Corporate Average Fuel Economy (CAFE) and emissions standards. On August 7, 1977, when he signed the bill, Carter announced that the bill provided automakers with a "firm timetable for meeting strict, but achievable emissions standards."[27]

CAFE standards were far from perfect.[28] They did, however, represent a historic effort to stimulate the manufacture of more efficient autos in hopes of reducing American dependence on foreign oil.[29] Each auto manufacturer was required to attain government-set mileage targets for all of its car and light trucks sold in the United States. In a compromise with manufacturers, the complex standards were calculated as a total for the entire fleet of autos and trucks made by each company. Thus, the manufacture of a few fuel-efficient models could offset an entire line of light trucks that fell below the standards. In fact, the standard is set separately for two classes of vehicles, cars and light trucks.[30] As a supplement to the original act, the 1978 CAFE standards required 18 miles per gallon (mpg) for cars (in 1974, American cars averaged 13.2 mpg; those built elsewhere, 22.2 mpg).[31] To spur innovation, the law authored by Congress increased the standard each year until 1985. With the car standard at 27.5 mpg in 1985, lawmakers expected that manufacturers would willingly surpass this goal. This standard had to be achieved for domestically produced and imported cars separately. If manufacturers failed to meet these standards, they were to be fined $5 per car per 0.1 gallon that the corporate average fuel economy fell below the standard.[32]

Given the degree of regulation and the immensely new expectations placed on vehicles, American auto manufacturers came out of the 1970s feeling under siege. Each leader in the industry forecasted expensive shifts that would raise vehicle prices and put American laborers out of work. In fact, some openly speculated about whether automobiles could hope to still be manufactured in the United States in the twenty-first century. They would apply their considerable creativity to extending the American tradition of car making into the next century; however, American manufacturers obviously directed this creative design toward circumventing new regulations. The reform of the 1970s was viewed as a pitched battle to be weathered instead of indication of consumer and cultural change.

The Consumer's Hand Made Visible

The taste of American auto consumers came out of the 1960s as nearly the exact antithesis of what Nader and others urged. For most American drivers, their automobile was not a utilitarian device; in fact, sitting in the driveway in front of their house, cars for many Americans were one of the most noticeable expressions of themselves—or, at least, who they wanted the world to think they were. Historian of automobile design David Gartman writes, "In America's automotive culture, size had always connoted the importance and luxurious excess in consumption that countered the degradation and frugalities imposed in production."[33] Was it possible that some consumers' basic priorities could be changed by the urgency of petroleum scarcity? Brock Adams, secretary of transportation, spoke to manufacturers in 1978 and left no doubt about his hope: "I bring you a challenge because I think it is time to create a car that is new from the inside out—a car that represents a commitment, not a concession, to a world short on energy and concerned about the future."[34]

With a reconsideration of the basic need for vehicles to perform transportation more efficiently—to use less petroleum for their task—Americans immediately altered their view of the few small vehicles already being sold in the American market, such as the Volkswagen Beetle. Thanks to the desire for Beetles, used models from the late 1960s sold for more than new models had just prior to November 1973. The industry reported that standard-size cars outsold subcompacts by two to one just prior to the autumn of 1973. By December, smaller cars were being sold at the same rate as larger ones, and throughout 1974 their sales jumped while the guzzlers remained in the showrooms. American manufacturers simply could not immediately step in and fill this new demand.[35]

Stepping into this breach in the market, Japanese manufacturers made a niche for themselves with inexpensive, small vehicles. The initial forays by these companies had actually sought to appeal to existing American tastes. When Toyota began marketing in the United States in 1965, it sold a standard sedan called the Corona. Datsun/Nissan similarly began selling its 310 series in the late 1960s.[36] By 1970, Datsun had 640 U.S. distributors, and by 1972 the Japanese manufacturers had overtaken German manufacturers to sell nearly 700,000 new vehicles to become the largest overseas seller in the United States. Word spread of the reliability of these models, and when many Americans sought small vehicles Toyota and Datsun were ready with the Corolla and 1200, respectively. The Corolla, for instance, sold for $1,953, which undercut the prices of the American competitors (Ford Pinto, Chevy Vega, and AMC Gremlin) as well as the Beetle. By the mid-1970s, the Japanese market would be joined by Mazda and Subaru.

Each of these companies focused on smaller vehicles while continuing to produce an entire line of autos. However, Honda specifically organized itself to take control of the emerging subcompact market. Beginning with motorcycles, Soichuro Honda entered the auto market in 1962. Honda created the Civic, modeled after the British Mini, and released it in the United States in 1973–74. With fuel economy near 40 mpg, the Civic took the American market by storm. It was followed by a Honda sedan, the Accord, in 1976—the same year that Japanese imports to the United States passed one million. By the end of the 1970s, each of the three "Big Three" U.S. auto manufacturers had entered into agreements of some type with foreign auto manufacturers to sell small models.

The idea of efficiency was not a smooth fit for Detroit. In addition to these new models, Chevrolet proclaimed in 1975, "It's about time for a new kind of American car," when it released its new Chevette, a subcompact capable of 35 mpg on the highway. AMC sought to get around the cheap, unsafe image of the subcompact by introducing the Pacer, which was billed as "the first wide small car." In its design, however, it also circumvented the efficiency of smaller vehicles and ran the heavy Pacer with a V-6 engine. Other companies created slightly downsized models, including Ford's Granada and Cadillac's Seville. Overall, though, the sales of American cars dropped precipitously in the mid-1970s. In 1974, Chrysler's sales dropped by 34 percent. To combat this freefall, Chrysler introduced the Cordoba in 1975, which, although it presented itself as a luxury vehicle, was the shortest Chrysler since World War II.

Although small was much more acceptable, the greatest impact on the overall fleet was the gaining status for small sedans that resembled Honda's Accord. Previously, as Finch explains, the U.S. manufacturers had resisted homogeneity and ideas such as efficiency and safety. The American car was, despite the pleading of Nader and others, about style. Finch writes, "Until 1973, the one thing that insulated the American car industry from this tendency was cheap gasoline, which permitted every man a grandiosity of expression that was forbidden to all but the rich elsewhere in the world. After 1973 many Americans began to play by the same rules as Asians and Europeans, and with this came the sameness of product that afflicts the automobile marketplace today. As fins and grinning chrome radiator grilles slipped into the past, they quickly became objects of nostalgia and veneration."[37]

"Detroit was far from dead in 1981," writes Finch, "but to a large extent it dominated the least interesting areas of the marketplace; its constituency had become the conservatives of the motoring world."[38] The future of the American industry ironically lay in creating a mechanism that would exploit opportunities left available by CAFE legislation and enduring consumer taste, manufactured over the previous eras of petroleum decadence.

Of Loopholes and Consumer Passion

Cultural preferences die hard. In fact, they often ebb and flow without actually passing away entirely. American consumers had shopped for, purchased, and driven automobiles for more than half a century without giving primary consideration to the machine's efficiency. The first oil shock changed all of this for a time. But American consumers' appetite for size and weight in their machines would flow again. In terms of auto preference, a significant portion of American consumers would (with minor adjustments) always expect certain basic qualities in their vehicles. In yet another example of petroleum culture's legacy of "intellectual paroxysms," as soon as their fears of scarcity abated, so did consumers' reformed priorities for vehicle selection. Although some consumers had permanently reprioritized their view of the automobile, others fell under the enticing spell of an entirely new breed of mass-consumed vehicle: the light truck.

In the decade after 1978, the average weight of domestic and imported cars dropped nearly 1,000 pounds, from 3,831 to 2,921. Although there are many variables to factor in (such as shifts in design and materials), we can at least say that, overall, the weight of the cars on American roadways has decreased since the 1970s.[39] In addition, the data clearly demonstrate that the percentage of cars on American roads has decreased while the share of light trucks/vehicles has exploded. The standards, in short, penalized large car purchases and manufacturing, subsidized that of small cars, and left light trucks largely unregulated. Therefore, writes economist Paul Godek, one could reasonably expect that over the ensuing decades *both* small cars and light trucks would grow in popularity.[40] However, Godek's data demonstrate that, in fact, the small-car share dropped from its 1980 level. The only growth, defying logic on a number of fronts, was in the light truck/vehicle category. That is where we wade into the murky terrain of the cultural taste of the American consumer.

Initially, the primary issue for manufacturers was vehicle weight. This is measured as "gross vehicle weight," which is the truck's weight when fully loaded with the maximum weight recommended by manufacturers. Instead of the 10,000 pounds used for trucks, light trucks were initially set at 6,000 pounds. Automakers realized that they could escape the light truck category all together by increasing the weight of their vehicles, so, as journalist Keith Bradsher writes, they "shifted to beefier, less energy-efficient pickups even in a time of rising gasoline prices rather than try to meet regulations that they deemed too stringent."[41] In 1977 the maximum for light trucks was raised to 8,500 pounds. In 1981, Ronald Reagan took office and made one of his first priorities to freeze most auto regulations where they stood at that time.

The growth in this category of vehicle is uncanny. The data demonstrate that by creating the category of light truck American manufacturers had found their safety valve within the very legislation that sought to restrict their designs. After the release of the Jeep Cherokee in the mid-1980s, the form of the SUV emerged to appeal to the desires of many consumers. Also in the light truck category, minivans offered families an alternative to the station wagons that were now being phased out, and pickup trucks matured from their status as a commercial, work vehicle. Together, this new class of vehicles revolutionized the American fleet of vehicles in exactly the opposite way that CAFE standards intended. Of course, this new category of vehicles contained very few vehicles when the standards were set.[42] The light truck share of the passenger vehicle fleet rose to 20.9 percent in 1975 and to 30 percent in 1987. By 1995 it had risen to 41.5 percent. And, remarkably, by the year 2001 there were almost an equal number of cars and light trucks on the road (approximately 8.5 million of each). By 1999 the Ford F-Series and Chevrolet Silverado pickups were the best-selling vehicles in the nation.[43] Other trucks, Dodge Ram and Ford Ranger, were the fifth- and seventh-best selling, respectively, meaning that pickups had taken four of the top ten slots in America's most popular vehicles.

Together, this segment of the American fleet is one of the remarkable developments of American technology in the twentieth century. Data for the weight of light trucks are available only as of 1984. From this start around 3,500 pounds, average weights of light trucks rose over the next decade and topped 4,000 pounds in 1995. With the addition of the category of the large SUV, vehicle weights topped 5,000 pounds by 2006. The SUV allowed American consumers to manage the people and materiel of their busy lives, which were lived in a world connected by roads, driveways, and parking lots. A vehicle served as a device of family management and maintenance; however, the SUV always allowed its driver the potential outlet of a rugged drive through deep snow or mud.[44]

In a bitter irony, the CAFE standards and ensuing legislation had created the opportunity to build large, heavy, inefficient vehicles. And, to the shock of manufacturers such as Jeep's AMC and others, Americans wanted such vehicles.[45] What began as a gimmicky, small-selling vehicle for a specific purpose morphed into ubiquity through the odd convergence of consumer taste and auto manufacturers' interest in exploiting a specific niche in new vehicle regulations.

Conclusion: Unintended Consequences of the 1970s and Our Energy Transition

When DaimlerChrysler looked for a way to make its Ram standout in the full-size truck category in 2002, it stared into the face of increased gas prices, increasing competition for existing gasoline supplies, and even the need—by some estimates—for our nation to go to war to ensure petroleum supplies. Instead, the auto marketers peered into the soul of American consumers. With no possible explanation other than cultural "drive," Chrysler reached into its vault of outmoded technology and played its Hemi card. And, thanks to the Hemi option, U.S. sales of Ram pickups were up 13 percent in 2003 to a record 450,000. More than half of these buyers opted for the Hemi. Now Chrysler offers the Hemi—an engine designed for the racetrack and the battlefield, not the highway—in its 300C sedan, Magnum wagon, Durango, and Jeep Grand Cherokee.

A sector in denial, American vehicle consumers completed their twentieth-century petroleum binge by staring into the face of scarcity and pressing the pedal to the metal—driving more miles in heavier cars than at any other time in history.[46] The petroleum moment had not changed; in fact, scarcity has become even more pronounced in the first decade of the twenty-first century and gasoline prices cost American record amounts. For this reason, writer Paul Roberts in the *End of Oil* may be stating the obvious when he writes: "The SUV represent the height of conspicuous energy consumption. The extra size, weight, and power of the vehicles are rarely justified by the way their owners drive them. Even though owners and carmakers counter that the SUV's greater size, weight, and capabilities provide an extra margin of safety, studies indicate that SUVs not only are more likely to kill people in cars they hit but, because they roll over more easily, can actually be more dangerous to their occupants as well."[47]

However, just as we assume that the conservation efforts initiated by policy initiatives of the 1970s have been lost—roadkill against the bumper of a super-sized SUV—we find that simultaneous with Americans' fetish with larger, heavier trucks and SUVs a clear cultural initiative has forced alternatives into the mainstream. As mass Americans veered to largesse in the 1990s, the State of California worked with manufacturers and scientists to create the first serious alternative vehicles on American roadways. The EV fate, immortalized in the film *Who Killed the Electric*

Car, proved a warm-up to prepare Americans for the first release of hybrid and electric vehicles to mass consumers at the turn of the twenty-first century.

The hybrid rebirth began when the Toyota Prius went on sale in Japan in 1997, making it the world's first volume-production hybrid car. Today, the five-passenger Prius is the world's most popular hybrid, but it has been joined by the Honda Insight, first sold in the United States as a 2000 model, and the Honda Civic Hybrid, which came to market in the United States as a 2003 model. In addition, ethanol and other biofuels have now been released as additives and supplements in every state. Thus, we are forced to admit that the environmental ethic that emerged in the 1970s has not faded. Instead, it has persisted and even expanded from roots in the fringe to include mainstream initiatives available to all consumers. This also must be claimed as a legacy of the 1970s.

The reaction of lawmakers and consumers to the initial shock of scarcity in the 1970s was a remarkable moment in the history of policymaking as well as that of petroleum. When the shock subsided, American consumers returned to their nostalgic passion for muscle, albeit in the form of a utilitarian, multidimensional vehicle. Manufacturers accommodated them by enhancing the beefy vehicles with the luxurious styling of the large vehicles of the past. Vehicle design in the twenty-first century is unfolding in a profoundly different climate, which appears to make the passion for Hemis of the 1990s even more striking—the dichotomy of our rides a bit less pronounced.

Industry figures in 2002 speculated that by 2010 the number of SUVs on the market would increase while hybrids would make scant impact on the market. Instead, the reality has been much more dramatic. A confluence of historical and economic factors saw the American auto industry—vested in large, heavy vehicles—entirely collapse in 2009 amid additional pressures regarding petroleum supplies, climate change, and national security. Leveraging the moment of economic panic and resource conservation, federal stimulus funds saved American manufacturers and demanded that they remode for an international market and a new era of vehicle design.

Although such changes appear as cataclysmic shifts in the existing American culture of consumption, they also encourage historians to focus new emphasis on the 1970s. From our current vantage point, the cultural process stretching from 1973 to the present—with all of its befuddling, inexplicable intellectual paroxysms—is our energy transition away from petroleum. The 1970s initiated a transition away from petroleum that today is marked, simply, by extreme bifurcation. Love it or hate it, this, in a nutshell, is what our energy transition looks like. Each end of the petroleum-use spectrum has swollen: sales of some of the largest, heaviest domestic vehicles in human history have soared, while hybrid and alternative technologies are, at last, reaching American consumers as viable, realistic options.

There is no denying the unique portion of petroleum use through which we have lived in the past few decades. With low-priced gasoline, the automobile of the mid-1900s was more decorative than efficient. The wake-up of the embargo of the 1970s made consumers reappraise the roles of automobiles in their lives, and lawmakers responded with the first efforts to force manufacturers to prioritize efficiency. Carmakers complained bitterly and consequently exploited loopholes in the efficiency and pollution standards that created a generation of oversized, overpowered vehicles. It is tantalizing to stop there—to place all the blame at the feet of American auto manufacturers; however, Americans bought them. Detroit provided American drivers, for a host of cultural reasons, just what they wanted.

Why did American consumers want these vehicles? It is a cultural thing. A large segment of the population is in the throes of denying our energy transition from petroleum. In doing so, we reach back to the simpler days of the Dodge Hemi Charger and its equivalent in a super-sized SUV or pickup. Within the next few decades, though, such denial will inevitably emerge as folly.

Notes

This chapter derives from portions of my book under contract with the University of Chicago Press, tentatively titled "Declaring Our Dependence: Petroleum in 20th Century American Life." Readers may also wish to see the special June 2012 issue of the *Journal of American History* on "Oil in American Life," for which I served as coeditor.

1. Aldo Leopold, *A Sand County Almanac: And Sketches Here and There* (New York: Oxford University Press, 1949).

2. J. R. McNeil, *Something New under the Sun: An Environmental History of the Twentieth-Century World* (New York, W. W. Norton, 2000), 298; David E. Nye, *Consuming Power: A Social History of American Energies* (Cambridge, Mass.: MIT Press, 1999).

3. As societies attempt to facilitate and integrate specific resources into a cultural and economic energy regime, we move from the systematic level and begin to see the stories that occur when people, communities, and governments attempt to make an energy source part of their everyday life. Though policy, available supplies, and geopolitics each play an important role in determining energy usage patterns in the United States, consumer preferences must also be factored into the mix.

4. *New York Times*, Jan. 21, 1974.

5. See Karen R. Merrill's collection, *The Oil Crisis of 1973–1974: A Brief History with Documents* (New York: Bedford/St. Martins, 2007).

6. This is a common term used for the post–World War II era in which the American middle class expands and cold war rhetoric helps to stimulate consumption of nonessential goods and services.

7. Eugene A. Rosa, Gary E. Machlis, and Kenneth M. Keating, "Energy and Society," *Annual Review of Sociology* 14 (1988): 149–72. This survey of approaches by sociology to understand energy's role in human society is fascinating, particularly for its emphasis of public reaction to the shifts during the 1970s. This quote (168) specifically refers to sociologists' interest in consumer/public reaction to energy concerns.

8. Daniel Yergin, *The Prize: The Epic Quest for Oil, Money and Power* (New York: Simon and Schuster, 1991), 612. Yergin's seminal work on oil history remains the leading source for reconstructing the scope of the timeline of petroleum's emergence in world affairs.

9. The formation of OPEC was precipitated by changes in the oil market after World War II. Lacking exploration skills, production technology, refining capacity, and distribution networks, oil-producing countries were unable to challenge the dominance of the oil companies prior to World War II. Although Mexico wrestled control of its oil industry from foreigners in 1938, it quickly receded from the lucrative international market because of insufficient capital for investment. Other nations also attempted to set up their own arrangements for oil development: In 1943, Venezuela signed the first "fifty-fifty principle" agreement, which provided oil producers with a lump sum royalty plus a fifty-fifty split of profits. Iran passed a law demanding the termination of previous agreements with Anglo-Iran (referred to as Anglo-Persian prior to 1935 and British Petroleum after 1954) and then nationalized its oil operations in May 1951 when such an agreement failed to occur. A new British-Iranian agreement was signed the following year; the newly restored shah of Iran became a pillar of American Middle East policy until the Iranian revolution in 1979.

With the good business acumen that Rockefeller had installed in them at the turn of the century, oil companies, of course, disliked such arrangements. The control of price and profit was pulled out of their corporate hands and placed in those of the nations beneath which the oil happened to exist. Soon nations with these independent agreements were shunned by oil companies. As the new business model took shape during the late 1950s, oil-producing nations set out to leave the oil companies no choice but to work with all of them. They formed the first large-scale, international political group framed around a single resource—a cartel.

OPEC's founding members in 1960 were Iran, Iraq, Kuwait, Saudi Arabia, and Venezuela. Eight other countries joined later: Qatar (1961), Indonesia (1962), Libya (1962), United Arab Emirates (1967), Algeria (1969), Nigeria (1971), Ecuador (1973), and Gabon (1975) (Ecuador and Gabon withdrew from the organization in 1992 and 1994, respectively). What these nations had in common was oil. To varying degrees, though, they also shared small size and little political influence. Together, though, OPEC's purpose was obvious: to limit supplies in the hope of keeping prices high.

It seems ironic, today, to talk about oil producers needing to manipulate markets in order to keep the price of petroleum up; yet major oil companies colluded from the 1920s to the 1960s to prevent prices (and profits) from falling. As their influence waned, other methods were used. One of the biggest difficulties was that, as prices fell, domestic producers simply could not compete. Moreover, the Eisenhower administration concluded (as the Japanese had prior to World War II), dependence on foreign oil placed the country's national security in jeopardy. The United States responded with an import quota. The quota kept domestic prices artificially high and represented a net transfer of wealth from American oil consumers to American oil producers. By 1970 the world price of oil was $1.30 and the domestic price of oil was $3.18; see Albert Danielsen, *The Evolution of OPEC* (New York: Harcourt, 1982), 150.

10. The primary rationale for this action was politics. Israel's military aggression outraged its Arab neighbors throughout the late 1960s. Israel's attack on Egypt in 1967 resulted in an earlier embargo, which proved unsuccessful because of oversupply of crude on the world market. In October 1973, President Richard Nixon agreed to provide more military jets to Israel after a surprise attack on the nation by Egypt and Syria. On October 19, the Arab states in OPEC elected to cut off oil exports to the United States and the Netherlands.

In petroleum circles, the embargo is often referred to as the "first oil shock." Accordingly, it combines new market features of the early 1970s: first, production restraints that were ultimately supplemented by an additional 5 percent cutback each month; and, second, a total ban on oil exports to the United States and the Netherlands and eventually also Portugal, South Africa, and Rhodesia.

11. Although the embargo had economic implications, it had begun as a political act by Arab OPEC members. Therefore, the Nixon administration determined that it needed to be dealt with on a variety of fronts, including, of course, political negotiation. These negotiations, which actually had little to do with petroleum trade, needed to occur between Israel and its Arab neighbors, between the United States and its allies, and between the oil-consuming nations and the Arab oil exporters. Convincing the Arab exporters that negotiations would not begin while the embargo was still in effect, the Nixon administration leveraged the restoration of production in March 1974. Although the political contentions grew more complex in ensuing decades, the primary impact of the embargo came through the residual effects it had on American ideas of petroleum supply.

12. In reality, Nixon's energy czar, William Simon, took only restrained action. Rationing was repeatedly debated, but Nixon resisted taking this drastic step on the federal level. Although he had rationing stamps printed, they were kept in reserve. In one memo, Nixon's aid Roy Ash speculated that "in a few months, I suspect, we will look back on the energy crisis somewhat like we now view beef prices—a continuing and routine governmental problem—

but not a Presidential crisis." Nixon's notes on the document read "absolutely right" and, overall, his actions bore out this approach. Yergin, *Prize*, 644–45.

13. See, for instance, Stephen Fox, *The American Conservation Movement: John Muir and His Legacy* (Madison: University of Wisconsin Press, 1986).

14. "The history of transportation policy through the late 1960s can be best described as one of accommodation to the automobile. Politicians and planners alike viewed the rise of the automobile and the concomitant highway construction as signs of progress, as clear indications of the superiority of the American political and economic system." Michael D. Meyer and Marvin L. Manheim, "Energy Resource Use: Energy, the Automobile, and Public Policy," *Science, Technology, and Human Values* 5, no. 31 (1980), 25. This article offers a superb overview of the incredible shift of priorities toward the automobile that was brought by 1970s legislation.

15. Paul S. Sutter, *Driven Wild: How the Fight against Automobiles Launched the Modern Wilderness Movement* (Seattle: University of Washington, 2002).

16. See, for instance, Adam Rome, *The Bulldozer in the Countryside: Suburban Sprawl and the Rise of American Environmentalism* (Cambridge: Cambridge University Press, 2001).

17. In Gary Cross, *An All-Consuming Century: Why Commercialism Won in Modern America* (New York: Columbia University Press, 2000).

18. The most useful survey of the evolution of environmental policy in the 1970s is Richard N. L. Andrews, *Managing the Environment, Managing Ourselves: A History of American Environmental Policy* (New Haven, Conn.: Yale University Press, 1999), esp. 225–30; see also Robert Gottlieb, *Forcing the Spring: The Transformation of the American Environmental Movement*, 2nd ed. (New York: Island Press, 1994).

19. Amory B. Lovins, "Energy Strategy: The Road Not Taken?" *Foreign Affairs* 55 (Oct. 1976): 65–96; and Lovins, *Soft Energy Paths* (New York: HarperCollins, 1979), 121–22.

20. Daniel Horowitz, Jimmy Carter and the Energy Crisis of the 1970s: The "Crisis of Confidence" Speech of July 15, 1979 (Boston: Bedford/St. Martin's, 2005), 20–25.

21. Ibid., 43–46.

22. Christopher Finch, *Highways to Heaven: The Auto Biography of American* (New York, HarperCollins, 1992), 321.

23. Unknowingly caught in Richard Nixon's Oval Office recordings in 1971, Ford Motor Company president Lee Iacocca called the inclusion of safety features such as head rests and seat belts as standard items in American autos "complete wastes of money." Jim Motavelli, *Forward Drive: The Race to Build "Clean" Cars for the Future* (San Francisco: Sierra Club Books, 2001), 41.

24. Quoted from Jack Doyle, *Taken for a Ride: Detroit's Big Three and the Politics of Pollution* (New York: Four Walls Eight Windows, 2000), 64.

25. Quoted in Hugh S. Gorman, *Redefining Efficiency: Pollution Concerns, Regulatory Mechanisms, and Technological Change in the U.S. Petroleum Industry* (Akron, Ohio: University of Akron Press, 2001), 315–17.

26. Ibid.

27. Quoted in Doyle, *Taken for a Ride*, 148.

28. There is a healthy scholarly debate over whether CAFE standards were an appropriate response to the energy crisis. For instance, I have consulted Andrews, *Managing the Environment.* For the purposes of this article, though, it is not necessary to discuss their merit or outcomes.

29. The battle over how far CAFE standards and emissions controls would extend required the auto industry to flex its political muscle like never before. Very quickly, the health and safety concerns morphed into threats of inflated prices on American cars and the economic threat of foreign autos encroaching on the American market. After meeting with Nixon during 1972–73, industry leaders altered their approach. When they met with President Gerald Ford in 1975 the auto industries offered to accept a 40 percent improvement in mileage standards if Congress would ease standards on emissions. Ford agreed and presented this policy to American consumers in his State of the Union address. Although Congress protested, this division (accepting CAFE while relaxing emissions) became the rallying point for the auto industry during the 1970s.

30. Doyle, *Taken for a Ride,* 240.

31. Rudi Volti, *Cars and Culture: The Life Story of a Technology* (Westport, Conn.: Greenwood Press, 2004), 124.

32. One other initiative begun in the 1970s was a federally mandated speed limit. In the 1970s, federal safety and fuel conservation measures included a national speed limit of 55 miles per hour (currently, consumers have led some states to loosen such restrictions); Doyle, *Taken for a Ride,* 251–62. Journalist Paul Roberts writes that perhaps "the most discouraging example of how developed nations misspend their efficiency dividend is transportation—and nowhere more so than in the United States"; Roberts, *The End of Oil: On the Edge of a Perilous New World* (New York: Houghton Mifflin, 2004). Prior to the early 1970s, energy costs were "trivial," and, therefore, "carmakers made no effort to build cars that were fuel-efficient." This changed with the implementation of CAFE standards.

33. David Gartman, *Auto Opium: A Social History of American Automobile Design* (New York: Routledge, 1994), 193.

34. Quoted in Meyer and Manheim, "Energy Resource Use," 24–30.

35. Volti, *Cars and Culture,* 125.

36. Finch, *Highways to Heaven,* 298–99.

37. Ibid., 318–19.

38. Ibid., 306.

39. See for instance, data from *Transportation Energy Data Book: Edition 26–2007* (Washington, D.C.: Department of Transportation, 2007). I have also consulted data collected in Paul E. Godek, "The Regulation of Fuel Economy and the Demand for 'Light Trucks,'" *Journal of Law and Economics* 40, no. 2. (1997), figure 2.

40. Godek, "Regulation of Fuel Economy," 503.

41. Keith Bradsher, *High and Mighty—SUVs: The World's Most Dangerous Vehicles and How They Got That Way* (New York: Public Affairs, 2002), 29.

42. Volti, *Cars and Culture,* 143.

43. Data findings are derived from *Transportation Energy Data Book* and *Automotive News Annual Reports.*

44. In addition to data collected by the DOT and manufacturing groups, I have consulted a scholarly literature concerning the effectiveness of CAFE standards. Most of these articles are based in economics and sociology. In the emergence of light trucks/vehicles in particular, I have used Godek, "Regulation of Fuel Economy," 495–509.

45. Bradsher, *High and Mighty*, 28–30.

46. Tables 3.4–3.5 in the *Transportation Energy Data Book 2007* lists total highway miles traveled in the United States as follows: 1970, 10.8 million vehicles traveling 1,109,724 miles; 2001, 24 million vehicles traveling 2,989,807.

47. Roberts, *End of Oil*, 154.

Chapter 11

Environmentalism and the Electrical Energy Crisis

Robert Lifset

In May 1974, Charles Luce stood in the glass-walled visitor's gallery and gazed down at the floor of the New York State legislature while nervously rubbing his hands. Luce was the chairman of Consolidated Edison of New York, the utility servicing New York City and Westchester County, and had spent the night lobbying legislators. Now, at 6 A.M., after a twenty-hour session, he watched as the Republican-controlled legislature passed a half-billion-dollar bailout of his company amid jeers from Democrats.[1]

This event marked the first time that a state had taken such action to preserve a major public utility, and it likely prevented the nation's largest private utility company from having to declare bankruptcy. How was it possible that a company with a monopoly on the sale of gas and electricity in the nation's largest city and permitted to create a rate structure designed to guarantee a rate of return on its investment could find itself on the brink of bankruptcy?[2] The answer is that Con Ed was painfully experiencing a series of problems endemic throughout the American utility industry during these years. These problems constitute what I refer to here as the "electrical energy crisis."

American utility companies found themselves in financial difficulty because the business model they had relied upon for the first six decades of the twentieth century was falling apart. The historian Richard Hirsh has argued that utility companies had long been successful in encouraging demand while expanding supply and reducing the price of electricity by building increasingly larger power plants, thereby taking advantage of economies of scale. But this model broke down in the 1970s for three reasons: technological stasis, the energy crisis, and the environmental movement.[3] In this essay I quickly summarizes the first two factors and focus on investigating how and why environmentalism became a factor in spurring the electrical energy crisis. I then argue that this crisis transformed the industry by enshrining conservation and efforts to reduce demand as essential components of thriving in a post–energy crisis environment.

The ability to build larger, more efficient plants hit a technological wall in the 1970s. For decades, greater efficiencies had been possible by building larger plants, thereby gaining from increasing economies of scale. These advances can be seen in the successful efforts to improve thermal efficiency (the percentage of a fuel's energy content actually converted into electricity). Thomas Edison's first generating station, built in 1882, had a thermal efficiency of 2.5 percent. By 1965 the average thermal efficiency was 33 percent. In the 1960s, manufacturers began to discover that improving thermal efficiency was producing diminishing returns, with metallurgical problems appearing at around 40 percent. Manufacturers and utilities learned that less-efficient plants could be run more reliably.[4]

Hoping to overcome the decline of thermal efficiency improvements and meet increasing demand, utility companies tried to build larger power plants. Lacking the time to test and slowly introduce larger turbines, manufacturers extrapolated from existing designs and produced equipment that frequently broke down. Utility companies were in a race to keep up with demand, and they were losing.[5]

In the 1960s and early '70s, utility companies were beginning to use ever larger quantities of oil to generate electricity. Oil was a cheap fuel; domestically produced oil dropped 30 percent in price from 1957 to 1970. It was also cleaner burning than coal, an important consideration for urban utility companies striving to meet new air pollution requirements. While oil consumption increased, its domestic capacity diminished, making Americans more reliant on imported oil from the Middle East. The timing here is important.[6] At the very moment when American utilities are beginning to switch away from coal to oil, they were confronted with the Organization of Arab Petroleum Exporting Countries (OAPEC) oil embargo of 1973, which drastically raised fuel costs. With these kinds of fuel costs, prices would no longer decline. As a result of these developments (technological stasis, the energy crisis), utility companies lost the ability to meet demand while lowering prices. In Con Ed's case, simply meeting demand became a significant challenge.[7]

Short on Power

Con Ed's inability to meet demand can be seen in the nearly annual blackouts that afflicted New York in the late 1960s and early 1970s. City-wide blackouts in 1965 and 1977 attracted national attention. But more typical were the smaller periodic blackouts that contributed to what many felt was a declining quality of life in the city. One example can be found in the summer of 1972 when a blackout left over half a million people in Brooklyn without power. Those unaffected by the power failures experienced three voltage reductions during a week in which the temperature at Kennedy International Airport reached 100 degrees.[8]

Sections of Brooklyn buildings had no elevator service, no air conditioning, no water, and limited lighting. Residents were said to be sleeping in lobbies because they could not climb the stairs. The beach at Coney Island saw over a million visitors that weekend, a record for a nonholiday. Supermarkets, butcher shops, restaurants, and small mom and pop stores were hit hard by the unreliable power, losing thousands of dollars in stock and business. Merchants along Brighton Beach Avenue in Brooklyn blockaded the street with their cars for several hours to dramatize their plight. They were given parking tickets by the police. Blackouts of this type, limited to a part of the city, typically did not attract a great deal of national attention. An editorial in the *New York Post* musing over what had just happened speculated that some "of Con Edison's afflictions may be incurable, as we have noted before, given the present state of technology, the environment, the inability of mere mortals to cope with megalopolis."[9] These episodes helped contribute to a growing sense that the city was becoming unlivable and ungovernable.

Environmentalism

To be sure, environmentalism exacerbated the crisis. Before the 1960s, utility companies enjoyed almost complete autonomy in choosing production and transmission technologies, the fuel used in their system, the location of generating plants, and the type of air and water pollution control systems employed. The historian Joe Pratt has noted that by the 1970s utilities had lost this autonomy as new environmental laws and regulations scrutinized and slowed down the ability of utility companies to build new plants and determine the fuels used to generate electricity. Perhaps most important, environmentalism questioned the growth-centered ideology and business model of the industry. Increasing power consumption had long been viewed as critical to the growth of the overall economy; as the historian David Nye has argued, it literally fueled the American way of life. Environmentalists were attracted to those critics who called into question this ideology of growth. Paul Ehrlich, E. F. Schumacher, Denis Meadows, Barry Commoner, and Amory Lovins all provided the intellectual ammunition with which environmentalists sought to focus the attention of utility companies on the demand for energy.[10]

But it should be noted that environmentalists attacked utilities because quite a large number of environmental problems are energy related. And though expanding energy production did garner the attention of conservationists looking to preserve a particular stretch of river or the sanctity of the national park system in the first half of the twentieth century, it was in the postwar years during an intensifying interest in pollution that utility companies found themselves in the crosshairs of environmentalists.

This increasing interest in pollution can be seen as both an inspiration for and result of Rachel Carson's *Silent Spring*. Focused on pesticides, this book spoke to a larger concern for the toll humans were taking on the environment. The message of *Silent Spring* was that humans were endangering their own lives through an arrogant and manipulative attitude toward other forms of life. This argument, cast in anthropocentric terms, resonated powerfully in a nation where the evidence for it appeared to be everywhere.[11]

Air

When New Yorkers began to think seriously about how their air and water were befouled, Consolidated Edison emerged as a significant culprit. This was made clear when in the spring of 1966 the Mayor's Task Force on Air Pollution reported its findings. Directed by Norman Cousins, editor of the *Saturday Review*, the report found that New Yorkers suffered from some of the worst air pollution in the country. Although many apartment buildings throughout New York still used coal-burning furnaces, Con Edison, as late as 1965, was still responsible for more coal smoke than any other source. New York had more sulfur dioxide gas in its air than any other American city, and Con Edison was cited as the single biggest contributor of this deadly poison.[12]

Widespread disaster had been averted only by a topography that enhances the cleansing effects of the prevailing winds. If New York had the same sheltered topography as Los Angeles, the city would be uninhabitable. In an interview, Cousins said that unless something was done the city would face a "disaster of substantial proportions," that under certain conditions "it is quite possible for New York to become a gas chamber."[13]

The report noted that at least three times in recent years stagnation of air loaded with gases and particles had resulted in a sudden rise in deaths (a statistical analysis revealed that heavy pollution for a two-week period in November 1953 caused the death of 240 New Yorkers). As if to emphasize the point, there was heavy smog during the Thanksgiving weekend that year, spurring the mayor, deputy mayor, and hospitals commissioner to all assure the public that the air had not killed anyone (168 deaths would later be blamed on the bad air that week). Mayor John Lindsey appeared hesitant to call for strict new standards, consistently referring to the fact that significant efforts to clean up the air would most likely involve a tax increase. It did not help that Cousins's report faulted the city's incinerators and buses as the worst violators of its own air pollution laws. But a significant amount of air pollution was created by Con Ed. The chief of the Air Pollution Division of the U.S. Public Health Service noted that sulfur dioxide, largely emitted from coal and oil the company used in its power plants, was found in New York at levels ten times above that which affect health. The Ralph Nader Study Group on Air Pollution concluded in 1970 that the city's air was responsible for the premature deaths of one to two thousand New Yorkers every year.[14]

The company's response was largely to deny to New Yorkers that a serious problem existed while working behind the scenes to reduce pollution emissions. One outside attorney working for Con Ed remembers that the one thing the company would not allow its lawyers to put into evidence, in their effort to license a controversial proposal for a pumped-storage hydroelectric plant at Storm King Mountain, was the company's private records on the state of New York City's air pollution.[15]

This record of publicly denying that a problem existed served to weaken the company when the political will for addressing the air pollution problem gained momentum in the 1960s. A series of local, state, and federal laws passed in the 1960s and '70s forced the company to adopt more advanced pollution control measures.[16]

Water

Water pollution was another serious problem the company was reluctant to admit publicly. In the winter of 1962/63, Hudson River fisherman began to notice large numbers of crows concentrating around a dump near Indian Point, a small peninsula jutting out into the river about 30 miles north of Manhattan. The crows were feasting on tremendous numbers of rotting fish. The fish were attracted to the hot water discharge of the new Indian Point nuclear power plant. This attraction intensified in the winter months as the fish became trapped in a nearby pier constructed to protect debris from the intake pipes. Some fish would swim into the nuclear plant, where they would meet their death. Most crowded under the pier, densely packed and trapped by the more recent recruits behind them, where they would die of starvation. The company installed a wire basket elevator to remove the dead and dying fish and trucked them to a nearby dump.[17]

On June 12, 1963, three Hudson River fishermen made a visit to Indian Point. The three men were moved and angered by what they had seen. As one recalled, "We saw ten thousand dead and dying fish under the dock. We learned that Con Ed had two trucks hauling dead fish to the dump when the plant was in operation." Pictures were taken by local fishermen and state Conservation Department officials of the dead fish rotting at the dump. In response, the company erected a fence and posted guards. The Conservation Department began quietly confiscating all the pictures it could acquire.

The December 1963 issue of *Southern New York Sportsmen*, a magazine devoted to hunting and fishing, published a photograph that had escaped the Conservation Department. It was captioned with the following commentary: "Enclosed is a photo taken one evening in early March and showing just one section of the dump. The fish seen here were supposed to be about 1 or 2 days' accumulation. They were piled to a depth of several feet. They covered an area encompassing more then a city lot." *Sportsmen* estimated the fish kill at about one million.

The assistant commissioner of New York's Conservation Department complained in a letter to the editor of *Sportsmen* that his numbers were unrealistic. He wrote that the peak kill was only eight hundred per day and of mostly smaller fish. Some fishermen later visited the regional office of the Conservation Department in Poughkeepsie to inquire about the fish kill and to find out what was being done about it. Officials denied ever having seen any pictures. They offered to open their files, then later claimed the files could not be made public while the matter was under investigation. *Southern New York Sportsmen* did not have a particularly large circulation, and Con Ed's efforts to suppress the story of what was happening at Indian Point remained largely successful until the fishermen found Bob Boyle, a young reporter at *Sports Illustrated* who was working on a piece about the Hudson River.[18]

Boyle's environmental consciousness had been growing in the early 1960s. He found an outlet for this concern in his work. "I started doing a series of articles like that for the magazine [*Sports Illustrated*]. We had a very large franchise at the time, and the reason we did it is that half the magazine was devoted to hunting and fishing or outdoor sports: sailing and skiing, whatever, people went outside and they ran into smog or they went into the water and it was polluted and they should know."[19]

Boyle finally managed to obtain a picture of the Indian Point fish kill, and he featured it in a *Sports Illustrated* article titled "A Stink of Dead Stripers." The article exposed the New York State Conservation Department for denying not only that it possessed pictures of the fish kill but that such pictures even existed.

The exposure helped to foster a growing opposition to a power plant Con Ed was then proposing for the west bank of the Hudson River a few miles north of Indian Point: a pumped-storage hydroelectric plant at Storm King Mountain. The company's expert witness on the impact of electricity production on the river's fish had testified (before the Federal Power Commission), under oath, that no studies had been conducted on Hudson River fish since the 1930s. Boyle discovered that this scientist had personally oversaw a 1956 study that concluded that the Hudson River striped bass spawned precisely where Con Ed wanted to place this new plant.[20] Boyle exposed these discrepancies in *Sports Illustrated* and in testimony before a congressional committee. But the Federal Power Commission had already completed taking testimony on the new license application, consistently refused to reopen the hearings, and issued Con Ed a license to build the Storm King plant in March 1965.[21]

Although these concerns among New Yorkers for the quality of the air and water were not new, the power that environmental organizations would soon wield and their ability to encroach upon the traditional authority of utility companies did change. That change took a dramatic turn when environmentalists sought to overturn Con Ed's license for its Storm King plant.[22]

Tools

Prior to the 1960s, activists struggled to highlight the environmental problems with energy projects. In the first half of the twentieth century many of the most prominent environmental struggles were waged against federal efforts to build giant hydroelectric dams across the American West. The struggles over Hetch Hetchy (1901–13) and Dinosaur National Monument (1950–56) have been well documented, for they stand as defining moments in the evolution of American preservationism.[23] Let us consider the manner in which that opposition was expressed. Because these were projects being constructed directly by the federal government with taxpayer money, the environmental organizations leading the struggle focused their attention first on the executive branch and then on Congress in their efforts to kill a project. This congressional lobbying was accompanied by a media strategy that sought to persuade the electorate that these projects were bad policy.

Although the successes and failures were felt more acutely, there were real benefits and drawbacks to this strategy of environmental conflict. Among the benefits can be counted the fallout from the publicity generated in the struggle itself. Each one of these struggles served to raise environmental consciousness and refine the skill with which environmental advocates might influence the public and their elected officials. Among the drawbacks was the intense investment of time, energy, and resources necessary to have a chance at stopping a single project.

The nature of environmental conflict significantly changed in the 1960s. At the heart of this change stands a legal decision that granted environmental advocates, for the first time, access to the federal courts. This change in American jurisprudence would come to have a powerful influence on environmental advocates and the nature of environmental conflict from the 1960s onward. In 1965, the 2nd Circuit Court of Appeals overturned the Federal Power Commission (FPC) license to build the hydroelectric plant at Storm King Mountain. A local environmental organization, Scenic Hudson, had sued the FPC, claiming that it had not fully developed the record, not satisfactorily developed the facts necessary to make informed judgments. Perhaps more important for our purposes, the FPC countered in its brief to the 2nd Circuit that Scenic Hudson had no standing to sue in federal court.

Environmental advocates had long attempted to use the federal courts to stop the government from licensing or building projects that promised to invade the national parks, destroy wilderness, or otherwise degrade the environment. These lawsuits inevitably failed. They failed because the federal courts had consistently denied environmental advocates standing to sue. Before you can successfully argue a case in court, you have to convince the court that you have a right (or "standing") to be there.[24]

In the Scenic Hudson case, in briefs filed with the court, the FPC argued that Scenic Hudson lacked a personal economic interest and therefore had no standing to sue. For some time, this had been the test for plaintiffs lacking a legislatively created interest. But the court disagreed with the FPC. The court found that (1) there was nothing in the Constitution that required that an aggrieved or adversely affected party have a personal economic interest; and (2) that the Federal Power Act protects noneconomic as well as economic interests.[25] The court held that in order to protect the public interest in "the aesthetic, conservationist, and recreational aspects of power development, those who by their activities and conduct having exhibited a special interest in such areas, must be held to be included in the class of 'aggrieved' parties."[26]

The decision was groundbreaking because it appeared to grant environmental causes access to the federal courts; when the court expanded the zone of public interests by including the environment, it strengthened environmental litigants. The court rejected the argument that this would open the floodgates to thousands of interveners, citing the time and expense necessary to intervene. The court found Scenic Hudson to be aggrieved, that it had a special interest in this area and legal right to protect that interest.[27]

A decision handed down in federal district court soon thereafter described the effect of this decision as being that neither economic injury nor a specific individual right were necessary adjuncts to standing. A plaintiff now need only demonstrate being an appropriate person to question the agency's alleged failure to protect a value specifically recognized by federal law as "in the public interest." Thus, the Scenic Hudson decision allowed citizen groups seeking to protect the public interest the ability to seek judicial review of federal agency decisions. This option gave these groups participation in the decision-making process.[28]

Yet this would do very little good if public interest representatives were to shoulder the burden of developing the environmental record. The Scenic Hudson decision admonished the FPC for failing to develop a full and complete record; if the commission is to act in the public interest, the court found, the FPC's role as representative of the public "does not permit it to act as an umpire blandly calling balls and strikes for adversaries appearing before it; the right of the public must receive active and affirmative protection at the hands of the Commission."[29]

This decision cast a long shadow over both the drafting and interpretation of new environmental legislation in the late 1960s and 1970s. The role prescribed for federal agencies under the National Environmental Policy Act (NEPA) of 1969 was premised on the role the court prescribed for the FPC in the Scenic Hudson case. The agencies had a duty to see that the record was complete and had an affirmative responsibility to inquire into and consider all relevant facts. The government could not burden litigants with the task of doing this job for it; rather, it should tap the expertise available to it within the federal bureaucracy and shift some of this burden to the applicant. As a result, the production of environmental impact statements would be the responsibility and financial burden of the government or applicant and not of those interest groups representing the public.[30]

This change in the jurisprudence of standing was ratified by the Supreme Court in *Sierra Club v. Morton* (1972). In this case, the Sierra Club sued the secretary of the interior for violating several laws and regulations when he approved a plan by the Disney Corporation to build a ski resort at Mineral King Valley in the Sierra Nevadas. The Sierra Club argued that its long-standing concern with and expertise in environmental matters granted it sufficient standing to act as a representative of the public. The Supreme Court actually rejected this view, holding that a mere interest in a problem is not sufficient to render the organization "aggrieved" or "adversely affected," a finding necessary to have standing.[31]

This minor setback was overcome when in the very same decision the Supreme Court embraced the reasoning of an earlier nonenvironmental case by ruling that environmental and aesthetic interests could be sufficient to constitute an injury in fact. When, in this decision, the Supreme Court held that "aesthetic and environmental well-being, like economic well-being, are important ingredients of the quality of life in our society," it was expanding the zone of interests available to potential litigants by writing into federal law a concern for the environment that can first be seen in the Scenic Hudson decision. The federal judiciary had handed environmental advocates a powerful new tool to influence land-use decision making. This tactic, the lawsuit, initiated a new style of environmental conflict.[32]

A New Landscape

Lawsuits were used to buttress the enforcement of the environmental legislation of the 1960s and 1970s. NEPA and the "little NEPAs"—versions of this national bill passed by several states—mandated environmental impact statements for all projects funded with federal (or state) money. The newly established Environmental Protection Agency (1970) enforced emission standards for several pollutants specifically detailed in the Clean

Air Act of 1970 and the Water Pollution Control Act of 1972. These laws were revisited and strengthened by Congress later in the decade. This served to encroach further upon the prerogatives of the utility industry.[33]

Indeed, we can trace specific provisions of these laws to forms of pollution relatively unique to power plants. The Clean Water Act of 1972 deliberately included language that defined thermal discharges as a form of pollution. Thermal water pollution was a problem primarily experienced by the utility industry, which typically sited power plants near large bodies of water so as to use the water as part of a plant's cooling process.[34] But in addition to regulating power plant discharges and forcing utilities to ponder environmental considerations in siting new power plants, the new legislation expanded the zone of interests and legal rights environmentalists could sue to protect (as well as creating new forums in which environmentalists could oppose proposed power plants). Furthermore, many of these laws explicitly contained citizen suit provisions that effectively incorporated this change in the jurisprudence of standing. Environmentalists now had a statutory right to sue in federal court.

This served to empower the grassroots of the nation's environmental movement at the very moment it was growing exponentially. Consider the types of organizations that tended to mobilize in opposition to the environmental consequences of energy production in the first half of the twentieth century. The Sierra Club and the Wilderness Society worked to prevent federal dams from being built at Hetch Hetchy and Dinosaur National Monument. These were large, regional and national organizations, recognized as leaders of the American conservation movement. When we examine the types of groups opposing energy production along environmental grounds in the second half of the century, especially after 1965 (the year of the Scenic Hudson decision), we see relatively obscure grassroots coalitions, often temporary organizations put together for a specific fight.[35]

The change in jurisprudence certainly empowered those opposed to a power plant at Storm King. After an additional fifteen years of hearings and court cases, Con Ed dropped its plans for the plant.

Conclusion

The historian Richard Vietor has written that it was not a coincidence that the environmental movement and energy crisis occurred together, for they had common origins in increasing energy demand. This served to tax the environment's ability to absorb new energy production while depleting cheap reserves of domestic oil and natural gas. Vietor argues that this led to an increasing reliance on coal and the search for oil and natural gas in more environmentally sensitive environments (offshore Alaska). In restricting the expansion of energy production, environmentalism contributed to the electrical energy crisis.[36]

What needs to be added to this story, as illustrated here, is an understanding of how radical changes in the jurisprudence of standing served to push power and agency down toward the grassroots. While local, state, and federal governments increasingly concerned themselves with the business of power generation (and especially its environmental impacts), both the government and the utilities were held to account by a newly energized and empowered environmental movement. This produced a more vigorous enforcement of environmental regulations while changing the ambitions and possibilities for grassroots environmentalism.

I think it can be argued as well that this change contributed to the decline of the nuclear power industry in the 1970s. It has also served as a brake on the market share of coal while adding to the attractiveness of natural gas. Even without a carbon tax, the environmental consequences of energy production, the externalities of the business, began to affect the bottom lines of energy companies the moment environmentalists were first empowered with access to the courts. As a result, it has changed the energy landscape in America, shaping the politics and economics of energy production and serving as a factor in determining how the United States generates power.

This essay focuses on how environmentalists acquired the power to influence land-use decisions, but the movement was not content to simply play defense, to block projects that threatened the environment. Environmentalists argued that the destructive impact of power generation could be reduced by embracing conservation and energy efficiency. Con Ed's problems were so severe that the company began to take seriously this criticism and adopt conservation measures. In the early 1970s, the company was facing both a capacity shortage and, after the OAPEC embargo, a fuel shortage. In the teeth of a strong and energized local environmental movement (as well as the challenges discussed at the beginning of this essay—technological stasis and the energy crisis), Charles Luce had no choice but to abandon a business model premised on encouraging the growth of electricity consumption. In 1971, Con Ed disbanded its sales force and launched the "Save-a-Watt" campaign. This new marketing effort sought to exhort the public to use less electricity and educate consumers on how to use energy more efficiently.[37]

These efforts would come to be embraced by state regulators and consumer advocates across the nation who saw in conservation a means to avoid the persistent rate hikes required by utilities bent on encouraging and meeting new demand.[38] Firmly in place thirty-five years later, national conservation efforts accounted for a reduction in electricity consumption of 78 billion kilowatt hours of electricity in 2009, saving utilities and their customers $3.6 billion dollars.[39] Environmentalism played a role in producing the electrical energy crisis, but it also pointed the way toward a long-term, sustainable solution.

Notes

1. "Albany Approves $500-Million Aid To Rescue Con Ed," *New York Times*, May 17, 1974; "Area Lawmakers Oppose State Buying Con Ed Plants," *Times Herald Record*, May 14, 1974; the bill passed 42–17 in the Senate and 85–63 in the Assembly; "Con Ed Sale of Plants Scored," *New York Times*, June 6, 1974. The bailout was controversial because Consolidated Edison had grown increasingly unpopular since the early 1960s; Democrats represented the bulk of Con Ed's downstate customers. Part of the company's unpopularity was due to the constant rate increases during these years. The company's average electricity consumer lived in an apartment and used 250 kilowatts of electricity per month. Between 1945 and 1970, the monthly bill for 250 kilowatts of electricity rose from $7.95 to $11.05. Between 1971 and 1974, the average monthly bill rose from $10.95 to $20.63; "Con Edison's Money Problems Are Serious and May Get Worse," *New York Times*, Mar. 31, 1974. This represents, adjusted for inflation into 2008 dollars, an increase from $57.59 in 1971 to $89.11 in 1974.

2. Consolidated Edison was caught in a vicious cycle in which rising interest rates and increasing demand for electricity led to high-priced construction requiring new financing at higher interest rates, which required higher rates to consumers to cover the interest and attract new investment. The Public Service Corporation, its regulator, generally gave the company a percentage of its requested rate hikes, producing a gradual financial erosion. Higher fuel prices in the 1970s drove the company to the brink of bankruptcy. See Joseph Pratt, *A Managerial History of Consolidated Edison, 1936–1981* (New York: Consolidated Edison Company of New York, 1988), 91.

3. This description of the factors that contributed to an energy crisis in the utility sector is taken from Richard Hirsh, who described the crisis as the breakdown of a "utility consensus"; see Richard Hirsh, *Power Loss: The Origins of Deregulation and Restructuring in the American Electric Utility System* (Cambridge: MIT Press, 1999). On how utility companies encouraged demand, see Mark H. Rose, *Cities of Light and Heat: Domesticating Gas and Electricity in America* (University Park: Pennsylvania State University Press, 1995), and Harold L. Platt, *The Electric City: Energy and the Growth of the Chicago Area, 1880–1930* (Chicago: University of Chicago Press, 1991).

4. Efficiencies were gained by increasing steam temperatures and pressures. Yet thermodynamic theory limited steam systems to a top efficiency of 48 percent.

5. Hirsh, *Power Loss*, 58. Under these circumstances one can see why nuclear power appeared so attractive to utility executives in the 1960s and 1970s. Nuclear power generated no air pollution, the chief environmental complaint utility companies faced; indeed, it was not until well into the 1970s that environmental opposition to nuclear plants could be counted upon. Nuclear plants could generate large amounts of power from relatively small amounts of domestically mined uranium. As fuel costs skyrocketed in the 1970s, this made nuclear plants, with virtually no fuel costs, look very attractive. Nuclear energy was the last hope of the growth strategy.

6. In 1973, 21 percent of the nation's electricity was generated by burning petroleum; Table 8.2a: Electricity Net Generation: Total (All Sectors), 1949–2010, in *Annual Energy Review 2011* (Washington, D.C.: U.S. Energy Information Administration, 2012).

7. Hirsh, *Power Loss*, 60–61.

8. "New Power Cuts Plague Brooklyn with 'Brownouts,'" *New York Times*, July 20, 1972; "Heat Equals '72 Record; Relief Is Due Tomorrow," *New York Times*, July 22, 1972; "Overload in Brooklyn; 300,000 Left in Dark," *Times Herald Record*, July 18, 1972.

9. "Aftermath of the Nightmare," *New York Post*, July 28, 1972; "New Power Cuts Plague Brooklyn with 'Brownouts,'" *New York Times*, July 20, 1972; "Heat Equals '72 Record; Relief Is Due Tomorrow," *New York Times*, July 22, 1972; "More Blackouts in Two Boroughs; 400,000 Affected," *New York Times*, July 25, 1972; the *Times* reported that 1,500 people slept on the beach that Tuesday night, July 23. The title of a political history of the era reflects this pessimism: Vincent Cannato, *The Ungovernable City: John Lindsay and the Struggle to Save New York* (New York: Basic Books, 2001).

10. See Pratt, *Managerial History*, 255; David Nye, *Consuming Power: A Social History of American Energies* (Cambridge: MIT Press, 1999); Paul Ehrlich, *The Population Bomb* (New York: Ballantine Books, 1968); E. F. Schumacher, *Small Is Beautiful; Economics As If People Mattered* (New York: Harper and Row, 1973); Denis Meadows and the Club of Rome, *The Limits to Growth: A Report for the Club of Rome's Project on the Predicament of Mankind* (New York: Universe Books, 1972); Barry Commoner, *The Poverty of Power: Energy and the Economic Crisis* (New York: Knopf, 1976); Amory Lovins, "Energy Strategies: The Road Not Taken?" *Foreign Affairs* 55 (Oct. 1976).

11. Rachel Carson, *Silent Spring* (Boston: Houghton Mifflin, 1962); see also Linda Lear, *Rachel Carson: Witness for Nature* (New York: Henry Holt, 1997).

12. "Air Study Finds Pollution Here Worst in Nation," *New York Times*, May 10, 1966. An account of New York City's struggle with air pollution can be found in Scott Dewey, *Don't*

Breathe the Air: Air Pollution and U.S. Environmental Politics, 1945–1970 (College Station: Texas A&M University Press, 2000), 113–74.

13. "City Panel Urges Steps to Clean Air," *New York Times*, Apr. 11, and "Air Study Finds Pollution Here Worst in Nation," *New York Times*, May 10, 1966. This is very charged language coming only twenty years after the end of the Shoah in a city where nearly a third of the population had Jewish heritage.

14. "Hospitals Report Public Unharmed by 3 Days of Smog," *New York Times*, Nov. 29, 1966; Neil Fabricant and Robert Marshall Hallman, Toward a Rational Power Policy, Energy, Politics and Pollution: A Report by the Environmental Protection Administration of the City of New York, Jerome Kretchmer, Director (New York: George Braziller, 1970), 16, 27; John Esposito and John C. Nader, Vanishing Air: Ralph Nader's Study Group Report on Air Pollution (New York: Grossman, 1970).

15. Sheila Marshall Interview, August 29, 2002. Con Ed was both a leading contributor to New York City's air pollution and a national leader in pollution control technology; see Pratt, *Managerial History*, 263. For insight into how other utilities faced similar challenges, see Marc J. Roberts and Jeremy S. Bluhn, *The Choices of Power: Utilities Face the Environmental Challenge* (Cambridge, Mass.: Harvard University Press, 1981).

16. In December 1966, Consolidated Edison signed a memorandum of understanding with the City promising to undertake additional research programs, to shut down the oldest generating plants in the city, to use more natural gas, and to try to build additional power plants outside the city. Though not legally binding, the memorandum was a sign of increased political interference with the company's prerogatives; see Pratt, *Managerial History*, 273–75.

17. A good portion of this account can be found in Robert Boyle, *The Hudson: A Natural and Unnatural History*, 2nd ed. (New York: W. W. Norton, 1979), 159–68. Many of the fish being killed were striped bass; for additional insight into the elevated status of striped bass among amateur anglers, see Dick Russell, *Striper Wars* (Washington, D.C.: Island Press, 2005).

18. Boyle, *The Hudson*, 161. Fish kills and thermal pollution would become significant problems for the advancement of nuclear power and a front in the struggles between nuclear proponents and environmentalists; see J. Samuel Walker, "Nuclear Power and the Environment: The Atomic Energy Commission and Thermal Pollution, 1965–1971," *Technology and Culture* 30, no. 4 (1989): 976–79.

19. Robert Boyle interview with author, Aug. 22, 2001.

20. Voluntary reports to the Federal Water Pollution Control Administration estimated 735,000 fish killed by thermal pollution for the period August 1962 to June 1969; see Fabricant and Hallman, *Toward A Rational Power Policy*, 53.

21. *Sports Illustrated*, Apr. 26, 1965, 84; Hearings before the Subcommittee on Fisheries and Wildlife Conservation, of the Committee on Merchant Marine and Fisheries, on Hudson River Spawning Grounds, House of Representatives, 89th Congress, 1st Session, May 10–11, 1965; Opinion and Order Issuing License and Reopening and Remanding Proceeding for Additional Evidence on the Location of the Primary Lines and the Design of Fish Protective Facilities, FPC Opinion 452, Mar. 9, 1965 (33 F.P.C. 428).

22. Widespread concern for the impact of pollution was clearly evident during the Progressive era; see Martin Melosi, *Effluent America: Cities, Industry, Energy, and the Environment* (Pittsburgh, Pa.: University of Pittsburgh Press, 2001), 209–24.

23. In recent years two new books have been added to a growing literature on Hetch Hetchy: Robert Righter, *The Battle over Hetch Hetchy: America's Most Controversial Dam and the Birth of Modern Environmentalism* (Oxford: Oxford University Press, 2006); and John W. Simpson, *Dam! Water, Power, Politics, and Preservation in Hetch Hetchy and Yosemite National Park* (New York: Pantheon, 2005). The most comprehensive account of the struggle over Dinosaur National Monument remains Mark Harvey, *A Symbol of Wilderness: Echo Park and the American Conservation Movement* (Seattle: University of Washington Press, 1994).

24. Ross Sandler, "Environmental Law," *Brooklyn Law Review* 42 (1976), 1040; see *National Parks Association v. Udall* (civil no. 3904–62, D.D.C. 1962).

25. See *FCC v. Sanders Bros. Radio Station*, 309 U.S. 470 (1940), and John L. and Eva Hanks, "An Environmental Bill of Rights: The Citizen Suit and the National Environmental Policy Act of 1969," *Rutgers Law Review* 24 (1970), n. 32. The Federal Power Act (1920) was the legal authority for the issuance of permits by the federal government for hydroelectric plants. In deciding that the Federal Power Act protected noneconomic interests, the court was reading an expansive clause of the Act (Section 10 (a)) which states:

That the project adopted, including the maps, plans, and specifications, shall be such as in the judgment of the Commission will be best adapted to a comprehensive plan for improving or developing a waterway or waterways for the use or benefit of interstate or foreign commerce, for the improvement and utilization of water-power development, for the adequate protection, mitigation, and enhancement of fish and wildlife (including related spawning grounds and habitat), and for other beneficial public uses, including irrigation, flood control, water supply, and recreational and other purposes referred to in section 4(e) [16 USC §§ 797(e)] [; and] if necessary in order to secure such plan the Commission shall have authority to require the modification of any project and of the plans and specifications of the project works before approval." 16 USC §§ 803.

26. *Scenic Hudson Preservation Conference v. Federal Power Commission*, 354 F.2d 608 (2nd Cir. 1965) 8; this decision overturned the FPC license and remanded the case back to the FPC. Karl Brooks downplays the significance of the Scenic Hudson decision by arguing that conservationists in Idaho had been granted intervener status by the FPC and had expanded the Federal Power Act's test for determining the public interest by making (between 1954 and 1957) scientific, recreational, and aesthetic evidence part of the record many years before Storm King became an issue. Putting aside the fact that this ignores the FPC's 1953 Namakagon decision (Namekagon Hydro Co., 12 F.P.C. 203 (1953), aff'd, 216 F.2d 509 (7th Cir. 1954), in which the FPC outright rejected an application based on conservationist grounds, the importance of this case has never rested on the fruits of Scenic Hudson's efforts before the FPC or on the FPC's willingness to grant Scenic Hudson standing, but on the production of a federal appellate decision that changed standing before the federal

judiciary. This provided environmental litigants a good deal more leverage. Brooks does offer a necessary corrective to the perception that environmental law emerged in the late 1960s and '70's by tracing the expansion of environmental law in the postwar period; see Karl Brooks, *Before Earth Day: The Origins of American Environmental Law, 1945–1970* (Lawrence: University of Kansas Press, 2009).

27. Hanks and Hanks, "Environmental Bill of Rights," 235–36. There was at the time an active debate as to whether environmental rights might be granted constitutional status. Lawyers were uniformly pleading environmental rights as constitutional rights under the Fifth, Ninth, and Fourteenth amendments. See David Sive "Some Thoughts of an Environmental Lawyer in the Wilderness of Administrative Law," *Columbia Law Review* 70 (1970), 642.

28. *Powleton Civic Home Owners Assoc. v. HUDD*, 284 F. Supp. 809 (E.D. Pa. 1968). The plaintiffs in this case are indicative of the wide-reaching effects that a change in the rules of federal standing can have. See Hanks and Hanks, "Environmental Bill of Rights," n. 40, 237; Sive, "Some Thoughts," 649. The ability for such groups to protect the public interest became far easier with the passage of the National Environmental Policy Act (NEPA, 1969). Many legal scholars commenting soon after the act was passed found that it created a cognizable interest in environmental values enforceable by suit from representative groups of citizens. NEPA became an environmental bill of rights by creating a legislative interest in a clean and healthy environment. This helped to push further open the courtroom doors first cracked by Scenic Hudson. The Administrative Procedures Act (1946) governs both how administrative agencies of the federal government propose and establish regulations and how the federal government reviews these agency decisions.

29. *Scenic Hudson v. FPC*, 354 F.2d 608 (1965), 5. NEPA was effectively given teeth by the Federal Court of Appeals in the District of Columbia in *Calvert Cliffs Coordinating Committee v. AEC*, 449 F.2d 1109 (1971). The court found that the Atomic Energy Commission was subject to NEPA requirements and that the AEC's environmental regulations violated the spirit and substance of the law. The court wrote that the AEC could not "simply sit back, like an umpire, and resolve adversary contentions at the hearing stage." This language was directly paraphrased from the Scenic Hudson decision.

30. Hanks and Hanks, "Environmental Bill of Rights," n. 40, 267.

31. This decision produced Justice Douglas's famous dissent in which he argues that environmental resources ought to be granted standing on their own with a guardian ad litem appointed to represent their interest; see Sierra Club v. Morton, 405 U.S. 727 (1972). This argument was further advanced by Christopher Stone, *Should Trees Have Standing? Toward Legal Rights for Natural Objects* (Los Altos, Calif.: W. Kaufmann, 1974).

32. *Sierra Club 405 U.S.* at 734; the earlier decision was *Association of Data Processing Service Organizations v. Camp* 397 U.S. 150 (1970). For insight in how standing has changed for environmental litigants since the early 1970s, see Ann E. Carlson, "Standing for the Environment," *UCLA Law Review*, April (1998), 45 UCLA L. Rev. 931; Philip Weinberg, "Are Standing Requirements Becoming a Great Barrier Reef against Environmental Actions," *New York University Environmental Law Journal*, 7 N.Y.U. Envtl. L.J. 1, 1999.

33. Although there was no federal legislation requiring the licensing of fossil fuel plants (unlike hydro or nuclear), if a federal decision was required the appropriate agency had to

prepare an environmental impact statement. The same is true of states with their own NEPAs when a state action was involved; additionally, many states passed power plant siting laws while expanding the power of their utility commissions; see Al H. Ringleb, "Environmental Regulation of Utilities," in *Electric Power, Deregulation and the Public Interest*, ed. John C. Moorhouse (San Francisco: Pacific Research Institute for Public Policy, 1986), 186. New York passed a siting law in 1972 that has since expired; see David Markell and Robert Nakamura, "The Effectiveness of Citizen Participation in the Article X Power Plant Siting Process: A Case Study of the Athens Project," report submitted to the Hudson River Foundation, Oct. 2002, 10–13, www.hudsonriver.org/ls/results_handler2_test.asp.

34. See Clean Water Act U.S. Code Title 33, chapter 26, subchapter III. For a book on the postwar history of the issue of water pollution, see Paul Milazzo, *Unlikely Environmentalists: Congress and Clean Water, 1945–1972* (Lawrence: University of Kansas Press, 2006).

35. Some examples are the Scenic Hudson Preservation Conference, which was assembled to stop Consolidated Edison from building the power plant at Storm King Mountain; the Hudson River Conservation Society (later renamed Riverkeeper), organized to clean up the Hudson River; the Clamshell Alliance, formed to stop the construction of a nuclear power plant in New Hampshire; and Citizen's Action for Safe Energy (CASE) group, organized to oppose a proposed nuclear power plant in Oklahoma. This essay is arguing that these types of organizations proliferated and came to define environmentalism by the early 1970s.

36. Richard H. K. Vietor, *Energy Policy in America since 1945* (Cambridge: Cambridge University Press, 1984), 229.

37. Pratt, *Managerial History*, 117–18, 123–27.

38. See Hirsh, *Power Loss*, 133–204.

39. Figure 8.13, Electric Utility Demand-Side Management Programs, *Annual Energy Review 2010* (Washington, D.C.: U.S. Energy Information Administration, 2011).

Contributors

Bruce Beaubouef is the managing editor of *Offshore* magazine, a monthly trade periodical published by PennWell that covers the worldwide offshore oil and gas industry. Prior to that, he served as editor of *Pipeline and Gas Technology*, a Hart Energy trade periodical that covered the oil and gas pipeline transportation industry. He has more than thirteen years of experience writing about the oil and gas industry and earned his Ph.D. at the University of Houston in 1997. Beaubouef's dissertation was published in book form by Texas A&M University Press in September 2007 as *The Strategic Petroleum Reserve: U.S. Energy Security and Oil Politics, 1975–2005*.

Brian Black is professor of history and environmental studies at Penn State—Altoona, where he currently serves as head of Arts and Humanities. His research emphasis is on the landscape and environmental history of North America, particularly in relation to the application and use of energy and technology. He is the author of several books, including the award-winning *Petrolia: The Landscape of America's First Oil Boom* (Johns Hopkins, 2003) and *Crude Reality: Petroleum in World History* (Rowman and Littlefield, 2012).

Jay E. Hakes is a the former administrator of the highly regarded Energy Information Administration at the U.S. Department of Energy. In recent years he has focused his research on American presidents and U.S. energy policy. He has also published the book *A Declaration of Energy Independence* (Wiley, 2008) as well as numerous articles on energy. He holds a Ph.D. from Duke University and began his career teaching political science at the University of New Orleans. In 2013 he stepped down after thirteen years as director of the Jimmy Carter Presidential Library in Atlanta, Georgia.

Robert Lifset (Ph.D., Columbia University) is the Donald Keith Jones Assistant Professor of Honors and History in the Joe C. and Carole Kerr McClendon Honors College at the University of Oklahoma. His research focuses on energy and environmental history. Lifset is the book review editor of H-Energy, an H-Net website devoted to the history of energy, and the author of *Power on the Hudson: Storm King Mountain and the Emergence of Modern American Environmentalism, 1962–1980* (forthcoming from the University of Pittsburgh Press).

Steve Marsh is a fellow of the Royal Historical Society and a reader in international politics at Cardiff University, United Kingdom. His research interests lie primarily in American foreign policy, Anglo-American relations, and European Union external relations. His latest books are the coauthored *The European Union in the Security of Europe* (Routledge, 2011) and the coedited *Anglo-American Relations: Contemporary Perspectives* (Routledge, 2013).

Yanek Mieczkowski is professor of history at Dowling College. His books are *Eisenhower's Sputnik Moment: The Race for Space and World Prestige* (Cornell University Press, 2013), *Gerald Ford and the Challenges of the 1970s* (University Press of Kentucky, 2005), and *The Routledge Historical Atlas of Presidential Elections* (Routledge, 2001). He has worked as a writing fellow for *The American National Biography*, to which he has contributed thirty-seven biographies. He has also served as an exam leader for the Educational Testing Service's AP U.S. History Exam grading.

David S. Painter teaches international history at Georgetown University. His publications include *Oil and the American Century: The Political*

Economy of U.S. Foreign Oil Policy, 1941–1954 (Johns Hopkins University Press, 1986), *The Cold War: An International History* (Routledge, 1999), and *Origins of the Cold War: An International History* (coedited with Melvyn P. Leffler; Routledge, 1994; 2nd edition, 2005), and articles on U.S. policy toward the third world, U.S. oil policies, and the cold war. His current project is a study of oil and world power in the twentieth century.

Tyler Priest (Ph.D., University of Wisconsin–Madison) is associate professor of history and geography at the University of Iowa. In 2010 he served as a senior policy analyst on the President's National Commission on the BP Deepwater Horizon Oil Spill and Offshore Drilling. He is the author of *Global Gambits: Big Steel and the U.S. Quest for Manganese Ore* (Greenwood Press, 2003) and *The Offshore Imperative: Shell Oil's Search for Petroleum in Postwar America* (Texas A&M Press, 2007) and coeditor of a special issue of the *Journal of American History* (June 2012) on "Oil in America."

Jason P. Theriot is an independent consultant and has a Ph.D. in history from the University of Houston. In 2011–12 he held a post-doctoral appointment at the Harvard Kennedy School of Government as an energy policy research fellow. His research interests and consulting practice focus primarily on energy development and environmental impacts in the Gulf Coast. He is the author of the forthcoming book *American Energy, Imperiled Coast: Oil and Gas Development* (LSU Press, 2014).

J. Samuel Walker is a former historian of the U.S. Nuclear Regulatory Commission. He is the author of *Three Mile Island: A Nuclear Crisis in Historical Perspective* (University of California Press, 2004), *The Road to Yucca Mountain: The Development of Radioactive Waste Policy in the United States* (University of California Press, 2009), and other books on nuclear power regulation.

Jeffery Womack is a Ph.D. candidate in the Department of History at the University of Houston, where he studies the historical impact of energy technologies on culture, politics, and medicine. His current work focuses on the adoption of ionizing radiation as a therapeutic medical technology.

Index

Acheson, Dean, 100, 101, 116n16
Adams, Brock, 270
Adams, Richard, 221–22
adAstra, 203
AEC (Atomic Energy Commission).
    See nuclear power industry
Afghanistan, 77, 79, 80–81
Agrifuels Refining Corporation:
    closure process, 193–94; failure
    factors, 190–91, 192–93; production
    expectations, 188–90, 200nn20–
    21; purchase of, 191–92; subsidy
    guarantees, 190, 197, 200nn23–24.
    See also ethanol industry
AIOC, 116n23
air pollution, 226, 267–68, 287–88,
    293–94, 298n15
Akins, James, 72
Alaska, 30, 141–43, 145–48, 167
Albert, Carl, 27, 29, 32
alcohol fuels industry. *See* ethanol
    industry

Algeria, 71, 73, 278n9
allocation programs, during Nixon
    administration, 22–23. *See also* price
    control *entries*
alternative energy, advocacy:
    Carter administration, 54–55,
    57nn11–12, 186–87, 199n8;
    Ford administration, 39; Nixon
    administration, 54, 206. *See also*
    ethanol industry; nuclear power
    industry; solar power, space-based
AMC, 269, 271–72
American Energy, Inc., 195
American Enterprise Institute, 54,
    57n7
American Hegemony and World Oil
    (Bromley), 62
American Nuclear Society, 242–43
American Petroleum Institute, 126,
    136–37
American Society of Civil Engineers,
    135

Amoco, 132–33, 146, 187
Andrus, Cecil, 144, 146
Anglo-Iranian Oil Company, conflict,
    95–102, 107, 117n26, 117n30
appliances, efficiency standards, 26, 173
Arab-Israeli conflicts, 64–65, 73, 102,
    165, 279n10
Arab oil embargo. *See* embargoes
Aramco, 98, 99
ARCO, 142
area-wide leasing plan, 152–53, 158–59
arms spending, 70–71, 74, 102–103,
    105
Arthur D. Little, 204, 207–208
artificial island concept, California
    coast, 139
Ash, Roy, 279n12
Ashland Oil, 195
Asimov, Isaac, 204
Assessment of Solar Energy (NASA/
    NSF), 206, 208, 212
Astounding Science Fiction, 204
Atlantic coast, offshore oil exploration,
    136–38, 153–54
Atomic Energy Act, 223
Atomic Energy Commission (AEC).
    See nuclear power industry
Atomic Industrial Forum, 228
Auger prospect, Gulf of Mexico, 157,
    158
automobile industry: conspicuous
    consumption theme, 261, 278n6;
    consumer culture role summarized,
    272–73, 277, 278n3; gasoline
    shortages, 3–4, 23, 47, 260–
    71; policymaking tradition, 171,
    280n14; regulatory changes, 267–
    70, 280n23, 281n29; sales patterns,
    270–72, 273–77. *See also* CAFE
    standards

bagasse, advantages, 189, 200n20,
    200n23

Bahrain, 73
Balogh, Brian, 208
Baltimore Canyon trench, 137–38, 153
Bankston, Gene, 127
Bayh, Birch, 30
Bea, Bob, 128
Beaubouef, Bruce: biographical
    highlights, 303; chapter by, 163–83;
    comments on, 10–11, 14
Beaufort Sea, Shell's offshore
    exploration, 146–47
Beckmann, Doug, 154, 156
Berard, Dailey, 188, 189, 190, 191, 192,
    193–94, 200n18
Berman, Ilan, 112
Beta field, Shell's, 139–41
Bethe, Hans A., 238–39, 242, 245, 248
bidding systems, offshore leases, 130–
    32, 143–44, 150–53. *See also* Shell
    Oil Exploration & Production
Biomass Energy and Alcohol Fuels Act,
    186
Black, Brian: biographical highlights,
    303; chapter by, 257–82; comments
    on, 12–13
Blackburn, Charlie, 126, 149, 152, 155
blackouts, New York City, 285, 297n9
BLM. *See* offshore oil exploration
Bodega Bay nuclear facilities, 229–30
Boeing Aerospace Company, 134–35,
    210
Bookout, John, 126, 136, 145, 153, 155
Boyle, Bob, 289–90
Bradsher, Keith, 273
Bray, Richard, 147
Brazil, ethanol industry, 196
breeder reactors, 53–54
Brezhnev, Leonid, 78–79
Bridenbaugh, Dale C., 243
Bridges, Harry, 126, 130
Bridges, James, 222
bright spot technology. *See* Shell Oil
    Exploration & Production

Britain: emergency petroleum
    reserve, 173–74; Iranian oil
    conflict, 95–102, 107, 117n26,
    117n30, 278n9; Libya's
    nationalization action, 71; Middle
    East military disengagement, 62,
    63–64; oil interests' power, 64
British Petroleum, 71, 76, 147
Broder, David, 31
Broman, Bill, 151, 157
Bromley, Simon, 62
Brook, J. H., 101
Brooklyn, blackout impact, 285
Brooks, Karl, 299n26
Brown, Edmund G., 139
Brown, Harold, 82
Brzezinski, Zbigniew, 75–78, 80–81
Bullwinkle prospect, Gulf of Mexico,
    152, 154–55
Bumpers, Dale, 33
Bundy, William, 62
Burns, Nicholas, 103
Bush, George W., 42
Business Week, 27, 28, 142, 229

CAFE standards, 7, 34, 42–43, 173,
    268–70, 281n29, 281n32
California: air pollution regulation,
    268; nuclear power measures,
    244–45, 248; offshore drilling,
    138–41; right-on-red traffic law,
    33
California Gas and Electric, 222
Calvert Cliffs nuclear facilities, 232–
    33, 300n29
Cannon, Jim, 183n18
Carroll, Phil, 140–41
Carson, Rachel, 286
Carter, Jared G., 171
Carter, Jimmy (and administration):
    overview, 47–49, 55–56, 57n12;
    alternative fuels support, 54–55,
    57nn11–12, 186–87, 199n8;

conservation initiatives, 7, 82,
    265–66, 269; energy department
    creation, 50–51, 57n6, 186; Iran
    relations, 48, 75–82, 264–65;
    nuclear power policy, 53–54, 241;
    Persian Gulf policy formalization,
    81–83; personnel continuity,
    49–50; presidential campaign,
    48–49; price regulation actions,
    52–53; solar power policies,
    210–12, 220n24; Soviet Union
    relations, 77–82
Carter Doctrine, 81–83
Chain Reaction (Balogh), 208
Cheney, Dick, 24, 103
Cherry, Myron M., 232, 235
Chevrolet, 271
China, 80–81, 104–105
China Syndrome, The, 222–23, 248–49
Chrysler, Hemi design, 257, 258, 275
Chukchi Sea, Shell's offshore
    exploration, 147–48
CIA, 70, 79–80
Circuit Court of Appeals, 2nd, 291–
    92, 299nn25–26
Citizens Energy Project, 237
Clarke, Arthur C., 204
Clean Air Act, 268, 293–94
Clean Atlantic Associates (CAA), 137
Clean Gulf Associates (CGA), 137
Clean Water Act, 293–94
CNPC of China, 110
coal industry: environmental impact,
    226; during Ford administration,
    26, 40, 172, 174; during Nixon
    administration, 54; and nuclear
    power, 53, 228
Cognac project, Shell's, 125, 130–35
Cohen, Murray, 260
Comey, David Dinsmore, 232, 235
complacency factor, energy
    legislation, 30–31, 38
Conable, Barber, 24, 35

Congress, energy-related activity:
during G. W. Bush administration,
42; during Carter administration,
50–53, 55–56, 57n10, 138, 269–70;
during Eisenhower administration,
223; during Nixon administration,
23, 209; during Reagan
administration, 54, 196–97. *See also*
Ford, Gerald (and administration)
Conoco, 155
Conservation Department, New York
State, 289
conservation initiatives: during Carter
administration, 7, 82, 265–66, 269;
Consolidated Edison's, 296; during
Ford administration, 7, 25–27,
33–34, 182n14; during Nixon
administration, 167, 263
Consolidated Electric of New York:
demand shortages, 285, 297n9;
financing crisis, 283, 296nn1–
2; hydroelectric plant litigation,
288, 290, 291–92, 294, 299nn25–
26; pollution problems, 287–90,
298nn15–18. *See also* electrical
energy crisis, environmentalism's
role
conspicuous consumption era, 261,
278n6
conspiracy theories, energy crisis, 170
consumer culture, influence, 278n3. *See
also* automobile industry
Cook Inlet, Kachemak Bay leases, 141
corporate average fuel economy
(CAFE) standards, 7, 42–43, 268–
70, 281n29, 281n32
costs, energy projects: ethanol
plan construction, 186, 200n26;
petroleum reserve, 167, 176; solar
power development, 209, 212–13
costs, Shell's offshore projects: Alaska,
142, 143, 146, 148; Atlantic Basin,
138, 154; California coast, 139, 141;

Gulf of Mexico, 132, 135, 148, 149,
154, 157
Cottam, R. W., 98
Cougar tract, Gulf of Mexico, 150
Cousins, Norman, 287
Creative Initiative Foundation, 248
Crude Oil Windfall Profits Tax, 186

Datsun, 271
Davies, Robert L., 172
decontrol initiatives. *See* price control
entries
deep-tow technology, 154
Deepwater Horizon oil spill, 124, 158
Defense, Department of, 82, 172
Defense Production Act, 99–100
demand perspective, overview, 12–13.
See also automobile industry;
electrical energy crisis *entries*
Department of Energy (DOE):
creation of, 50–51, 57n56, 186;
ethanol program development, 190,
193–94, 195, 197, 199n8, 200n24
Department of Interior. *See* offshore
oil exploration
Détente and Confrontation (Garthoff), 62
Diapir basin, Beaufort Sea, 146
DiBona, Charles, 22, 24
Dick Cavett Show, 234
Dingell, John, 29, 33
Dinosaur National Monument, dam
project, 290–91
Discovery Seven Seas, 154–55
Disney Corporation, 293, 300n31
distillate oils, 22, 39
Doha conference, OPEC's, 75
Domenici, Pete, 57n10
Douglas, Michael, 222
Douglas, William O., 300n31
Downey, Marlan, 142
Dubos, Rene, 247
Dulles, John Foster, 113
Duncan, Charles, 49–50

Dunn, Mike, 157
Dunn, Pat, 124, 129
Dykstra, Bouwe, 124

Early Storage Reserve program, 174
Earth Day celebration, 264, 267–68
Eckhardt, Bob, 27
Economist, 31
ECPA (Energy Conservation and Production Act), 39, 40
Ecuador, 278n9
Edgington Oil, 191–93
Edison, Thomas, 284
Edwards, Edwin, 188, 192
Egypt, 65, 73, 79, 262, 279n10
Eisenhower, Dwight D. (and administration), 5, 94, 278n9
Eizenstat, Stuart, 49
electrical energy crisis: Consolidated Electric bailout, 283, 296nn1–2; oil embargo factor, 284–85, 297n6; technology factors, 4, 284, 297n4
electrical energy crisis, environmentalism's role: overview, 286, 295–96, 301n35; air pollution problem, 287–88, 298nn15–16; hydroelectric litigation, 290–92, 299n26, 300n29; regulatory legislation, 291–94, 299n25, 300n28, 300n33; Sierra Club suit, 293, 300n31; water pollution problem, 288–90, 298nn17–18, 298n20
electric vehicles, 275–76
Ellen platform, Beta field, 141
Elly platform, Beta field, 141
embargoes: Carter administration actions, 48, 264–65; electrical utilities impact, 284–85, 297n6; Nixon administration actions, 23, 129–30, 227; as response to Arab-Israeli conflict, 15n2, 22–23, 73–74, 165–66, 279nn10–11; as response to

Mossadegh's threats, 99–101, 107, 113
Emergency Petroleum Allocation Act (EPAA), 6, 23, 171
Emergency Preparedness report, NPC's, 167, 175
emissions standards, automobile industry, 267–70, 281n29. *See also* air pollution
End of Oil (Roberts), 275
Energy Conservation and Production Act (ECPA), 39, 40
energy crisis. *See specific topics, e.g.,* embargoes; price control *entries*; solar power, space-based
Energy Independence Act, 172–74, 182n14
Energy Policy and Conservation Act (EPCA), 33–38, 39, 43, 52, 173–76, 183nn17–18
energy policymaking, overview, 5–15. *See also specific topics, e.g.,* automobile industry; ethanol industry; Iran-U.S. relations; Nixon, Richard M. (and administration)
energy regime, McNeil's, 258
Energy Research and Development Administration, 50
Energy Security Act, 186, 195, 199n8
"Energy Strategy" (Lovins), 265
Energy Tax Act, 186
Environmental Protection Agency (EPA), 268, 293–94
environmental rights, constitution debate, 300n27. *See also* electrical energy crisis, environmentalism's role
EPAA (Emergency Petroleum Allocation Act), 6, 23, 171
EPCA (Energy Policy and Conservation Act), 33–38, 39, 43, 52, 173–76, 183nn17–18
Ericcson, John, 205

ethanol industry: benefits debate,
57n11, 194–95, 197, 201n38,
201nn40–41; Carter administration
initiatives, 55, 57n11, 186–87;
feasibility factors, 184–85, 189;
Louisiana's subsidy patterns, 187–
89, 192, 193, 201n31; success
barriers, 11, 191, 195–98, 201n27;
trade deficit impact, 200n26
Etzler, John Adolphus, 205
Eureka drillship, deep-water cores,
150–51
Eureka platform, Beta field, 141
Exxon, 73, 132, 133, 155
Exxon Valdez, 147

Federal Energy Administration (FEA),
50. *See also* Zarb, Frank
Federal Energy Office: during Ford
administration, 39, 40, 167–68, 172;
during Nixon administration, 22,
167
Federal Gasohol Plan, 186, 199n8
Federal Power Act, 291, 299n25
Federal Power Commission,
Storm King plant, 290, 291–92,
299nn25–26
fees, oil import, 25–26, 27–28
Ferguson, Niall, 62
Ferriter, John Pierce, 183n19
Fesharaki, F., 117n30
fifty-fifty agreement, 278n9
Finch, Christopher, 272
finished petroleum products, decontrol,
39
fish kills, electrical industry, 288–90,
298nn17–18, 298n20
Flowers, Billy, 126–27, 150–52, 154,
156
Fonda, Jane, 222
Forbes, 231
Forbes, Ian A., 243
Ford, Daniel F., 233–34

Ford, Gerald (and administration):
alternative fuels, 39; anti-inflation
campaign, 24–25; coal mining
support, 40; congressional relations,
20, 27–28; conservation initiatives,
7, 25–27, 33–34, 281n29; decontrol
proposals, 25–27, 32–33; economic
philosophy, 7, 21–22, 23–24, 172;
energy department proposal, 50;
Energy Independence proposal,
172–73, 182n14; EPCA bill,
33–38, 173–74, 175, 183nn17–
18; leadership philosophy, 31;
legacy of energy policymaking,
7, 41–43; nuclear power policy,
227; and OPEC's price increases,
75; personnel continuity, 49–50;
presidential campaign, 35–36,
38, 48–49; Project Independence
program, 167, 172; *Sequoia* cruise
purpose, 19–20
Ford, Gerald (as vice-president), 19, 21
Foreign Affairs, 237
foreign policy's role, overview, 8–9,
61–63, 83–85. *See also specific topics,*
e.g., Iran-U.S. relations; Nixon,
Richard M. (and administration)
Forrest, Mike, 130, 145, 155, 157
Fortune, 226, 248
Frederick, Don, 156, 158
Freeman, David, 49–50
Friedersdorf, Max, 20, 29
Friends of the Earth, 235
Frosch, Robert A., 211–12, 216
Fuel Alcohol report, 187, 199n12
fuel efficiency, gasohol performance,
201n38, 201n41. *See also* CAFE
standards

Gabon, 278n9
Gaddis, John Lewis, 61
Gaitskell, Hugh, 116n13
Garthoff, Raymond, 62

Gartman, David, 270
gasohol. *See* ethanol industry
gasoline: decontrol attempt, 39;
    shortages, 3–4, 23, 47, 260–71.
    See also CAFE standards; ethanol
    industry
Gazprom, 110
General Agreement on Participation,
    71
General Electric, 224–25, 243
geography factors, energy programs,
    29, 35, 48
geopolitics of oil, overview, 61–63,
    83–85. *See also* Iran-U.S.
    relations; Nixon, Richard M. (and
    administration); oil economy,
    transition factors
Georges Bank trough, 137, 138
geothermal energy, 237
Gephardt, Richard, 57n10
Gingrich, Newt, 54
Glaser, Peter E., 204, 207–208, 209,
    212, 216–17
Global Cold War (Westad), 62
Godek, Paul, 273
Godfrey, Dan, 134
Goldsmith, Paul, 207
Goodell, Jack, 221–22
Gore, Al, Jr., 57n10
Grassley, Charles, 38
Gray, Mike, 222
Great Britain. *See* Britain
Green Canyon area, Gulf of Mexico,
    154–55, 158
Greenspan, Alan, 23, 35
Gruman Aerospace Corporation, 208
Gulf Corporation, 116n23, 149
Gulf of Alaska, Shell Oil's explorations,
    141–43
Gulf of Mexico: Nixon administration
    leasing, 129–30, 132; petroleum
    structure statistics, 123. *See also*
    Shell Oil Exploration & Production

Hakes, Jay E.: biographical highlights,
    304; chapter by, 47–67; comments
    on, 8, 14
Hamilton, Pat, 193
Hammer, Armand, 68
Hanley, James, 21
Hart, Gary, 57n10
Hart, Tom, 127
Hemi design, automobiles, 257, 258,
    275
Hetch Hetchy, dam project, 290–91
Hill, John, 27, 35, 39
Hirsh, Richard, 284
Hodel, Don, 152
Holdren, John P., 246
Honda, 271, 276
Hondo platform, Exxon's, 133
Honston, Bennett, 52
Hoover, Herbert, Jr., 102
hostage crisis, Iranian, 79–80, 81,
    264–65
House of Representatives, U.S., 21, 22,
    106, 209. *See also* Congress, energy-
    related activity
Hubbard, Richard B., 243
Hurricane Katrina, 42
Hutton field, North Sea, 155
hybrid vehicles, 275–76
Hydrocarbons Reversion Law,
    Venezuela, 71
hydroelectric dams, 290–93

I, Robot (Asimov), 204
Iacocca, Lee, 280n23
import fees. *See* tax policies
import quotas, United States, 5, 171,
    278n9
India-Iran relations, 105
Indian Ocean, Soviet Union
    deployments, 64
Indian Point nuclear facilities, 288–89,
    298nn17–18

Indonesia, 278n9
inflation, 24–25, 27, 33, 36. *See also*
    price control *entries*; pricing, oil-
    producing countries
Interior, Department of. *See* offshore
    oil exploration
International Energy Agency, 112
International Energy Program, 173–
    74, 175, 177–78
Iran: arms spending, 70–71, 102–
    103; civil conflict, 75–76, 144;
    conflict with Britain, 95–102, 107,
    117n26, 117n30; nationalization
    actions, 71, 98, 99, 102, 117n30,
    278n9; and OAPEC's embargo,
    15n2; oil production patterns,
    76, 107; oil revenues, 70, 74; in
    OPEC's founding, 278n9; partition
    possibility, 97, 116n13; Seven-Year
    Plan, 96, 116n9
Iran Counter-Proliferation bill, 106
Iran-U.S. relations, 93–95, 113–15;
    during G. W. Bush administration,
    93, 103–107, 113–15; during Carter
    administration, 48, 75–82, 264–65;
    during Eisenhower administration,
    101–102, 113; during Nixon
    administration, 65, 66, 69–71,
    74–75, 102–103; during Truman
    administration, 95–101, 117n26,
    117n30
Iraq: in Anglo-Iranian oil conflict,
    99; nationalization actions, 71,
    73; in OAPEC, 73; oil revenues,
    70; in OPEC's founding, 278n9;
    production decline, 68; U.S.
    invasion, 85
iron triangle phenomenon, space-based
    solar power, 208–209
Irwin II, John, 69–70
Israel, 64, 73, 102, 279n10

Jackson, Henry, 28, 30–31, 32, 50–51

Japan: automobile industry, 271,
    276; in Carter administration
    policy considerations, 69, 78, 82;
    emergency petroleum reserve
    system, 173–74; and Iran's oil
    bourse, 104; oil imports and
    consumption, 67, 84, 85, 109–10
Jeffords, Jim, 57n10
Jersey Central Power and Light
    Company, 225
Johnson, Lyndon B. (and
    administration), 63
Justice Department, U.S., 68–69

Kachemak Bay leases, 141
Katrina, Hurricane, 42
Kazakhstan, 112
Kemp, Jack, 54
Kendall, Henry W., 233–34, 235
Kennedy, Paul, 61
Khamenei, Ayatollah Ali, 112
Khashoggi, Adnan, 192, 201n28,
    201n33
Khomeini, Ayatollah, 76, 80
Kingdon, John, 215
Kissinger, Henry, 62, 63, 69, 71, 74–75,
    113
Knight, Frank, 3
Kobrin, S., 108
Koch Oil, 146
Korechoff, Mary, 3
Kraft, Christopher, 207, 209–12, 216
Kraft, Joseph, 30
Kuwait, 70, 73, 85, 99, 278n9

Laborde, Doc, 124
Laird, Frank, 215–16
Laird, Melvin, 57n7
Lance, Bert, 211
Lapp, Ralph E., 240, 242
leaded gasoline, 268
Leffler, Melvyn, 62
Lemmon, Jack, 222

Leopold, Aldo, 257–58
Libya: nationalization actions, 71, 107; in OAPEC, 73; oil production statistics, 68; oil revenues, 70; in OPEC, 278n9; pricing demands, 67–69, 70
Lifset, Robert: biographical highlights, 304; chapters by, 3–23, 283–301; comments on, 13
light truck category, CAFE standards, 7, 269, 273–74
light-water reactors, 53–54
Limited Test Ban Treaty, 229
Lindsey, John, 287
Little, Jack, 148
loan guarantees, coal industry, 40
Long, Russell, 29, 52
Lott, Trent, 54
Louisiana, 187–89, 192, 193, 199n14, 201n31
Love, John, 24
Lovins, Armory B., 237, 240, 265
Lowell, Elliot, 221
Luce, Charles, 283
Lynn, James, 24

MacArthur, Douglas, II, 65
Macondo/*Deepwater Horizon* oil spill, 124, 158
Madison, Ed, 195
Malibu nuclear facilities, opposition, 229–30
Mansfield, Mike, 32, 42
Mapco, 37
Marathon, Shell partnership, 146
Marsh, Steve: biographical highlights, 304; chapter by, 93–119; comments on, 9
Massachusetts, 137, 138
McCain, John, 31
MccGwire, Michael, 64
McCloy, John J., 68
McCracken, Paul, 41–42, 57n7

McDermott, J. Ray, 134
McNeil, John, 258
Mead, Margaret, 247
Mensa prospect, Gulf of Mexico, 158
Mexico, 278n9
Michel, Bob, 24
microwave radiation risks, space-based solar power, 213–15
Mieczkowski, Yanek: biographical highlights, 304; chapter by, 19–46; comments on, 7, 14
mileage statistics, driving, 275, 282n46. *See also* CAFE standards
Mineral King Valley, Sierra Club litigation, 293, 300n31
Minerals Management Service (MMS), 153
Mineta, Norm, 57n10
mining industry. *See* coal industry
Minnesota, ethanol industry, 195–96
Minor, Gregory C., 221, 243
Mobil, 73
Mondale, Walter, 220n24
Montana, nuclear power ballot measure, 245
Morrison, Herbert, 96–97
Morton, Rogers, 129–30
Morton, Sierra Club v., 293, 300n31
Mossadegh, Mohammad, 97–99, 101, 116n16
Muluk field, Beaufort Sea, 146–47
Muskie, Edwin, 267–68

Nader, Ralph, 234–35, 237, 240, 242, 243, 248
Namakagon decision, Federal Power Commission, 299n26
Nanz, Bob: on Alaska exploration, 143; on bidding system competitiveness, 144; Cognac prospect presentation, 131; estimate of offshore oil, 145; leadership value, 127; offshore development lobbying, 136–37, 146

NASA. *See* solar power, space-based
Nasser, Gamal, 65
National Academy of Sciences, 42–43
National Alcohol Fuels Commission,
    U.S., 187, 189, 190–91, 195, 199n12,
    201n38
National Council of Churches, 247
National Energy Act, 7, 16n12
National Environmental Policy Act
    (NEPA), 292–93, 300nn28–29
National Highway Traffic Safety
    Administration, 267
National Iranian Oil Company
    (NIOC): during Anglo-Iranian crisis,
    98, 99, 100, 102, 117n26; during
    G. W. Bush administration, 110;
    international partnerships, 104, 115
nationalization programs. *See* ownership
    conflicts
National Ocean Industries Association
    (NOIA), 137
National Petroleum Council (NPC),
    126, 167, 174–75
National Science Foundation, solar
    energy research, 206–207
"National Security and U.S. Energy
    Policy" (NSC), 72–73
National Security Council (NSC):
    during Carter administration,
    75–82; during Nixon administration,
    65, 69, 72–73; during Truman
    administration, 95
National Security Space Office, U.S.,
    203
National Traffic and Motor Vehicle
    Safety Act, 267
natural gas, 34, 39, 52, 55, 148, 181n9
Natural Resources Defense Council
    (NRDC), 233
naval petroleum reserves, 26, 34, 172,
    174, 175
Naval Petroleum Reserves Production
    Act, 38–39

Navarin basin, Bering Sea, 146
Nelson, Gaylord, 267
NEPA (National Environmental Policy
    Act), 292–93, 300nn28–29
Nessen, Ron, 41
Netherlands, 73, 166
New Energy, alternative fuel grant, 197
Newgard, Mark, 193
New Hampshire, presidential
    campaigns, 35
New Mexico, ethanol subsidies, 192
New York, Consolidated Energy
    legislation, 283, 296nn1–2
New York City: air pollution, 287–
    88, 298nn15–16; Earth Day
    demonstration, 267; electricity
    shortages, 206, 285, 297n9
New York Post, 285
New York Times, 40, 205, 207, 260
Nigeria, 67–68, 70, 278n9
NIOC. *See* National Iranian Oil
    Company (NIOC)
Nippon Oil, 104
Nixon, Richard M. (and
    administration): alternative energy
    research, 54, 206; Arab-Israeli
    conflict, 279nn10–11; congressional
    relations, 20; conservation
    initiatives, 167, 263; on embargo
    threats, 107; energy personnel, 49;
    Iranian arms sales, 70–71, 102–103;
    offshore oil leases, 129–30, 132; oil
    import quota, 5; overview of energy
    policies, 6; price controls, 22, 23,
    41, 51–52, 57n7, 164–65; Project
    Independence initiation, 50, 167–
    68, 227, 263, 279n12; twin pillar
    strategy, 62, 63–66, 73–75, 102–103
Nixon Doctrine, 62, 63–66, 73–75,
    102–103
nominations process, offshore tracts,
    130–31, 150, 151
Nordhaus, Robert, 28

North Dakota, 192, 195
North Sea platforms, 129
Norton basin, Bering Sea, 146
NPC (National Petroleum Council), 126, 167, 174–75
NRDC (Natural Resources Defense Council), 233
NSC. *See* National Security Council (NSC)
Nuclear Industry, 227, 228, 232, 234
Nuclear News, 244
nuclear power industry: ballot measures, 244–45, 248; Carter administration policy, 53–54, 241; construction statistics, 16n14; debate polarization factors, 12, 245–47; energy crisis impact, 227–29; Ford administration policy, 227; growth of, 39, 223–27; movie portrayal, 222–23, 248–49; opposition arguments, 229–38, 241–44, 247, 298n18, 300n29; public opinion polls, 244, 245, 246; support arguments, 238–41, 242–43, 297n5
Nucleonics, 224
Nucleonics Week, 226, 227, 228, 230, 235
Nye, David, 258, 286

OAPEC (Organization of Arab Petroleum Exporting Countries), 15n2, 73, 165–66, 285
Oasis Petroleum Corporation, 188, 189, 200n18
Obama, Barack, 47
Occidental Petroleum Company, 68
OCSLAA (Outer Continental Shelf Land Act Amendments), 138, 144. *See also* offshore oil exploration; Shell Oil Exploration & Production
Odom, William E., 85
offshore oil exploration, 123–24, 158–59; bidding/leasing systems,

130–32, 143–44, 150–53; Nixon administration lease statistics, 129–30. *See also* Shell *entries*
oil bourse, Iran's, 103–104, 105
oil consumption, statistics: European, 67; global, 66, 76; Japan, 67; United States, 4, 67, 76, 85, 164, 166
Oil Crisis, The (Venn), 14
Oil Diplomacy in the Twentieth Century (Venn), 61–62
oil economy, transition factors, 62–63, 66–75, 108–109. *See also specific topics, e.g.,* embargoes; Iran *entries*; pricing, oil-producing countries
oil imports, statistics: European countries, 64; Japan, 109–10
oil imports, U. S. statistics: during Bush administration, 94; during Carter administration, 76, 94; Ford administration goals, 26; growth patterns, 4, 165; during Nixon administration, 6, 72, 85, 94, 165, 166
oil production, statistics: British companies, 64; Libya, 68; Middle East countries, 67, 165, 166; Seven Sisters share, 110, 116n23; in United States, 4, 67; Venezuela, 67
oil revenues/exports, statistics, 66, 70, 74, 104
oil spills, 124, 139, 147, 158
oil weapon. *See* Iran *entries*; OPEC; ownership conflicts
O'Leary, Jack, 49
Omang, Joanne, 241
O'Neill, Paul, 24
O'Neill, Tip, 21, 29
OPEC: embargo incentives, 22–23, 227, 279nn10–11; formation of, 262, 278n9; influence pattern, 110, 262; ownership negotiations, 71–72; price increases, 68–70, 75, 76, 165, 166

Organization of Arab Petroleum
  Exporting Countries (OAPEC),
  15n2, 73, 165–66, 285
Organization of Petroleum Exporting
  Countries. *See* OPEC
Otteman, Lloyd, 152
Outer Continental Shelf Land Act
  Amendments (OCSLAA), 138, 144.
  See also offshore oil exploration;
  Shell Oil Exploration & Production
Overseas Consultants, Inc., 116n9
ownership conflicts: during Anglo-
  Iranian crisis, 95–102, 117n26,
  117n30; nationalization actions, 71,
  73, 107–108; OPEC negotiations,
  71–72; significance of, 108. *See also*
  pricing, oil-producing countries
Oyster Creek nuclear facilities, 225

Pacesetter II, 132
Pacific Gas and Electric Company, 203
Paine, Sam, 133
Paine, Thomas, 207
Painter, David S.: biographical
  highlights, 304–305; chapter by,
  61–102; comments on, 8–9
Pakistan, 79, 80
Palisades nuclear facilities, opposition,
  232
Payne, Jon, 244
PDVSA of Venezuela, 110
Percy, Charles, 57n10
Persian Gulf: Britain's military forces,
  63–64; Carter administration policy,
  82–83. *See also* Iran *entries*; Saudi
  Arabia
Pesonen, David, 245
Petrobras of Brazil, 110
Petroleum Reserves Corporation,
  176–77
Petronas of Malaysia, 110
pipelining approach, offshore
  wellheads, 155

Placid Oil, 155
political leadership, overview of energy
  policies, 5–8. *See also* Carter, Jimmy
  (and administration); Ford, Gerald
  (and administration); Nixon,
  Richard M. (and administration)
Portugal, 166
"Power from the Sun" (Glaser),
  204–205
power-pooling arrangements, nuclear
  power growth, 225–26
power utilities, energy needs, 166,
  181n5. *See also* electrical energy
  crisis
Pratt, Joe, 286
presidential campaigns, impact on
  energy programs, 48–49
price controls, disadvantages
  summarized, 176–77. *See also* tax
  policies
price controls, during Carter
  administration, 7, 48
price controls, during Ford
  administration: congressional
  resistance, 27–28, 30; legislation,
  32, 33, 34–38, 173; president's
  philosophy, 7, 23–24, 52; production
  impact, 37, 39; proposals for, 25–27,
  32, 42
price controls, during Nixon
  administration: impact of, 22, 41,
  164–65; imposition of, 6, 51–52;
  legislation, 23; opposition studies,
  57n7
price controls, during Reagan
  administration, 42
pricing, oil-producing countries:
  Iranian revolution impact, 76,
  144; Iran's strategies, 69–70, 103–
  104; Libyan demands, 67–68, 70;
  OPEC-led actions, 68–69, 70, 75,
  76, 165–66, 278n9; profit-sharing
  negotiations, 98–101, 117n26,

117n30; Saudi Arabia's actions, 74–75; Soviet Union benefits, 84. *See also* embargoes; oil economy, transition factors; ownership conflicts

Priest, Tyler: biographical highlights, 305; chapter by, 123–62; comments on, 10, 14

Pró-Álcool, Brazil, 196

production statistics. *See* oil production, statistics

profit-sharing arrangements, oil-producing countries, 98–101, 117n26, 117n30, 278n9

Progressive, The, 222–23

Project Independence program: energy efficiency standards, 182n14; Ford administration approach, 167, 172; Nixon administration initiation of, 50, 167–68, 227, 263, 279n12; nuclear power recommendation, 227; oil lease mandate, 129; petroleum stockpiling recommendation, 168–70, 175; tax proposals, 173

Proposition 15, California, 244–45

Prospect Powell, Gulf of Mexico, 157

Proxmire, William, 50

public interest standard, environmental litigation, 291–93, 299n26, 300nn28–29

Pylan, Bob, 210

Qaddafi, Muammar, 65, 67

Qatar, 73, 278n9

radiation risks: atomic bomb testing, 229; nuclear power plants, 230–31, 236–37, 239, 246–48; space-based solar power, 213–15

radical Arab nationalism, 64–65

Rankin, John, 149

ranking systems, presidential, 41

Rapid Deployment Joint Task Force, 82

Ravenswood nuclear facilities, opposition, 229–30

Raytheon Corporation, 208

Reagan, Ronald (and administration): automobile regulation, 273; decontrol actions, 42, 52; Energy Department budget, 51; ethanol industry opposition, 191, 196–97; nuclear power policy, 54; oil leases, 10, 152; presidential campaigns, 35, 38, 48

"Reason" (Asimov), 204

Redmond, John, 126, 131, 132

Reece, Ray, 208–209

residual oils, 39, 169

Reuss, Henry, 31

Rhodes, John, 29

Rhodesia, 166

Rickover, Hyman, 53

right-on-red traffic law, 33–34, 42

Rise and Fall of the Great Powers (Kennedy), 61

Roberts, Paul, 275, 281n32

Roemer, Buddy, 192, 193

Roisman, Anthony Z., 232–33, 235–36

Rossin, A. David, 239

Ruckelshaus, William D., 268

Rumsfeld, Donald, 24

Russell, Don, 127

Russia, Iran's oil agreement, 104–105. *See also* Soviet Union

Sadat, Anwar, 65, 262

safety regulations, automobile industry, 267–68, 280n23

salt dome storage, FEC's proposal, 167–69, 181n9

sanctions regimes, 85, 93, 103–105

Santa Barbara Channel, oil production, 129, 133, 138–39

Santa cartoon, Ford administration, 36

satellite conversion system. *See* solar
    power, space-based
Saudi Arabia: Afghanistan conflict, 79;
    Ford administration relations, 75;
    Nixon administration relations,
    64–66, 73–75; in OAPEC, 73; oil
    revenues, 70; in OPEC, 75, 278n9
Saudi Aramco, 110
Sawhill, John, 24, 49–50
Scenic Hudson, 291–92, 299n25
Schleede, Glen, 25
Schlesinger, James R., 49, 50, 57n6, 211
Schulz, George, 5
Schumacher, E. F., 246–47, 248
Science, 204, 217, 233–34, 242
Scowcroft, Brent, 24
Seabrook nuclear facilities, 243–44
Segura, Donald P., 189–90
Seidman, William, 24, 36
seismic surveys. *See* Shell Oil
    Exploration & Production
Senate, U.S., 21, 22, 50, 51. *See also*
    Congress, energy-related activity
September 11 attacks, 85
Sequoia cruise, Ford's, 19–20
Seven Sisters companies, listed,
    116n23. *See also* ownership conflicts
Seven-Year Plan, 96, 116n9
Shell America, 156
Shell Oil: Iranian revolution response,
    76; offshore development lobbying,
    136–37; in Seven Sisters list,
    116n23
Shell Oil Exploration & Production:
    Alaska activity, 141–43, 145–48;
    Atlantic Basin activity, 137–38, 153–
    54; bidding processes, 130–32, 149–
    50, 154, 156–57, 158; California
    coast activity, 138–41; deepwater
    development challenges, 150–
    57; leadership personnel, 126–27;
    natural gas leases, 148; platform
    technology development, 128–29,
132–35, 140–41, 155–56; tradition
    of offshore development, 10, 124–
    26, 127–28
Shirley, O. J., 137
Shuman, Frank, 205
Sierra Club, 233
Sierra Club v. Morton, 293, 300n31
Silent Spring (Carson), 286
Simon, William E., 22, 24, 34–35,
    74–75, 260
Small Is Beautiful (Schumacher),
    246–47
Smith, Howard K., 241
Socony Vacuum, 116n23
Sohio, 147
Solar Caucus, 55, 57n10
Solaren Corp., 203
Solar Energy Panel, 206–208
solar power, space-based: aerospace
    industry's influence, 207–209,
    210; budget context, 207; Carter
    administration positions, 210–12,
    220n24; conceptual beginnings,
    204–205; congressional interest,
    209; energy crisis context, 205–
    206; environmental risks, 213–15;
    infrastructure requirements, 210;
    Kraft's early role, 209–12; NASA/
    NSF's research report, 206–207,
    208; Nixon administration policy,
    206; policy community schism,
    215–16; technological barriers,
    212–13; technological expectations,
    11–12, 203–204, 216–18
SONAT, 153
sorghum, in ethanol production, 190,
    200n23
South Africa, 166
Southern New York Sportsmen, 289
Soviet Union: Afghanistan conflict, 77,
    79, 80; and Arab-Israeli conflicts,
    64; Carter administration relations,
    77–82; Indian Ocean arrival, 64;

Iranian relationship, 77, 95–96; oil production, 77, 84, 91n81; space-based solar power, 214, 220n30; Truman administration relationship, 95–96, 97–98

space-based solar power. *See* solar power, space-based

Spanke, Bill, 181n9

specialty oils, decontrol, 39

speed limits, automobile, 6, 167, 266, 281n32

Sporn, Philip, 224

Sports Illustrated, 289–90

SPR. *See* Strategic Petroleum Reserve

Standard Oil, 116n23

Stang, William, 97

State Department actions, during Truman administration, 100–101, 117n26

State of the Union addresses: Carter's, 81; Ford's, 25–26, 227, 281n29; Nixon's, 63

Sterling, Gordon, 133

St. George basin, Bering Sea, 146

Stockman, David, 54

Storm King Mountain, hydroelectric plant, 288, 290, 291–92, 294, 299nn25–26

Strategic Petroleum Reserve: in congressional legislation, 34, 173–76; FEC's proposal, 167–70; in Ford's Energy Independence bill, 7, 172; implementation of, 174–76, 183n17; and political power patterns, 170–71; as positive supply-side management, 10–11, 42, 176–79, 183n19

Strategies of Containment (Gaddis), 61

strip mining, 40

stripper wells, 39

subsidies, alternative fuels. *See* ethanol industry

sugar cane industry, 187–88, 189–90, 196, 199nn13–14

Sun Betrayed, The (Reece), 208–209

supply approach, overview, 9–12. *See also* ethanol industry; nuclear power industry; offshore oil exploration; solar power, space-based; Strategic Petroleum Reserve

Supreme Court, U.S., 137, 138, 144, 293, 300n31

Sutter, Paul, 264

SUV category, automobiles, 274, 275

Swearingen, John, 142

sweet sorghum, in ethanol production, 190, 200n23

Syria, 68, 73

System Definition Study, Boeing's, 210

tariffs. *See* price controls *entries*; tax policies

tax policies: during Carter administration, 7, 52–53, 55, 186; disadvantages summarized, 178–79; during Ford administration, 24, 25–26, 27, 31–32, 38, 173, 174; during Nixon administration, 167. *See also* bidding systems, offshore leases; price control *entries*

tax policies, ethanol industry: complexity problem, 195–96, 202n48; Louisiana legislation, 188, 192, 201n31; revenue costs, 202n44

tax rates, oil-producing countries, 66, 68, 69–70. *See also* pricing, oil-producing countries

Tenneco, ethanol sales, 197

Tennol, alternative fuel grant, 197

Texaco, 187, 193, 197

Texas, presidential campaigns, 35

Textron, 208

Theriot, Jason P.: biographical highlights, 305; chapter by, 184–202; comments on, 11

thermal efficiencies, electrical plants, 4, 284, 297n4

thermal pollution, nuclear power
plants, 230, 232
Three Mile Island, 53–54
Threet, Jack, 127, 137–38, 145, 155
THUMS Group, California coast, 139
Time, 243
Toca, Carlos, 188, 189
Toyota, 271, 276
Trans-Arabian Pipeline, 68
Transportation, Department of, 267
Treen, David, 189
Triad America Corporation, 191, 192,
201n28
Tripoli Agreements, 70
truck categories, automobile industry,
7, 269, 273–74
Tsongas, Paul, 57n10
turnkey era, nuclear power, 225
twin pillar strategy, Nixon
administration, 62, 63–66, 73–75,
102–103

Udall, Morris, 40
Udall, Stewart, 40
Union of Concerned Scientists (UCS),
233–35, 242–43
Union Oil, 129, 146
United Arab Emirates, 73, 75, 278n9
United Methodist Church, 247
U.N. Security Council, 103, 105, 106
Ursa prospect, Gulf of Mexico, 158
USDA, 186, 195, 199n8

Velleca, Tom, 156
Venezuela, 15n2, 67, 71, 98, 278n9
Venn, Fiona, 14, 61–62
Ventana nuclear facilities, 221–22
Vienna Summit, Carter's statement,
78–79
Vietor, Richard, 295
Virginia, ethanol subsidies, 192

Voiland, Gene, 157
Volkswagen Beetle, 270–71
von Hippel, Frank, 236, 238

Walker, J. Samuel: biographical
highlights, 305; chapter by, 221–54;
comments on, 12, 14
Washington Post, 241
water pollution, electrical industry,
288–90, 294, 298nn17–18, 298n20
Watt, James, 152, 153
Weinberg, Alvin M., 238
Wells, Kimberly, 221–22
West, J. Robin, 152
Westad, Odd Arne, 62
West Germany, 173–74
Westinghouse Electric Corporation,
224–25
White, Lee C., 205–206
Wickizer, Carl, 153, 157
Wilkinson, Edward, 195
Wilmington Canyon, 153, 154
windfall profits tax, 52–53, 173, 186
Wireless World, 204
Womack, Jeffrey: biographical
highlights, 305; chapter by, 203–20;
comments on, 11–12
Woodcock, Gordon, 210

Yannoni, Nicholas F., 207
Yergin, Daniel, 61, 62, 262, 278n8

Zarb, Frank: appointment of, 24; on
complacency factor, 30–31; on
decontrol approach, 32, 42; on
EPCA bill, 35–36, 37, 183n18; on
gasoline taxes, 38; on right-on-red
traffic law, 33–34; on *Sequoia* cruise,
19–20
Zhuhai Zhenrong Corporation, 104

CPSIA information can be obtained
at www.ICGtesting.com
Printed in the USA
FFOW05n0118280314